THERMAL INSULATION

THERMAL INSULATION
Recent Developments

Joseph B. Dillon

NOYES DATA CORPORATION

Park Ridge, New Jersey, U.S.A.

1978

Library of Congress Catalog Card Number: 77-15218
ISBN: 0-8155-0687-2
Printed in the United States

Published in the United States of America by
Noyes Data Corporation
Noyes Building, Park Ridge, New Jersey 07656

FOREWORD

The detailed descriptive information in this book is based on U.S. patents issued since the early 1970s that deal with thermal insulation materials and their application.

This book serves a double purpose in that it supplies detailed technical information and can be used as a guide to the U.S. patent literature in this field. By indicating all the information that is significant, and eliminating legal jargon and juristic phraseology, this book presents an advanced, technically oriented review of thermal insulation technology as depicted in U.S. patents.

The U.S. patent literature is the largest and most comprehensive collection of technical information in the world. There is more practical, commercial, timely process information assembled here than is available from any other source. The technical information obtained from a patent is extremely reliable and comprehensive; sufficient information must be included to avoid rejection for "insufficient disclosure." These patents include practically all of those issued on the subject in the United States during the period under review; there has been no bias in the selection of patents for inclusion.

The patent literature covers a substantial amount of information not available in the journal literature. The patent literature is a prime source of basic commercially useful information. This information is overlooked by those who rely primarily on the periodical journal literature. It is realized that there is a lag between a patent application on a new process development and the granting of a patent, but it is felt that this may roughly parallel or even anticipate the lag in putting that development into commercial practice.

Many of these patents are being utilized commercially. Whether used or not, they offer opportunities for technological transfer. Also, a major purpose of this book is to describe the number of technical possibilities available, which may open up profitable areas of research and development. The information contained in this book will allow you to establish a sound background before launching into research in this field.

Advanced composition and production methods developed by Noyes Data are employed to bring these durably bound books to you in a minimum of time. Special techniques are used to close the gap between "manuscript" and "completed book." Industrial technology is progressing so rapidly that time-honored, conventional typesetting, binding and shipping methods are no longer suitable. We have bypassed the delays in the conventional book publishing cycle and provide the user with an effective and convenient means of reviewing up-to-date information in depth.

The Table of Contents is organized in such a way as to serve as a subject index. Other indexes by company, inventor and patent number help in providing easy access to the information contained in this book.

15 Reasons Why the U.S. Patent Office Literature Is Important to You —

1. The U.S. patent literature is the largest and most comprehensive collection of technical information in the world. There is more practical commercial process information assembled here than is available from any other source.

2. The technical information obtained from the patent literature is extremely comprehensive; sufficient information must be included to avoid rejection for "insufficient disclosure."

3. The patent literature is a prime source of basic commercially utilizable information. This information is overlooked by those who rely primarily on the periodical journal literature.

4. An important feature of the patent literature is that it can serve to avoid duplication of research and development.

5. Patents, unlike periodical literature, are bound by definition to contain new information, data and ideas.

6. It can serve as a source of new ideas in a different but related field, and may be outside the patent protection offered the original invention.

7. Since claims are narrowly defined, much valuable information is included that may be outside the legal protection afforded by the claims.

8. Patents discuss the difficulties associated with previous research, development or production techniques, and offer a specific method of overcoming problems. This gives clues to current process information that has not been published in periodicals or books.

9. Can aid in process design by providing a selection of alternate techniques. A powerful research and engineering tool.

10. Obtain licenses — many U.S. chemical patents have not been developed commercially.

11. Patents provide an excellent starting point for the next investigator.

12. Frequently, innovations derived from research are first disclosed in the patent literature, prior to coverage in the periodical literature.

13. Patents offer a most valuable method of keeping abreast of latest technologies, serving an individual's own "current awareness" program.

14. Copies of U.S. patents are easily obtained from the U.S. Patent Office at 50¢ a copy.

15. It is a creative source of ideas for those with imagination.

CONTENTS AND SUBJECT INDEX

INTRODUCTION

Only a few short years ago thermal insulation was largely known, and of concern, only to those working in industrial plants and the construction industry. However, with the advent of the energy crisis in 1973–1974, and the concomitant rapid increase in fuel prices, thermal insulation and the conservation of energy resources have been the subject of new national policies in many countries on a worldwide basis. Individual homeowners, faced with staggering increases in monthly fuel expenses, are installing insulation at a rapidly increasing rate.

In many countries, particularly those hard hit by oil shortages, government policies have been implemented to absorb some of the cost of improving the insulation systems in both the private and industrial segments of the economy. In a very significant way, the saving of fuel can now influence the standard of living in most developed countries, and, grossly affect the balance of payments position of a nation. Profitability of many industrial processes is highly dependent upon the energy/product relationship and the overall energy balance of an integrated producing complex.

Thermal insulation is used not only to conserve energy, but to provide year-round comfort for living and working spaces. Insulation is used for extremes of both heat and cold ranging from aerospace reentry problems and open hearth furnaces to the cryogenic storage of liquefied gases and refrigerated storage units.

Common materials which provide varying degrees of insulation (low thermal conductivity) are cork, mineral wool, perlite, vermiculite, clay, other ceramics, wood fibers, felts or boards, many foamed plastics, paper, reflective metal foils, silicates and cellular glass. Many of these materials are used in combination, and in recent years, foamed, low density inorganic and ceramic systems have been increasingly utilized.

One way of classifying insulation materials is by the temperature range in which they are most useful. The following classification may be used as a guideline.

Temperature Range	Applications	Insulation Type
Above 2000°F	Aerospace reentry vehicles, automotive catalytic converters	Refractory metals and oxides Aluminosilicate fibers
Above 1200°F	Furnaces, kilns, roasters, flues, direct flame	Diatomite, silica bricks, fire-clay brick
600° to 1200°F	Pipe boilers, high pressure steam lines, furnaces, turbines	Asbestos, special mineral calcium silicate
300° to 600°F	Ovens, boilers, stills, locomotives, steam lines, pipe wrapping	Magnesia-asbestos fiber, calcium silicate, felted asbestos, asbestos paper, mineral wool
50° to 300°F	Home insulation, roofs, commercial buildings	Mineral wool, macerated paper, insulating board, aluminum foil, expanded vermiculite, insulating cements, plastic foams
Below 50°F	Cryogenic chemical and food processes, liquefied gasses, pipe insulation	Cork, mineral wool blankets, cellular glass, polystyrene and polyurethane, perlite

Obviously, in actual use, considerable overlap exists as many insulation materials provide, alone, or in combination, the desired cost/performance benefits for a wide variety of applications.

In this book, some 190 processes, as presented in the U.S. patent literature of the past few years, are described in considerable detail with reference to specific applications wherever possible. The worldwide concern for improved and increasingly efficient insulation systems is immediately evident from the patent literature. Extensive development work is evident in Germany, Japan and Canada, with processes from England, Russia, Switzerland, Sweden, France and other countries also presented. Overall, some 30% of the processes described in this book were developed outside the United States.

This very timely book provides a detailed view of the research, product development and applications technology for insulation systems which will increasingly become an integral part of construction and industrial process planning over the next decade.

PANELS AND WALL UNITS

CONCRETE COMPOSITES

Anchoring System for Sandwich Slab

E. Haeussler; U.S. Patent 3,757,482; September 11, 1973 describes a system for anchoring a front concrete plate to a rear concrete plate in substantially juxtaposed and coextensive relationship, through an insulating layer interposed between these plates. The anchor comprises an elongated tubule or sleeve provided at opposite extremities received in the respective concrete plates with means engaging them. Such means are preferably located adjacent each edge of the sleeve.

According to an important feature of this process, the tubules or sleeves are tube or pipe sections of a regular cross section, i.e., a cross section corresponding to that of a conic section or a regular polygon, and preferably which is a constant over the entire length of the anchor although it has been found to be advantageous in certain circumstances to provide a constant convergence to the anchor. In the latter case, the anchor sleeve is conical, the conicity assisting in anchoring the concrete plates together.

The anchor sleeves are formed with rows of holes or perforations proximal to their ends while the aforementioned formations receive bars or rods traversing these holes and projecting laterally beyond each anchor to overhang or underlie a reinforcing rod or bar of the lattice. Advantageously, two longitudinally spaced rows of such holes are provided and at least one rod is fitted into each row of holes so that the transverse rods engage on opposite sides of the reinforcing mats. Each row of holes will generally include a number of pairs of diametrically opposite perforations which are aligned to permit the transverse rod to extend through them.

In addition or as an alternative, the means for locking the sleeves within the concrete plates can include tabs or lugs which may be bent outwardly to project laterally. Figure 1.1 illustrates the system.

3

The sandwich slab construction may, as shown diagrammatically in Figure 1.1d, comprise a rear concrete slab **103** having a mat **106** of reinforcing rod embedded therein and of generally rectangular configuration. The concrete plate **103** is coextensive with a slab or plate of cellular synthetic resin as represented at **102**, this insulating layer of foamed polystyrene, for example, being overlain by the upper plate **101** of concrete. The plates are interconnected by anchor sleeves generally represented at **104** at spaced locations, the anchor sleeve **104** being constituted as described in connection with Figures 1.1a, 1.1b and 1.1c.

Referring to Figure 1.1a, in which the anchor assembly is shown in greater detail, it will be apparent that the reinforced concrete plate **1**, the insulating layer **2** of foamed synthetic resin, and the rear reinforced plate **3** constitute a sandwich slab. The concrete plates **1** and **3** are anchored together by tubules or sleeves **4** provided at respective extremities with means for locking the sleeve to the respective concrete plates. As illustrated in Figures 1.1a and 1.1c, the lower concrete plate **3** is provided with a lattice work or mat **6'** or reinforcing rod, the rod running parallel to the plane of the paper being represented at **6a'** while the rods **6b'** run perpendicular to the plane of the paper. The rods **6a'** and **6b'**, therefore, define a lattice work in an interstice of which the sleeve **4** is seated. The mat has an effective thickness **V** (Figure 1.1a).

FIGURE 1.1: SANDWICH SLAB CONSTRUCTION

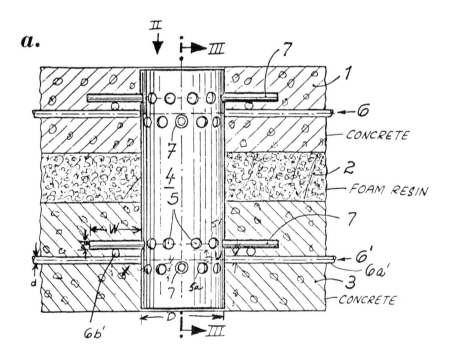

(continued)

b.

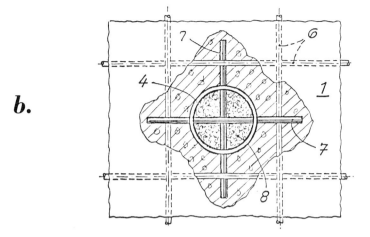

c.

d.

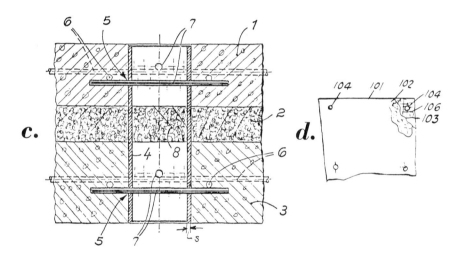

(a) Cross section perpendicular to the plane
    of sandwich slab illustrating the
    anchor assembly in elevation
(b) View taken in the direction of arrow II
    and partly broken away
(c) Cross section taken along the line
    III — III of Figure 1.1a
(d) Plan view of structural member

Source: U.S. Patent 3,757,482

The means at **5** at each end of the sleeve **4** may include rows **5a** and **5b** of perforations which are angularly staggered and are spaced axially by a distance $U > V$, each hole being paired with a diametrically opposite hole so that rods **7** may be passed through the holes and project laterally to engage the rods of the mats **6**. The rods **7** may be lengths of reinforcing rod and project to a distance $W > X$ where X is the spacing of a rod of the mat from the tube **4**.

It has been found to be advantageous to have the projecting portions of the rods **7** approximate the diameter **D** of the tube. In practice, $0.5 D \leqslant W \leqslant 1.5 D$ The rods **7** have diameters d'' which are slightly less than the diameters d' of the holes **5a** and **5b**. A similar anchorage of the tube **4** to the upper concrete plate is provided. The sleeve **4** may have a thickness S (Figure 1.1c) which is less than d' and a minor fraction of the diameter **D** of the tube, this diameter corresponding approximately to the thickness T or T' of the concrete plates.

In general, insulating layer **2** will have a thickness R ranging between 0.3T and 1.5T. The interior of the sleeve **4** is provided with a plug **8** of the foamed synthetic resin constituting the layer **2** and punched out when the tube **4** is driven into the insulating layer. To facilitate the punching of this plug, an edge of the tube can be chamfered.

According to a related process described by *E. Haeussler; U.S. Patent 3,996,713; December 14, 1976* a sandwich panel assembly or slab structure has a pair of steel reinforced concrete plates generally coextensive with one another and a cushion of insulating foamed synthetic resin therebetween. A tubular anchor is located at the center (centroid) of at least a portion of the panel and is embedded in the concrete plates.

At least one slightly flexible elongated auxiliary anchor plate or membrane sheet provides additional fixation of the assembly. This membrane is spaced from the tubular anchor and traverses the resin cushion and is embedded in each of the concrete plates. Metal bars extend through and from the membrane and are embedded in the concrete plates.

Polyvinyl Acetate Emulsion-Inorganic Foam

C.C. Sachs; U.S. Patent 3,775,351; November 27, 1973 describes the production of polymer-inorganic hybrid foam construction materials. The process comprises whipping air into a resin emulsion, such as a polyvinyl acetate emulsion, establishing a prefoam or resin emulsion bubbles, adding an inorganic phase, which can be, for example, portland cement or gypsum, in the form of particles so as to suspend such inorganic particles on the foam bubbles without precipitating out the particles, pouring the resulting polymer-inorganic hybrid foam into suitable molds, permitting the foam to set to a solid self-supporting mass, usually taking place in a short period of the order of about 10 to 30 minutes, and hardening the inorganic phase to fully set the hybrid foam.

The product is a hard rigid polymer-inorganic hybrid foam which is load supporting, fire resistant, thermally insulating, and of low density. Such hybrid foam which is strong and lightweight, is particularly useful for production of wall, floor or roof panels in construction of buildings and homes.

The characteristics of typical polymer-inorganic hybrid foam constructions produced according to the process employing cement and gypsum, respectively, are as follows:

Properties	Cement	Gypsum
Density, pcf	43.31	37.32
Compressive strength, psi	997	792
Modulus	18.1×10^5	7.5×10^5
Flexural strength, psi	1,440	1,036.4
Resistance to water	*	**
Chemical resistance	Good	Good
Insulating property (K factor), Btu/in	0.35	0.28
Rockwell Hardness, R	109	90

*Excellent
**Poor

Example: A resin emulsion was provided by adding to 400 parts water, 75 parts of a 45% aqueous emulsion of polyvinyl acetate, 25 parts of a 45% aqueous emulsion of vinyl acetate-acrylic copolymer and 0.2 part triethanolamine. The resulting aqueous resin emulsion in an open vessel was air whipped by means of a Mixmaster type device equipped with standard beaters to aerate the polymer emulsion formulation noted above, in a manner similar to whipping egg whites.

When the emulsion was whipped to about 5 times the volume of the original emulsion, taking about 2 or 3 minutes, 600 parts by weight of portland cement powder was introduced onto the surface of the prefoam using an electric vibrator to advance the powder down a chute and allowing it to free-fall onto the surface of the prefoam while the latter was being vigorously agitated. Addition of the cement powder took place over a period of about 2 minutes.

The resulting hybrid foam mixture having the particles of cement powder suspended therein was poured into a mold or casting cell. In about 10 minutes the consistency of the foam in the mold was sufficiently rigid to be self-supporting, although not hard.

Following such initial set, the article was removed from the mold, and the resulting molded article permitted to set over a period of 28 days at ambient temperature to form the finally cured polymer-inorganic hybrid foam construction material. The cure could be accelerated to about 6 hours by placing the article in a steam bath at 200°F and ambient pressure. Final drying can be accomplished by oven drying at temperatures from 200° to 300°F. The resulting solid foam material had a density of about 37 lb/ft³, was strong and load-supporting.

Cast Reinforced Panels

A process described by *S.W. Shelley; U.S. Patent 3,671,368; June 20, 1972* relates to the casting of reinforced insulated building panels. Such panels generally comprise a layer of any suitable insulating material sandwiched between two external layers of reinforced concrete or other suitable moldable buildings material.

The process involves inserting into the openings of a single concrete reinforcement mat having a generally open wire mesh configuration, an insulating material in such a manner that the insulating material occupies a substantial portion, but not all, of the opening, this unoccupied portion, of the opening referred to as the "free space" of the opening. The insulating material is restrained within the opening so that it will not become dislodged when the mat is moved or placed in an upright or inclined position.

The mat containing the insulating material is disposed within a closed mold which is suitable for gravity pouring or pumping of concrete or other suitable moldable building material into the mold when the mold is disposed at an angle greater than zero degrees above the horizontal position. The mold has a configuration conforming to that desired for the panel. The mat is disposed at the desired position within the mold by the use of chairs, spacing devices or other known techniques. The insulating material can be emplaced within the openings of the open mesh reinforcement mat either prior to or after the mat is disposed within the mold.

Concrete is next placed into the mold while the mold is disposed at an angle which is greater than zero degrees, and typically either about 90°, or 30° to 60° above the horizontal. This angle is measured from the plane of the panel defined by the mold. As the concrete enters the mold, the free space of the openings of the reinforcement mat, which are not occupied by insulating material, provide channels of communication between the concrete flowing downwardly into the mold on opposite sides of the mat, thereby allowing the concrete on each side to seek a common level within the mold. This is desirable because it minimizes the damage which could result to the mat as a result of the pressure differential created within the mold in the case where one side of the mold fills with concrete more rapidly than the other.

Moreover, by providing communication between both external concrete layers, the free spaces of the openings also enable the formation of a multiplicity of bonding contacts between the two layers when the concrete which occupies these free spaces hardens.

An additional feature of the free spaces within the openings of the reinforcement mat is that they provide access for the concrete to the wires which comprise the mat, so that the mat is firmly attached to each layer of concrete when the concrete hardens. This results because at least a portion of the wires or rods defining the free space have their entire periphery exposed to the concrete and, become firmly embedded within the concrete at the interface of the two concrete layers.

Since the panel is prepared in closed molds, disposed at an angle greater than zero degrees above the horizontal position, both external surfaces of the panel exhibit a smooth finish which eliminates the need for additional finishing operations.

The reinforcement mat can take a variety of forms. For example, it can be a sheet of welded wire, mesh or a matrix of metallic rods such as, for example, steel rods.

The insulating material inserted into the openings of the mat can similarly take a variety of forms. For example, it can be a panel which is snugly insertable into the openings of the mat, or it can take the form of one or more continuous strips which are woven through the mesh of the reinforcement mat. Any suitable material such as, for example, polyurethane and polystyrene foams can be employed.

If effective insulating properties are to be imparted to the panel, it is apparent that the portions of the openings of the wire which are not occupied by insulating material cannot be excessively large. The percentage of the openings which can remain unoccupied will vary depending on a number of factors such as, for example, the nature and thickness of the concrete material. In general, the insulating material should occupy at least 65%, and preferably 65 to 95% of the area of the openings of the reinforcement mat.

The resulting panel is insulated, reinforced by a single mat, and has smooth external surfaces which require no additional finishing as a result of pouring the moldable building material into an inclined mold. Moreover, the two outer layers of building material, between which the insulated reinforcement is sandwiched, are bonded together and to the reinforcement mat by means of the building material within the unoccupied portions of the mesh openings.

Facade Composite

A process described by *E. Blum; U.S. Patent 3,750,355; August 7, 1973; assigned to Blum-Bau KG, Germany* provides a facade-forming element comprising an outer facade or facing slab, an inner slab preferably having an inwardly facing surface designed to form an interior wall and spaced from but generally parallel to the facade slab, at least one single-shear truss-like reinforcement lattice embedded in each of the slabs and having diagonal struts extending across the space between them, a layer of gravel-containing concrete, preferably cast in place, in the space between the slabs and embedding the struts, and an insulating layer, preferably of a foamed synthetic resin, elastomer or rubber, interposed between the outer slab and the gravel-concrete core.

The key elements of the facade-forming structure are thus an inner shell or slab, a core bonded to the inner shell or slab and an outer shell or slab bonded to this core. The core comprises a filler of the gravel concrete, preferably cast in place, the struts embedded in this gravel concrete and spanning the slabs, and a layer between the gravel concrete and the outer slab of a thermal and acoustical insulation. This insulating layer between the core and the outer slab consists preferably of a polyurethane foam although similarly moisture-resistant foamed elastomeric insulating materials may be used.

This sandwich construction provides not only the considerable structural strength desired in modern building structures but also affords a degree of insulation of the wall commonly associated with more massive structures in terms of both thermal and acoustical transmission. For example, an insulating layer having a thickness of 3 cm of polyurethane foam has the heat conductivity apart from diffusion processes, of 0.023 or approximately the same as a pumice-block masonry wall of 30 cm thickness.

Referring to Figure 1.2a, it will be apparent that each structural element comprises an inner shell or slab 1 or 1' of concrete or other mineral matter, an outer slab or shell of concrete or other hardenable mineral substance as represented at 3 and 3' and a common gravel-concrete core generally designated at 2 and bridging all of the structural elements of a particular facade or wall. Bridging the slabs of the structural elements **100** and **100'**, respectively, are reinforcing steel trusses **4** and **4'** which are described in greater detail for the truss **4**. Each truss **4** is generally planar, i.e., is a single-shear reinforcing lattice whose plane, as illustrated, is the plane of the paper in Figure 1.2a.

A plurality of such trusses can be seen in Figure 1.2d in which they lie parallel to one another in respective planes perpendicular to the plane of the paper. Each of the trusses also comprises an inner chord or reinforcing bar **4a**, preferably extending the full length or width of the facade-forming element **100** and parallel to an outer chord or reinforcing bar **4b** embedded in the outer slab **3** and likewise extending the full length or width thereof.

Spanning the coplanar bars **4a** and **4b** are a multiplicity of struts **11** which are inclined alternately in opposite directions and are welded, tied or otherwise anchored to the bars **4a** and **4b**. The struts **11** likewise are substantially co-planar with bars **4a** and **4b**. The bar **4a** is embedded in the inner slab **1** as are the junctions of the struts.

The inner slab or shell **1** is composed of a lightweight concrete, i.e., a concrete containing expanded slag or other expanded mineral as an aggregate or filler, and has its inner surface **5** finished to the desired esthetic state.

FIGURE 1.2: FACADE COMPOSITE PANEL ELEMENT

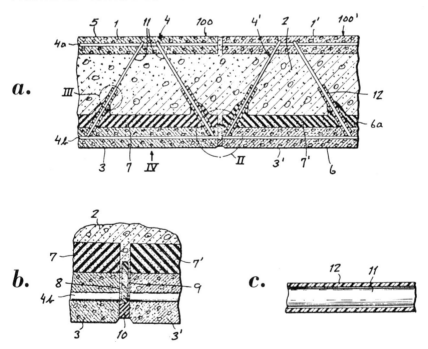

(continued)

FIGURE 1.2: (continued)

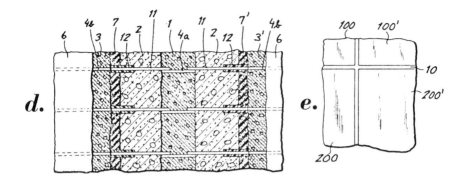

(a)   Transverse cross-sectional view through a pair of facade-
      forming elements at a junction
(b)   Detail view of the region II of Figure 1.2a
(c)   Detail view of the region III of Figure 1.2a
(d)   Elevational view taken in the direction of arrow IV of
      Figure 1.2a
(e)   Elevational view of a portion of a facade formed by the
      elements of Figure 1.2a

Source: U.S. Patent 3,750,355

If the surface 5, for example, is to have a textured appearance, this appearance
may be created by troweling the wet concrete during casting of the slab 1, by
sandblasting or shockblasting, by carving of the embedding texturing materials
in the slab during casting. Where a smooth appearance is desired, the surface 5
may be troweled smooth, provided with a plastic or other facing layer, polished
or ground.

The outer slab or shell 3 is likewise constituted as a finished structure so that
further treatment at the erection site is not required. In this case, the outer
slab 3 is composed of a gravel concrete which may be sandblasted or shock-
blasted along its outer surface 6 to provide a roughened texture. When a polished mineral appearance is desired, this surface can be ground by conventional
processes after hardening of the concrete.

Between the core 2 and the outer slab 3 or 3', there is provided an insulating
layer 7 or 7' which preferably consists of foam polyurethane and which is applied by casting, spraying or molding to the inner surface 6a of each outer slab
3 or 3' after the casting thereof. Prior to the casting of the outer slab, however, each strut 11 is surrounded by an insulating sheath 12 preferably of a
thermoplastic material, so that the sheath is partly embedded in the slab 3 or
3' (Figure 1.2a) and extends substantially midway into the free space remaining
for the casting of the concrete core 2.

The slabs. 1, 1' and 3, 3' with the interconnecting trusses 4, 4' are cast at a.pre-
fabrication site, provided with the polyurethane layers 7, 7' and given the desired
surface treatments at 5 or 6. They are then delivered to the construction site

and erected in, for example, an orthogonal array as shown in Figure 1.2e and the structural elements **100, 100'** described in connection with Figure 1.2a and the identical elements **200** and **200'**. Between each pair of outer slabs **3**, there is inserted a sealing band of foam rubber as represented at **9** for preventing the cemented liquor from passing on to the finished surfaces of the outer slabs.

It has also been found to be advantageous to provide between the mutually confronting edges **8** of the adjacent slabs, from the exterior in, a crack sealer **10** of a flexible thiokol in a thickness of about 1 cm, preferably by a caulking gun. The core **2** can then be cast from gravel concrete. The resulting structure has been found to be creep and deformation resistant to weathering and moisture diffusing, and can serve as a fire barrier. German Industrial Standards (DIN) 4102, 4108 and 4109 are met.

In Situ Foamed Core

A process described by *M.R. Piazza; U.S. Patent 3,984,957; October 12, 1976; assigned to Maso-Therm Corporation* provides a monolithic-like building module which is extremely light in weight as compared to precast concrete panels, for example, and which has greatly improved insulating and vapor barrier properties per se. Because the process utilizes an in situ foamed core, an adhesive interlock between core and shell is formed which is stronger than either material by itself.

The chemical foaming reaction that takes place, plus the fact that foaming takes place in an enclosed shell under retention, results in an overall intimate adhesive interlock and a prestressed structure wherein the shell is under tension and the core is under compression. This means that the shell and core are now united together into a monolithic-like structure that has far greater strengths (because of the overall adhesive interlock) than prior laminated panels using preformed foam plastic cores, and, at the same time, is light in weight and has excellent insulating and vapor barrier properties.

The composite building module comprises a rigid foam core (preferably urethane polymer foam) encased in an enclosed shell having a bottom half and a top half, each made of fiber reinforced cement, preferably glass fiber reinforced cement containing a filler such as sand. The bottom half of the shell has a peripheral ledge and inwardly adjacent thereto a rib member extending above the level of the ledge. The top half of the shell has side walls forming a channel with the rib when the top half of the shell is in place on the bottom half. The foam core is foamed within the enclosed shell filling the interior thereof including the channel formed by the two shell halves.

The process provides a surprisingly strong and even load supporting building module which is light in weight and has outstanding insulating and vapor barrier properties. The process is described with reference to Figure 1.3 and the preferred use of a rigid urethane foam polymer core and a cement/glass fiber shell.

Figure 1.3a shows a preferred building module in cross section having a rigid urethane polymer foam core **14** encased in a shell having a bottom half **12** and upper or top half **12'**, each made of a mixture of cement and glass fibers. The finished panel or module is indicated generally by the reference numeral **10**. The bottom half **12** has a peripheral ledge **38** and an inwardly adjacent rib **34**.

FIGURE 1.3: COMPOSITE BUILDING MODULE

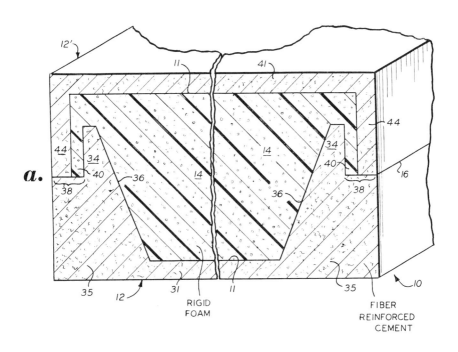

a.

RIGID
FOAM

FIBER
REINFORCED
CEMENT

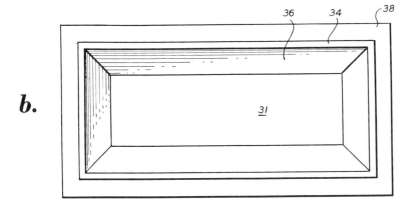

b.

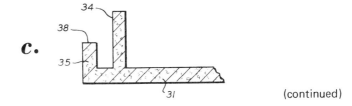

c.

(continued)

FIGURE 1.3: (continued)

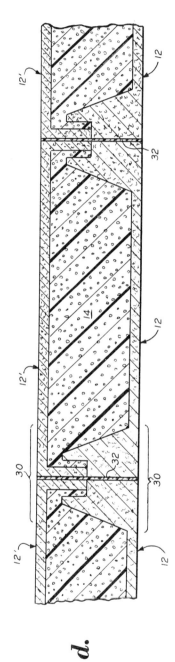

d.

(a) Cross-sectional view of a composite building module
(b) Top plan view of the bottom half of the shell **12** as shown in Figure 1.3a
(c) Fragmented cross-sectional view showing an alternate form for the bottom shell **12**
(d) Cross-sectional view of a series of composite building modules assembled edge-to-edge and forming a load bearing structure

Source: U.S. Patent 3,984,957

As shown in Figures 1.3a and 1.3c, the ledge **38** and the rib **34** are integrally formed and together make up the side wall of the bottom half **12** having the stepped configuration shown and indicated generally by the reference numeral **35**.

The top half of the shell **12'** has side walls **44** which are less thick than the width of the ledge **38**. This means that when the top shell **12'** is placed on the bottom shell half **12** the side walls **44** of the top half **12'** rest on the ledge **38** creating an edge seam **16** and, within the interior of the shell the side walls **44** form with the rib **34** a channel **40**. In the embodiment shown in Figure 1.3a, the channel **40** is actually formed by rib **34**, ledge **38** and side wall **44** of the top shell half **12'**. As shown in Figure 1.3c, the rib **34** can extend up from the bottom **31** of the shell half **12** which means that the corresponding channel will be larger in size.

The configuration shown in Figure 1.3a is preferred because the stepped configuration **35** provides additional bulk at the edge of the panel and when two panels are assembled and joined in an edge-to-edge assembly, a load-bearing column results.

The rib **34** preferably has a sloped wall **36** as shown in Figure 1.3a to facilitate mold removal and again to provide additional bulk at the edges of the panel. The rib **34** preferably is a continuous rib extending around the entire periphery of the interior of the panel shell but it can be discontinuous or interrupted depending on the intended use of the panel.

The rigid urethane foam core **14** is formed within the enclosed shell formed by shell halves **12** and **12'** and fills the interior of the panel shell including the channel **40** formed by the side walls **44** of the top shell half **12'** and the rib member **34** of the bottom shell half **12**.

The encapsulation of the rigid foam core **14** by the shell halves **12** and **12'** results in the formation of intimate adhesive bond **11** between the core **14** and the shell halves **12** and **12'** over the entire surface area of the core **14**. Because the rigid foam core **14** is formed in situ, the foaming urethane polymer enters and fills surface irregularities and the channel **40** to provide an intimate overall rigid interfacial adhesive interlock between the rigid foam core **14** and the shell halves **12** and **12'**. The channel **40** provides additional surface area within the interior of the panel shell and in effect increases the interfacial adhesive interlock in the important area of the edge seam **16** where the two shell halves **12** and **12'** come together.

Figure 1.3d shows finished modules joined in an edge-to-edge assembly whereby the joined edges form a load-bearing column **30**. The finished modules can be conveniently joined using an elastomeric material which forms a joint **32** between adjacent panels. Depending on the intended use of the modules, the side walls **44** and the flat portion **41** of the shell half **12** and the flat portion **31** of the shell half **12** can have thicknesses in the range of from about ⅛ inch to about 1 inch or more. If desired, this thickness can be greater or less. For curtain-wall construction, thicknesses in the range of from ¼ inch to ⅜ inch are preferable.

In a preferred case, the side walls **44** and the flat portion **41** of the shell half **12'** are ⅜ inch thick. The same thickness is employed for the flat portion **31** of the bottom shell half **12**. The width of the ledge **38** is ⅝ inch and the rib **34** extends 1 inch above the ledge **38**. The top of the rib **34** is ¼ inch thick and the channel **40** is ¼ inch wide. This means that the stepped portion **35** of the bottom shell half **12** is approximately ⅞ of an inch from the edge of the shell. When two panels are joined in edge-to-edge assembly, as shown at **30** in Figure 1.3d, this means that there are approximately 2 inches of glass reinforced cement extending across the joined edges of two panels and this assembly together with the side walls **44** of the top shell halves **12'** and the foam in place core **14** forms a load-bearing column.

The rigid foam core **14** can range in thickness from about 1 inch to 10 inches or more and this can be greater or less depending on the structure involved and the intended use. The building modules themselves can be made in almost any size ranging from small modular units up to large curtain wall units or roof deck members.

A preferred method for making the composite building modules will now be described. At a first station, metal or glass fiber/polyester molds, preferably with fold down ends to facilitate product removal, in the form of shell tops **12'** and bottoms **12** have applied thereto a mixture of cement and glass fiber preferably containing 35 to 40% by volume glass fiber. The mixture of cement and glass fiber can be premixed dry and water subsequently added to provide a viscous mixture. This mixture can then be sprayed into the mold interiors or applied by hand.

In the preferred case, hot wet cement (made with water at about 120° to 200°F, e.g., 180°F) with glass fiber is applied to or sprayed into the interior of the molds. The molds can then be vibrated to obtain uniform distribution of the glass throughout the entire volume of the cement and complete filling of the molds. Following this, or at the same time, the mixture of glass fiber and cement in the molds can be pressed with a forming member to distribute the cement/glass fiber mixture within the interior of the molds. Also, if desired, suction can be applied to the mold walls to remove water.

At the same time the glass is chopped and sprayed, a coating can be applied by spraying, for example, with a polyester in a water-miscible solvent such as alcohol, to impart alkali resistance to the fibers. The molds are then fed to a curing line. If hot cement is used, oven curing can be eliminated. Oven curing generally requires about 6 hours to produce a hardened shell for the glass fiber reinforced cement tops and bottoms, curing time for hot cement is generally 50% less.

Next, flowable, foamable rigid urethane polymer composition is poured into the open shell half **12** and the top half **12'** is then set in place before substantial foaming begins. The assembly is then held under pressure while the urethane polymer foams and sets. This can be accomplished using a hydraulic press or other restraint device. After the urethane polymer foam sets, filling the shell and providing an overall rigid interfacial adhesive interlock between the rigid foam core and the interior of the shell, the panel is removed from the restraint device.

The composite panel is now ready for use or can have a surface finish applied. Preferably, the surfaces of the panel have a sealer, such as a polyester type of sealer. While the panel itself is substantially waterproof, the application of a sealer insures that the panel will maintain its waterproofness. If desired, in addition to or in place of a sealer, the panel can be painted, stained or other types of coatings can be applied, for example, to provide for easy removal of graffiti. It has also been found that a sealer, when applied in a thick coating, can also be used to adhere aggregate to a surface of the panel to provide a surface finish.

The composite building module can be used and installed in the same manner as conventional building modules such as curtain-wall panels but with a great reduction in weight (and simplified installation procedures). Because of the greatly improved insulating and water vapor barrier properties of the modules, no further steps have to be taken to ensure these properties as is the case with conventional building modules.

Concrete-Polyurethane Foam Adhesion

R. Labrecque; U.S. Patent 4,010,232; March 1, 1977 describes a method of making a construction panel including a layer of urethane which is successfully adhered to a layer made of a cement mixture. The method of making the construction panel includes forming a dry cement mixture by mixing cement, silica and an aggregate with preferably a coloring agent to obtain a concrete of the desired color. Preferably, the cement is a white cement, the silica is of 70 mesh, and the aggregate is expanded mica, such as the heat expanded mica known as Zonolite.

Some of the dry cement mixture is mixed with water to form a wet cement mixture which is spread into a layer in a mold. Some dry cement is thereafter powdered onto the wet cement mixture until a dry blanket is formed onto the latter. As soon as such dry blanket has been obtained, an unset foaming urethane fomulation in liquid form is poured onto the dry cement blanket and the mold is closed by a cover. The setting operation is started at room temperature that is some 70°F, no heating being required.

When the foaming urethane formulation has set into a foam urethane layer, the construction panel is removed from the mold to allow the cement mixture to complete its setting into a concrete layer. The urethane formulation upon reacting in the mold, expands and develops heat and pressure whereby the wet cement mixture is heated and the resulting urethane layer is pressed against the cement layer. Water in the wet cement layer migrates into and wets the dry blanket whereby the latter turns into concrete. This water migration takes place in the closed mold where it is helped by the pressure and heat conditions therein. A firm bond is thus produced between the concrete layer and the foam urethane layer. The mold may be vibrated to pack the wet cement mixture into a compact layer before setting.

Example: The following dry cement mixture was prepared: 1 lb of white cement, about 24 wt %; 3 lb silica 70 mesh, about 73 wt %; 1 ounce of coloring agent, about 1.5 wt %; and 1 ounce of Zonolite aggregate, about 1.5 wt %.

The following wet cement mixture was prepared by adding 40 ounces of water to the above dry cement mixture giving then by weight about 38% water, 15% white cement, 45% silica, 1% coloring agent, and 1% Zonolite aggregate. The wet cement mixture was spread in the bottom of the mold, then dry cement mixture was sprinkled over the wet cement mixture just enough to form a dry blanket. Then a liquid urethane formulation was poured over the dry blanket and the mold was closed. During the setting in the mold, the temperature inside the mold reached 180°F. The setting of the foaming urethane composition took only 30 seconds after which the mold was opened to remove the panel therefrom. The foaming urethane formulation was the result of the isocyanate with a dark liquid containing thorough mixing of a clear liquid containing polyol, a suitable catalyst and a blowing agent in liquid form all in accordance with the conventional one-shot technique.

It is very important to ensure that the element blanket between the wet cement mixture and the foaming plastic composition prevent the ingress of water or moisture into the latter during the foaming action to allow the latter to take place. The construction panel thus formed includes a layer of concrete bound to a layer of foamed urethane.

The layer of concrete may for instance be ⅛ inch thick while the layer of foamed urethane may be ¾ inch thick with at least one edge of the latter extending laterally from the corresponding edges of the harder layer of concrete to allow nailing through that one edge. The construction panel may be used to produce floor surfaces, ceilings, as well as exterior and interior wall finishing panels or tiles.

METAL COMPOSITES

Continuous Elongated Laminar Strips

R. Klinkosch; U.S. Patent 3,785,917; January 15, 1974; assigned to Hoesch AG, Germany describes a process which permits the continuous production of laminar panels or structural units provided with sound-retarding, fire-retarding or analogous intermediate layers.

The use of conventional spacing elements in the manufacture of such laminar structural units is avoided and the manufacture of the unit is greatly simplified not only by this, but also by the generally simpler and more economical operation. The method and apparatus permit a continuous automated mass-production wherein the expenses involved in the material and in the manufacture of the separate spacing elements for the webs are eliminated, and the time-consuming and expensive handling and transportation of individual precut web panels is avoided. The danger of damage to special coatings for the webs is avoided or largely eliminated and there exists no need to readjust the entire apparatus whenever a change in the length of the finished individual structural units is necessary.

Referring to Figure 1.4a, it will be seen that supply means for a plurality of elongated webs is identified with reference numeral 1. This supply means comprises a pay-off 2 for an upper web 20 and a pay-off 3 for a lower web 30. A further pay-off 4 is provided for an intermediate web 40. Each of the pay-offs 2, 3 and 4 supports a coil of the respective web 20, 30 or 40 so that the webs can be withdrawn simply by exerting tension on them in a direction from the left towards the right-hand side of Figure 1.4a.

FIGURE 1.4: LAMINAR STRUCTURAL UNITS

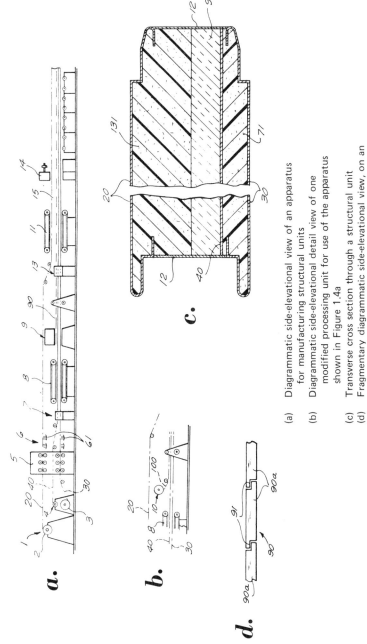

a.

b.

c.

d.

(a)   Diagrammatic side-elevational view of an apparatus for manufacturing structural units

(b)   Diagrammatic side-elevational detail view of one modified processing unit for use of the apparatus shown in Figure 1.4a

(c)   Transverse cross section through a structural unit

(d)   Fragmentary diagrammatic side-elevational view, on an enlarged scale, of one form of an intermediate layer

Source: U.S. Patent 3,785,917

ventional and well known to those skilled in the art, are usually made adjustable
so that the length of the individual structural units to be severed from the assembly **15** can be varied at will.

If desired, the intermediate layer **90 (100)** may be apertured, that is, it may be
provided with holes, slits, slots or the like, so that by the foam of the second
layer **131** the material of the layers **71** and **90** (or **100**) may directly adhere to
one another through the existing apertures and the web **40**. However, the layer
90 (100) may also be continuous and nonapertured, and suitable means, such
as any one of many well-known adhesives suitable for the purpose, may be provided to adhere the layer **90 (100)** through the web **40** to the layer **71**. In any
case there is a safe connection between the webs and layers.

Sectional Partitioning

A process described by *F.J. Howells; U.S. Patent 3,879,911; April 29, 1975;
assigned to GKN Sankey Limited, England* provides a partioning section comprising two metal side walls and a rigid heat insulating material which is interposed between the side walls so as to hold them apart and out of contact and which is
secured to the side walls to form an integral assembly, the side walls having at
at least one side edge of the section inturned flanges disposed in spaced parallel
planes and each of a width less than one-half of the thickness of the section.

When two sections of this form are fitted together in a partition so that the
sections are disposed in a common plane and their two side edges provided with
the flanges are disposed in abutting relationship, the two flanges at each side of
the partition are in face-to-face contact. Since the flanges of each section are
disposed in spaced parallel planes and each of a width less than one-half of the
thickness of the section, there is a substantial distance between the two pairs
of abutting flanges and also between each pair of flanges and the opposite side
wall, so that there is a good heat insulation at the location where the sections
fit together.

Since there is no metal-to-metal contact between the walls at the respective
sides of the partition, the transfer of heat from a fire at one side to the other
side to an extent sufficient to cause a fire on the other side is prevented, at
least for a period of time whereby the spread of fire is inhibited. Thus, it takes
a considerable time for the metal walls on the other side of the partition to
reach a temperature which is such as to set on fire any inflammable material in
contact with or close to the walls.

Since the side walls of the section are secured to the rigid heat insulating material between them, the section is of a unitary form and does not have to be assembled during installation of the partition. Preferably the flanges at the side
edge of the section are at right angles to the side walls. The insulating material
is preferably arranged so that it extends up to the inner surface of the outer
one of the two offset flanges at the side edge of the section so that a portion of
the material is exposed between the flanges. Thus, the (or each) side edge is of
rabbeted form.

The heat insulating material preferably comprises a rigid corrugated board of a
heat insulating material. Thus, there are provided at opposite sides of the board
channels the mouths of which are closed by the side walls and which may be

filled with air or mineral wool. These channels assist the heat insulation between the metal walls, improve the soundproof qualities of the partitioning, assist in reducing the weight of the partitioning, and also enable electrical wiring to extend through the partitioning.

The metal side walls are preferably secured to the insulating material by a suitable adhesive which is heat resistant at least up to high temperatures and nontoxic, i.e., does not give off fumes when heated. One conventional, commercially available adhesive of this type is Tretobond 848.

The heat-insulating material is preferably a mineral or a mineral-based material. It may be asbestos but is preferably a vermiculite-based material. Thus, the material may comprise exfoliated pretreated vermiculite bonded with an inorganic binder and pressed to produce a noncombustible board of the required corrugated shape. Such a material is available under the name Vicuclad. Both side edges of the section may be provided with inturned flanges, or flanges may be provided at only one side edge, the other side edge being planar. The section will usually be rectangular in shape and of planar form. If desired, the metal side walls can be painted or coated with a plastic material such as polyvinylchloride.

The process is primarily applicable to so-called demountable partitioning. Thus, the upper and lower parts of the sections may be removably received in channel section members secured, for example, to the floor and ceiling in the case of a building or to the deck and the deckhead in the case of a ship.

Vacuum Panels

A.W. Rowe; U.S. Patent 3,936,553; February 3, 1976; assigned to Rorand (Proprietary) Limited, South Africa describes an insulation material comprising a pair of generally parallel impervious surface sheets of material sealed together through thermally insulating material at their free edges and held in spaced relation by a series of transverse pins spaced apart over the area of the surface sheets, the space between the surface sheets being evacuated.

It has been found that by using pins made of a material having a high resistance to compression an effective evacuated panel can be constructed. Thus the pins are preferably made of pure epoxy, reinforced epoxy such as glass fiber filled epoxy, certain nylons and the like. With the correct selection of materials for the pins these may cover, in cross section, an area of only about 1% of the area of the surface sheets and thus they will transmit very little heat through an insulation material made according to this process.

The pins may also be split transversely along their lengths to enable a sheet of reflective material to be supported between the surface sheets of the panel. Such a sheet of reflective material would serve to prevent radiative heat transfer and would thus be of a suitable material for this purpose. Alternatively, the pins may extend through the intermediate sheet of reflective material with the latter being held in position by sleeves or washers located over the pins and supporting the intermediate sheet in its position parallel to the surface sheets.

The sheets may be preshaped to provide rigidity to a panel and so that the effect of evacuation will not deform the surfaces of the panel into concavities between the support pins. The sheets are of a concave shape between the internal support pins with spaced indentations. The surface sheets are sealed together around their free edges in any convenient way which allows for a vacuum to be created between them. Thus syntactic foam, being a mixture of epoxy or other synthetic plastics material and microscopic hollow glass or ceramic spheres is considered suitable. Alternatively, metal webs may be used wherein conduction is minimized by selection of the thickness and material of the webs. Preferably gettering material will be placed between the panels to maintain the vacuum applied when the panel is manufactured.

Coupling Technique

A process described by *P. Bellagamba; U.S. Patent 4,028,859; June 14, 1977* relates to heat-insulating panels, consisting of two thin metal sheets, being more or less ribbed, between which there is placed a suitable synthetic resin. The heat-insulating panels are generally used as walls of buildings.

The heat-insulating panel is characterized by the fact that the two covering sheets are in positions corresponding to each other, and at a regular distance reliefs in undercut project on a face where there are disposed coupling organs consisting of material having a low thermal conductivity; on each projecting relief there are movably fastened fastening means to the support structures.

Furthermore, the end portions of the two sheets are folded over and so shaped as to build a concave and a convex edge, which may be coupled by embedding between them and providing on their inner surface gaskets having a series of longitudinal lamellae placed in intercalated positions. The reliefs thus build between the two panels, when embedded therebetween, a labyrinth chamber. The coupling organs are preferably made from plastic material and formed by a V-shaped and a C-shaped piece, connected to each other by a spacer member. Of the pieces, the V-shaped piece is inserted by embedding it into the cavity delimited by the relief projecting; the C-shaped piece is fitted over the recessed relief.

The fastening means of the heat-insulating panel to the support structures consists of a small block provided with an undercut cavity, which may be prismatically coupled with the reliefs of the panel. The block has two vertical slides and thereto there is fixedly secured a small bar embedded into the vertical slides which serves as an anchoring member to the support structure.

Adjustable Height Panel

A process described by *W.T. Nungesser; U.S. Patent 3,986,315; October 19, 1976; assigned to Diamond Power Specialty Corporation* relates generally to reflective insulation assemblies and particularly to selectively extendable panels of such reflective insulation.

The panel assembly includes an outer panel having a number of spaced reflective insulation sheets and a pocket formed along one side to slidably accommodate an inner panel. The inner panel may be extended in and out of the pocket to any position and captured in that extended position by a latching assembly provided on the outside face of the pocket.

This latching assembly may be formed as an L-shaped clip which has one leg extending along the face of the pocket and the other leg extending through a wall of the pocket to pierce the wall of the inner panel and prevent any movement of the inner panel within the pocket. To prevent the removal of the L-shaped clip the leg extending along the pocket is retained by spring clips fastened to the pocket wall by spotwelding. The L-shaped clip may be removed by rotating the clip out of the spring clips and pulling the second leg out of engagement with the inner panel.

As may be seen with reference to Figures 1.5a through 1.5c the panel assembly 10 includes an outer panel assembly 12 forming a pocket 14 within which an inner panel 16 is telescopically moved to numerous positions. The outer panel assembly 12 also provides an enclosed area 18 which is lined with individual sheets 20 of spaced reflective insulation. The spacing between sheets 20 may be maintained in numerous known ways such as spacing clips or offsets formed in the sheets themselves. The inner panel 16 may be similarly lined with sheets of spaced reflective insulation.

FIGURE 1.5: ADJUSTABLE HEIGHT INSULATION PANEL

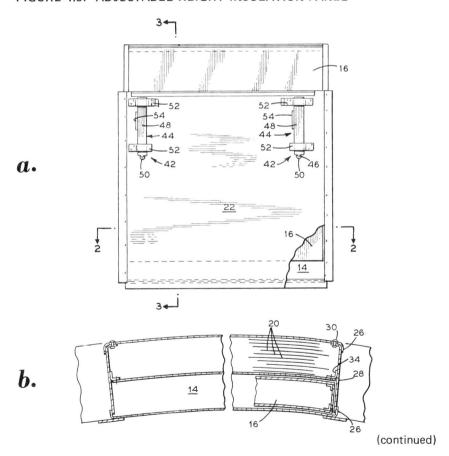

(continued)

FIGURE 1.5: (continued)

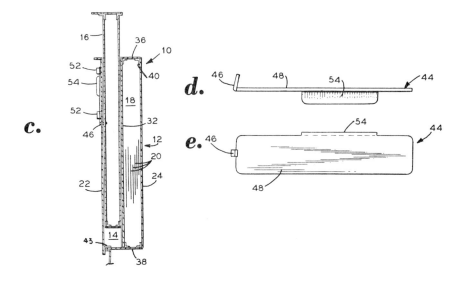

(a)   Panel assembly
(b)   Enlarged sectional plan view of the panel assembly taken along
      line **2–2** of Figure 1.5a
(c)   Sectional side view of the panel assembly taken along line
      3–3 of Figure 1.5a
(d) (e)   Side and top views of the L-shaped clip retainers for retaining
          the inner panel of the panel assembly of Figure 1.5a

Source: U.S. Patent 3,986,315

The outer panel assembly **12** is formed from heavy gauge material such as stain-less steel sheets approximately 0.025 inch thick to provide a rigid structure which is still lightweight. The outer panel assembly **12** is also assembled to in-sure that the pocket **14** is free from any obstructions such as pop-rivet heads which may jam the inner panel **16**. To assemble the outer panel assembly **12** an inner case **22** and an outer case **24** are formed to have lips **26** inside of which a pair of L-shaped side members **28** are located.

The short leg of members **28** is pop-riveted to the outer case **24** by pop-rivets **30**. The ends of the long leg of the members **28** are spotwelded to the lips **26** of the inner case **22**. To form the pocket **14** a central partition **32** is spotwelded to bracket clips **34** which extend into the enclosure **18** and may be either pop-riveted or spotwelded to the members **28**. The top of enclosure **18** is sealed by a top closure panel **36** which may be spotwelded to brackets **40** located in the enclosure **18**.

The bottom of the panel assembly **10** is sealed by a bottom closure panel **38** which may be formed as a flanged extension of the outer case **24** spotwelded to a flange **43** of the inner case **22** or which could also be formed as a flanged extension of the inner case **22**. Clearly the bottom closure panel **38** is easily formed as a separate sheet and spotwelded to flanges of both the inner and outer cases **22** and **24**. The foregoing assembly thus insures that the pocket **14** is free of any burrs or protrusions, such as pop-rivets, which may jam the inner panel **16** to move smoothly within the pocket **14**. This type of panel assembly **10** is especially useful as a skirt for a nuclear reactor wherein the inner panel may be slidably adjusted to fit snug against the bottom head of the reactor.

To maintain the extended position of the inner panel **16** a pair of identical clip assemblies **42** are located on the inner panel **22** at opposite sides of the pocket **14** opening. Each clip assembly **42** has an L-shaped member **44** having a short end **46** extending at substantially a right angle from a flat face **48**. Holes **50** are formed in the inner case **22** through which the inner case **16** may be drilled or otherwise punctured when the desired extension of the inner case **16** from the pocket **14** is achieved.

The short end **46** of the clip **44** is then extended through the hole **50** to further extend into the holes formed in the inner case **16**. The clip **44** is then rotated to engage the flat face **48** of the clip **44** within a pair of spring clips **52** spotwelded at one end to the inner case **22**. To ease the rotation of the clips **44** into the spring clips **52** a tab **54** is formed at substantially right angles to the flat face **48** of the clip **44**.

Wall Unit with Foil Reflector and Cavity

O.A. Becker; U.S. Patents 3,990,202; November 9, 1976; and 3,803,784; April 16, 1974 describes a wall unit comprising opposite sheet metal panels, seals arranged between the margins of the panels and insulation accommodated in the cavity between the panels. There is provided at least one insulating group extending parallel to the panels which is composed of two high gloss foils capable of reflecting thermal radiation and chambers arranged between the foils. The walls of the chambers consist of insulating material. Thus a highly effective insulation is attained. Heat absorbed, for example, by the outer panel is imparted to the foil on the inside.

From this high gloss foil only a few thermal rays are radiated, which penetrate through the chambers, impinge the opposite foil, and are almost completely reflected by the latter, so that this second foil is heated to a very low extent only. The walls of the chambers are made of insulating material, e.g., in the form of honeycombs, the webs of which are thin, so that heat conduction through these webs is low. The chambers are small and tightly sealed, so that likewise hardly any heat transfer takes place by convection through the air.

Further improvement of the insulation is attained by evacuating the air from the cavity between the sheet metal panels. For this purpose the individual wall elements may be provided with valves, or all the wall elements may be connected to a vacuum pump by a pipe-line. Evacuation prevents any heat transfer by convection. Moreover, the formation of condensate is prevented which condensate might reduce the reflectivity of the high gloss reflector foils.

In Figure 1.6 sections **1** of Z-shaped profile are attached by means of screws **1a** to an end face of the ceiling **2** of a building. The upper section **1** carries an outwardly directed flange on an upper wall unit **3** and the lower section **1** carries the upper edge of a lower wall unit **3**.

FIGURE 1.6: INSULATING WALL UNIT

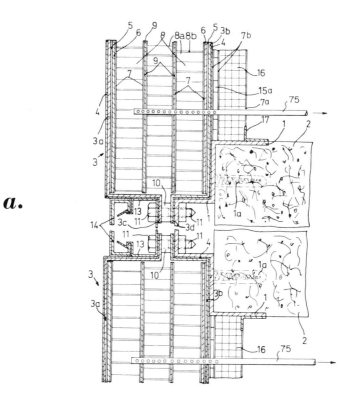

a.

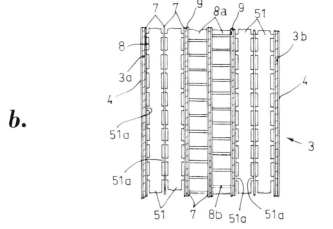

b.

(continued)

FIGURE 1.6: (continued)

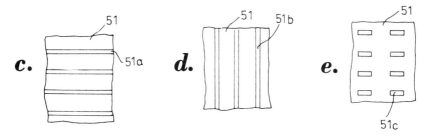

c.   d.   e.

    (a)   Vertical section of parts of two mutually abutting wall units
             constituting an external wall fixed to the ceiling of a
             building
    (b)   Vertical section of a wall unit with insulation by plates and
             by honeycombs
    (c)   Insulating plate having horizontal ribs
    (d)   Plate having vertical ribs
    (e)   Plate having interrupted ribs

Source: U.S. Patent 3,990,202

Each wall unit comprises an outer sheet metal panel **3a** and an inner sheet metal panel **3b**. Between the margins **3c, 3d** of the panels an insulating and sealing strip **10**, preferably of ebonite, is inserted. Screws **11** penetrate the margins, the sealing strips **10** and the sections **1**. They press the sheet metal panels in an airtight manner against each other, and at the same time support the wall units on the building.

In the cavity between the sheet metal panels an insulation is accommodated which comprises at least one highly effective insulating group extending parallel to the panels. This insulating group comprises a thin high gloss foil, e.g., of aluminum capable of reflecting thermal radiation, small cells **8** of about 8 mm width, the walls of which consist of insulating material, and a second reflecting foil, these foils sealing off the chambers.

These chambers are formed by honeycomb plates **8a**, the webs **8b** of which extend between the two opposite foils. Between the foils of the three insulating groups illustrated there are moreover inserted insulating plates **9**, e.g., of felt-board or solidified synthetic foam, for reduction of thermal conduction and for stiffening.

For improved appearance and for protection from corrosion each of the steel metal panels is provided outside with a coating **4** of lacquer or synthetic material, and inside with a foil **5** of a sound-deadening material and a stiffening sheet metal panel **6**.

Upon solar radiation of the outer sheet metal panel also the adjoining first foil is warmed. However, the high gloss foil radiates but little heat inwards, namely about one-fifth of the radiation emitted by a black body. The heat radiation emanating from the foil penetrates the chambers, impinges the opposite foil and is almost completely (93%) reflected by the high gloss surface of the latter and is absorbed only to a small proportion (7%), so that the opposite foil is warmed up but little and can discharge little heat (1.25%) only.

Heat transfer by radiation is accordingly very small. Moreover, heat conduction takes place through the edges of the webs of the honeycomb contacting the foils. However, since the cross-sectional area of the webs contacting the foils amounts to about 4% only of the whole area of a foil, and the webs consist of insulating material, the heat conduction is also very low. Finally heat could be transferred by convection through the enclosed air. However, since a great many and consequently small cells are formed by the honeycomb plates, an air flow and consequent heat transfer can hardly take place.

When several insulating groups are arranged in series, their effect is greatly increased. In a test with five insulating groups, the outer sheet metal panel was raised to a temperature of 100°C and kept at that temperature, while the inner sheet metal panel was exposed to a room temperature of 20°C. After four hours a steady condition was attained in which the temperature of the inner sheet metal panel had increased from 20° to 30°C. At an external temperature of 70°C the temperature of the inner sheet metal panel rose by 7°C and at an external temperature of 45° by 3°C only.

The walls of the honeycomb cells may be provided with highly reflective very thin metal surface layers (e.g., by deposition of aluminum from the vapor phase in vacuo). For reasons of fire protection the honeycombs may be made fireproof by impregnation or the like. Likewise mats of glass fibers may be arranged as an insulation on the internal surfaces of the wall panels as well as, for example, between two aluminum foils whereby the wall units, in conjunction with the reflecting metal foils and sheet metal panels (thermal reflection) and the impregnated and hardened honeycombs, are made extraordinarily fireproof.

The gaps between any two adjacent wall units may be outwardly closed by angle sections **13** held by screws **11** and by screens **14**, which engage behind resilient tongues in the angle sections. For further thermal and acoustic insulations from the interior space the inner sheet metal panels may each carry a reflecting foil **7** and spacer strips **15**. On the latter a plate **16** of plaster of Paris and covered on both surfaces with reflecting foils **7** is placed, which is attached to the buildings by angle sections **17** and forms the inner wall surface.

The cavity between the sheet metal panels **6** with foils **7** may be evacuated through perforated pipe **75** which is connected to a vacuum pump, and then sealed. Or the air in the cavity may be replaced by dry air or a dry gas such as nitrogen through the pipe **75** and then also sealed. Or perforated containers with a drying agent such as calcium chloride, silica gel, or the like may be placed into the cavity which is then sealed, so as to dry the air therein. Or dry air or a dry gas may be circulated through the cavities of superposed wall units. Or the cavity filled with dry air may be vented to the atmosphere.

Any means to provide dry air or a dry gas atmosphere in the cavity of the wall units may be employed provided they maintain a dry air or gas atmosphere under subatmospheric, atmospheric, or superatmospheric pressure in the cavity and thus prevent deposition of moisture on the walls of the wall unit and especially on the reflecting foils.

A second example of the process is illustrated in Figures 1.6b through 1.6e. The wall unit 3 comprises an outer sheet metal panel 3a and an inner sheet metal panel 3b. Each of them has a coating 4. The inserted insulation comprises six insulating groups. The two middle groups are made of honeycomb plates 8a as in the first embodiment. The outer groups are formed by insulating plates 51, e.g., of hardened synthetic foam, which have narrow ribs 51a about 5 mm wide on both faces and form strip-shaped cells 8. In Figure 1.6c an insulating plate is illustrated in elevation having horizontal ribs 51a, in Figure 1.6d a plate with vertical ribs 51b and in Figure 1.6e one with rows of projections 51c.

The first insulating plate is followed by a second reflecting foil 7 and by a second insulating plate, the ribs of which run at right angles to the first plate; thus thermal conduction can occur at the crossing points only of the ribs. A third reflecting foil 7 seals these cells off. Between the insulating groups sheet metal panels 9 may be arranged to prevent buckling of the thin reflecting foil under major pressures, e.g., upon evacuation, caused by the webs of the honeycomb plates or ribs of the insulating plates being juxtapositioned to each other on points only. When panels 9 themselves are highly reflective, the foils may be omitted. Units 9 may also be provided with dry air or a dry gas atmosphere or may be evacuated as described above.

Clapboard

E.R. Lewis; U.S. Patent 3,998,021; December 21, 1976 describes an insulated siding panel assembly comprising an elongated outer panel formed of deformable sheet material and having a body section providing inside and outside faces and a flange extending along at least one longitudinal edge. The siding panel assembly includes an elongated synthetic resin backing member of greater thickness than the sheet material of the outer panel disposed against the inner face. The backing member has at least one rib spaced intermediate the longitudinal edge thereof and projecting away from the inner face of the outer panel, the backing member providing resistance to deformation of the outer panel.

In another example of the siding panel assembly the outer panel and the backing member each have a transverse configuration providing two body sections extending in generally parallel planes joined by a web extending at an angle to the planes of the body sections in the same direction as the flange to space the planes of the body sections. Each of the backing member body sections has a longitudinally extending rib, and the siding panel assembly has a double clapboard configuration with each section of the siding panel supported for resistance to deformation.

In the preferred case the flange extends inwardly of the inner face of the outer panel and has a lip extending in spaced relationship to the inner face to provide a U-shaped channel along one longitudinal edge of the outer panel. The outer panel has a second flange extending adjacent the other longitudinal edge and outwardly of the outer face. The synthetic resin backing member preferably has a closed cellular structure and is of substantially uniform thickness. The rib is defined by a corrugation providing a recess in the opposite surface of the backing member.

A plurality of siding panel assemblies of either embodiment may be interengaged to provide an insulated siding assembly having a clapboard configuration. Each outer panel has flanges extending along both longitudinal edges thereof, flanges of adjacent panels being interengaged to provide the siding assembly. Preferably one flange on each outer panel extends inwardly of the inner face of the outer panel and has a lip as defined above to provide a U-shaped channel, the other flange extending outwardly of the outer face thereof. The U-shaped channel on one of the outer panels seats the other flange of the adjacent outer panel to secure the siding panel assemblies in a clapboard configuration.

Siding panel assemblies may be mounted to a plurality of spaced stringers extending transversely thereto. Each stringer has mounting means longitudinally spaced thereon which cooperate with the U-shaped channels and other flanges of outer panels to secure siding panel assemblies to the stringers.

The distance from the plane of the backing member face adjacent the outer panel to the opposite face of the rib at its maximum spacing exceeds the width of the flanges. The ribs abut a vertical support surface when the siding panel assemblies are mounted to provide resistance to deformation of the outer panels.

The outer panels may be of any conventional material including aluminum, steel and polyvinyl chloride. The insulating backing member is formed of any suitable synthetic resin of good insulating properties. To provide lightweight, highly effective insulation closed cell foam structures are desirable. Relative rigidity of the resin structure is also a factor to be considered in order to provide the desired support for the outer panel. Among the resins which may be employed are polystyrene, polyethylene, polyurethane and polyvinyl chloride.

The relative thicknesses of the outer panel and backing member are on the order of 1–3:15, typical thickness being $\frac{1}{32}$ inch for the outer panel and $\frac{3}{8}$ inch for the backing member. Of course, these dimensions may be varied according to installation requirements.

The process provides an insulated siding panel assembly having a synthetic resin backing member which is inexpensive and easy to manufacture and which provides improved support for the outer panel. The siding panel assembly may be easily interengaged to an adjacent panel and secured to a vertical support surface in a clapboard configuration. In one example a double clapboard configuration is provided by offset body sections in the outer panel and backing member.

PLASTIC COMPOSITES

Supporting Core Material for Sandwich Type Construction

A process described by *H. Fathi; U.S. Patent 3,940,526; February 24, 1976; assigned to Sintef, Norway* relates to insulating elements particularly intended for use as a supporting core material in wall structures and the like, e.g., of the sandwich type. The insulating element consists of an insulating body of a nonsupporting material such as mineral wool, which is intersected by a plurality of parallel, relatively thin sheet-formed bracing members of a stiffer material, such as polyester, polyethylene or the like, imparting to the element improved rigidity in one or more directions.

The method of producing such insulating elements mainly comprises placing mats of the nonsupporting material in layers with intermediate coatings of the stiffer bracing material, and then cutting or dividing the assembled layers into slices with a thickness corresponding to the desired thickness of the insulation.

Referring to Figure 1.7a, as basic material in the production process is used a rectangular sheet or a mat **1** of soft nonsupporting (i.e., having very low strength properties) insulating material, such as mineral wool. The insulating mat may be of a common type with standard dimensions, i.e., having a thickness up to about 10 cm.

FIGURE 1.7: BRACED INSULATING ELEMENT

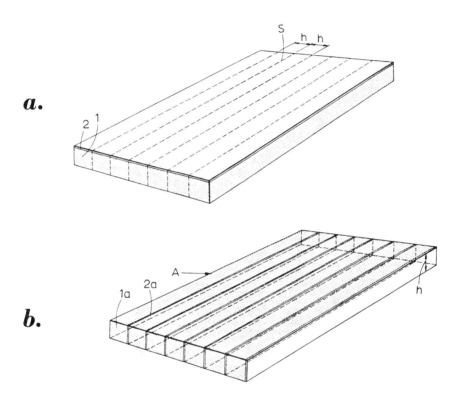

Source: U.S. Patent 3,940,526

On one of its flat sides the mat **1** in a known manner is provided with a rela- tively thin coating **2** of a stiffer, yet flexible material, for instance, polyester or the like, i.e., a coating which is rigid in the general plane of the coating and flexible in a direction perpendicular thereto. The plastic is applied preferably in liquid form, for instance, by spraying or spreading, such that the plastic coat- ing **2** after setting is fixedly adhered to the adjacent insulating material **1**. If de- sired preformed plates of the stiffer material may also constitute the coating **2** which then is bonded to the insulating material, e.g., by means of an adhesive.

The insulating mat **1** with applied coating **2** is then cut or divided normal to the coating **2** along parallel lines **S** into identical, parallelepipedal strips, each of which thus consists of an insulating part **1a** and a coating or bracing part **2a** of width **h** corresponding to the desired thickness of the insulating layer of the wall structure for which the element is intended.

A predetermined number of the thus cut out strips are then placed adjacent each other as shown in Figure 1.7b, in such a manner that the coating or bracing part **2a** of each strip forms a backing against an uncovered side face of an adjacent strip. The adjacent strips **1a**, **2a**, when provided with outer facings or panels for use in a wall structure or the like, thus constitute an insulating element **A** which is orthotropically braced, i.e., braced in one cross direction.

The products described have an advantageous lightweight structure which, as a result of the intersecting relatively stiff coating parts, is braced in a direction parallel to the plates of the coating ribs, the necessary rigidity in the general plane of the insulating element being provided by the areal extension of the mat. The insulating elements are thus self-supporting and easily handled during further fabrication of wall structures or the like.

Honeycomb Core Member for Double-Skin Panel

J.W. Anderson; U.S. Patent 3,998,023; December 21, 1976; and L. Frandsen; U.S. Patent 3,998,024; December 21, 1976; both assigned to H.H. Robertson Company describe a double-skin building panel having an insulating core and having an outer facing sheet laterally offset from and spaced apart from an inner facing sheet, each panel presenting an overlapping edge portion along one longitudinal side and an overlapped edge portion along its opposite longitudinal side. The panels are adapted to be erected in lapped relation without externally visible fasteners.

The insulating core comprises a honeycomb core member filling a major portion of the space between the facing sheets, and rigid spacing means, one positioned along each side of the panel and thermally insulating the facing sheets from each other. The honeycomb core member presents open ended cells which are substantially entirely filled with an insulating medium, e.g., expanded silicate.

Figure 1.8a illustrates a building structural framework **10** of which only vertically spaced horizontal subgirts **11** and a vertical column **12** are illustrated. The building structural framework **10** supports a wall structure **13** assembled from plural building panels **14** erected in edge overlapped relation and presenting plural joints **15**. Each of the panels **14** is secured to selected ones of the subgirts **11** by fasteners **16** (only one is visible in Figure 1.8a).

Each of the panels **14** (Figure 1.8b) comprises an outer facing sheet **17**, an inner facing sheet **18**, and an insulating core **19** disposed between and secured to the facing sheets **17**, **18**. The insulating core **19** comprises rigid spacing means **20**, one positioned along each longitudinal side of the panel **14**, and a honeycomb core member **21**. The facing sheets **17**, **18** may be formed from steel, anodized aluminum, stainless steel, weathering steel and the like, in thicknesses from 20 gauge (0.81 mm) to 24 gauge (0.61 mm). Each of the facing sheets **17**, **18** (Figure 1.8c) includes a central web **22** having first and second side walls **23**, **24** extending in the same direction from opposite longitudinal edges.

FIGURE 1.8: DOUBLE-SKIN INSULATED BUILDING PANEL

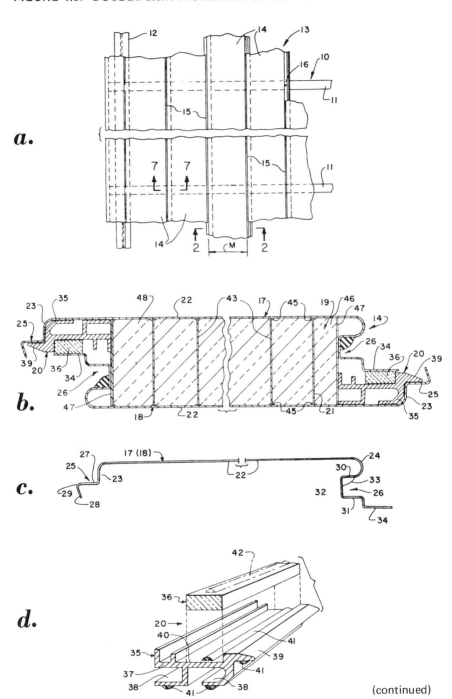

(continued)

FIGURE 1.8: (continued)

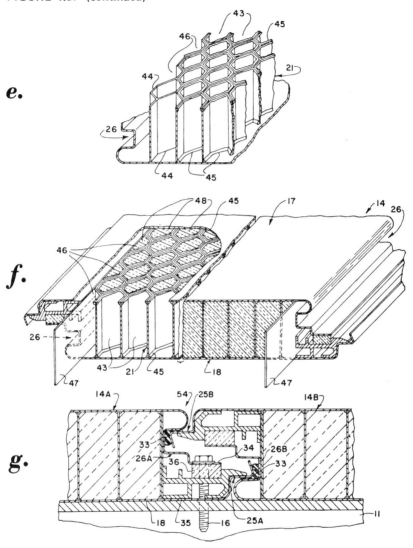

(a)   Fragmentary elevation view of the building wall panels erected on a building
      structural framework
(b)   Cross-sectional view taken along the line **2-2** of Figure 1.8a, illustrating the
      transverse profile of the building construction panel
(c)   End view illustrating the profile of the facing sheets used in the building
      panel of Figure 1.8b
(d)   Exploded perspective view of rigid spacing means incorporated in the panel
      of Figure 1.8b
(e)   Fragmentary isometric view of the honeycomb core member secured to a
      facing sheet
(f)   Broken, fragmentary isometric view of the panel of Figure 1.8b
(g)   Fragmentary cross-sectional view taken along the line **7-7** of Figure 1.8a
      illustrating the joint between the building panels

Source: U.S. Patent 3,998,023

A tongue 25 is formed along the first side wall 23 and extends laterally there-from away from the central web 22. A complementary groove 26 is formed in the second side wall 24 and extends inwardly and confronts the central web 22.

The tongue 25 includes first and second wall segments 27, 28. The first wall segment adjoins the first side wall 23 and extends outwardly generally parallel with the central web 22. The second wall segment 28 adjoins the first wall seg-ment 27 to provide a leading edge 29 remote from the first side wall 23. The second wall segment 28 extends away from the central web 22 and is inclined relative to the first wall segment 27 so as to form an acute angle.

The recess 26 comprises spaced apart wall segments 30, 31 which are generally parallel to each other and to the central web 22; and connected by wall segment 32. A bead 33 of sealant material is provided along substantially the entire length of the groove 26. Each facing sheet 17 (18) additionally presents a flange 34 which adjoins the second side wall 24 at a location remote from the central web 22 and which extends outwardly from the second side wall 24 gen-erally parallel to the central web 22.

It will be observed in Figure 1.8b that the facing sheets 17, 18 are laterally off-set from one another, the overall arrangement being such that the flange 34 of each facing sheet 17 (18) confronts the opposing central web 22 of the other facing sheet 18 (17) and is laterally spaced apart from the first side wall 23 of the other facing sheet 18 (17).

Each rigid spacing means 20 (Figure 1.8d) may comprise a profiled metal ele-ment 35 and an isolation strip 36. The metal element 35 which is extruded preferably from aluminum, includes a central plate 37 from which extend a pair of legs 38 and an arm 39. The metal element 35 additionally presents a recess 40 for receiving the isolation strip 36. Each of the legs 38 may be pro-vided with a bead 41 of adhesive material for securing each of the metal ele-ments 35 on the facing sheets 17, 18 in the position illustrated in Figure 1.8c. The adhesive material 41 may comprise any of the well-known urethane adhe-sives, or epoxy structural adhesives.

The isolation strip 36 may be formed from any suitable rigid thermal insulating material, such as glass fiber reinforced gypsum, cement asbestos board, and the like. An additional bead 41 of adhesive material, provided, for example, in the recess 40, secures the isolation strip 36 to the metal element 35. Alternatively, the isolation strip may comprise cast-in-place gypsum. In either instance, the isolation strip 36 receives a bead 42 of adhesive material, shown in dash-dot out-line in Figure 1.8d, by which each of the rigid spacing means 20 is secured to the flange 34 of the facing sheets 17, 18. It will be observed in Figure 1.8b that the arm 39 of each metal element 35 extends interiorly of the tongue 25 thereby strengthening the same.

The honeycomb core member 21 (Figure 1.8e) presents plural open-ended cells 43 having upper and lower faying edges 44 each provided with a bead 45 of ad-hesive material, such as an epoxy structural adhesive. The honeycomb core mem-ber 21 (Figures 1.8b and 1.8f) extends transversely between the complementary grooves 26 and the rigid spacing means 20, the open-ended cells 43 being capped by the facing sheets 17, 18. The honeycomb core member 21 is secured to the facing sheets 17, 18 by the beads 45 of adhesive material.

The honeycomb core member 21 may be made from Kraft paper impregnated with a fire resistant salt and with a phenolic resin, the impregnants rendering the honeycomb core member fire, moisture and fungus resistant. Paper honeycomb core members useful in the panel have a depth of up to 3 inches (26.3 mm) and have cell sizes of ¾ inch (19 mm) to 1 inch (25.4 mm). Alternatively, the honeycomb core member 21 instead may be formed from metals such as aluminum of 0.003 inch (0.076 mm) thickness having a depth of up to 3 inches (76.2 mm) and cell sizes of ¼ inch (6.4 mm) to 3.4 inches (19 mm).

It will be observed in Figures 1.8b, 1.8e and 1.8f that the opposite longitudinal edge sides of the honeycomb core member 21 present partial or half-cells 46 confronting the complementary grooves 26. As illustrated in Figures 1.8b and 1.8f, a longitudinal filler strip 47 is provided between each longitudinal side of the honeycomb core member 21 and the adjacent complementary groove 26. The longitudinal filler sheets 47 serve as sides which render the half-cells 46 capable of holding an insulation medium 48. The insulating medium substantially entirely fills the open-ended cells 43 and the half-cells 46. The insulation medium may comprise an expanded silicate, such as perlite, or other thermal insulating material, any one of which increases the heat-insulating properties of the panel 14.

Figure 1.8g illustrates a joint 54 between adjacent panels 14A, 14B. The panel 14A is secured to the subgirt 11 by the fastener 16 which extends through the flange 34, the isolation strip 36, the metal element 35, the facing sheet 18, into threaded engagement with the subgirt 11. With panel 14A thus secured to the subgirt 11, the building panel 14B is erected by introducing the tongue 25B thereof into the complementary groove 26A of the panel 14A. Simultaneously, the tongue 25A of the panel 14A is received in the complementary groove 26B of the panel 14B. The tongues 25A, 25B penetrate the sealant bead 33 presented in each groove 26A, 26B thereby to provide weather-tight seals.

Referring to Figure 1.8a, each of the panels 14 has a modular width indicated at M, which corresponds to the distance between the first and second side walls 23, 25 (Figure 1.8c). Panels 14 may be provided having modular widths M ranging from 24 inches (610 mm) to 60 inches (1,524 mm).

Woven Jute Layers and Glass Fibers

A.G. Winfield and B.L. Winfield; U.S. Patent 3,819,466; June 25, 1974; assigned to Care, Inc. describe a double-walled reinforced and insulating building panel. The inner skin is constructed from a plurality of woven jute layers saturated in polyester resin, and the outer skin constructed of a single base layer of woven jute with an exterior coating of chopped glass fibers saturated in resin and an intervening corrugated layer constructed of a plurality of woven jute layers and saturated in polyester resin and bonded between the inner and outer skin. The panel is characterized by its light weight and durability under extremes of temperature and wind, as well as the extreme economy of construction in using woven jute as the reinforcing element.

Reinforced Panels with Key Members

J.B. Ellingson; U.S. Patent 4,004,387; January 25, 1977 describes a simplified reinforced panel which may be readily mass produced to be set up on a building site in building construction. The improved panel is formed of a fiber glass and insulating material with steel rods spaced through the same to provide for the studs or bracing within the panel and with the fiber glass mats on either surface providing the interior and exterior surface of the panel. The rods extend beyond the height of the panel to enable the same to be readily interconnected by bolts to floor and ceiling joists or roof plates, and the panels include keyway members at the ends of the same to readily interconnect to adjacent panels to form exterior and interior walls of the building.

The panels include their own insulation with excellent thermal characteristics and the keyway spline included in the panels and the key members provide the means for interconnecting the panels and the means for compensating for differences in temperature, frost heaves, etc. The panels may be constructed of a size to provide for entire wall or exterior surfaces to facilitate simplified construction or erection of a building on the building site.

Figure 1.9a shows a floor plan view taken through the walls of a typical building construction utilizing the improved wall panels of the process and the interconnecting coupling means or key members which connect the same to provide for forming of the rooms and the entire structure which is connected to a floor frame and a ceiling frame with a roof structure on top.

FIGURE 1.9: REINFORCED PANELS

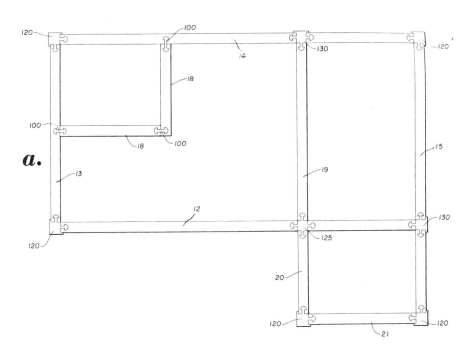

(continued)

FIGURE 1.9: (continued)

b.

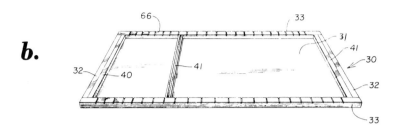

c.

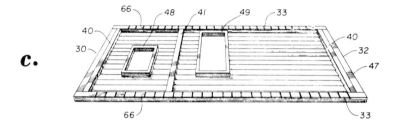

d.

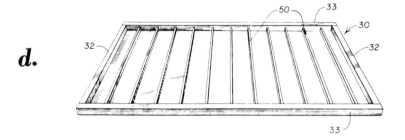

e.

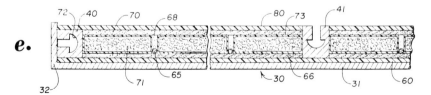

(continued)

FIGURE 1.9: (continued)

f.

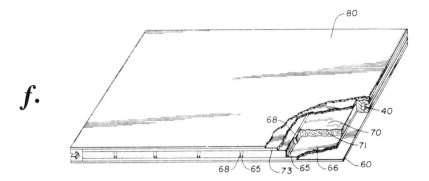

(a) Floor plan view of a building utilizing a new method of house construction and panel configuration
(b) Perspective view of a mold for the formation of a wall panel
(c) Perspective view of a mold in a modified form for exterior panel construction showing the placement of a window and door
(d) Perspective view of a mold for roof or ceiling panel construction
(e) Cross-sectional view of a mold of Figure 1.9b showing the construction of the panels with the mold
(f) Pictorial view of a panel with parts broken away

Source: U.S. Patent 4,004,387

Thus, in Figure 1.9a, the exterior walls indicated at **12**, **13**, **14**, and **15** are interconnected at the corners to form a generally rectangular configuration with interior petition walls **18** being connected to the exterior walls and to one another and to define another room within the enclosure. Additional wall panel **19** extends across the width of the enclosure to define still another room area and an addition to the external room as indicated by the exterior wall panels **20** and **21** which define still another enclosure attached to the exterior walls in a typical type of construction.

Figures 1.9b, 1.9c, and 1.9d show variations in a mold form for the construction of panels, such as exterior wall, partition wall and ceiling used in the building construction. Thus, in Figure 1.9b, a mold is shown at **30** having a solid base portion **31**, fixed ends **32** secured thereto and upwardly extending sides **33** which are hinged to the base portion and releasably interconnected with the ends **32** to define a recessed mold area in which a panel is to be constructed.

Positioned in the mold form and included as part of the panel are extruded key members, indicated at **40**, which are positioned at the ends of the molds and one is indicated at **41** intermediate the length of the mold for purposes to be later noted. In Figure 1.9c, a typical exterior wall is indicated in which the mold **30** with its base structure **31** and ends **32** and upstanding sides **33** has included in addition therein window and door frames indicated at **48** and **49** respectively which frames are inserted into the mold area as are the extruded keyway members **40** and **41** and located therein to be included as part of the panel being constructed. Similarly, exterior siding material, as indicated at **47**, may be included in the mold to be included as a part of the panel. Interior partitions or walls may include only the door frame member **49** in which a door will be mounted to divide rooms and provide access thereto. In Figure 1.9d, the mold form **30** with its upstanding sides **33** hinged to one another has positioned therein a rafter or stringer member, preferably 2' x 4' or 2' x 6' as indicated at **50**, and spaced along the length of the mold and across the width of the same so that as the panel is constructed the stringers **50** or rafters will become a part of the same.

Figure 1.9e shows a section of the mold of Figure 1.9b with a portion of the panel construction therein and with parts broken away. Thus, in the bottom of the mold **31** there is positioned a layer of fiber glass **60** or fiber glass mat which on an interior wall will extend continuously along the bottom of the same except in areas where it is to be broken by a window or door. The mold will be of a size to effect a panel construction sufficient to extend from ceiling to floor or a mold width equal to the height of the panel and a length equal to a part or all of the wall, for example, up to approximately 20 feet for ordinary house construction.

Positioned over the fiber glass mat and spaced along the extent of the same being supported by the sides **33** of the mold are a series of steel rods **65** which are positioned in slots **66** in the sides of the molds extending across the width of the same and in spaced parallel relationship and at a spacing of approximately 16 inches for conventional house construction to simulate the stud spacing in a conventional wall. The rods extend beyond the edges of the mold or greater than the width of the same and extend through the slots having threaded peripheral surfaces at the end of the same.

With the rods in place over the lower fiber glass mat, layer **66** of acrylic material such as Resonate, is positioned over the entire exposed surface of the fiber glass mat. In the areas of the rods, a filler is added to the acrylic such that there is a buildup of material **68** in the area of the rods. Next is positioned in the mold a series of plastic container members **70** filled with a plastic or foam polystyrene chips **71** as insulating material; the thickness of the container is substantially equal to the depth of the panel except for the mats on either side of the same. The plastic containers are of a width dimension sufficient to extend substantially between the pair of rods **65** in the molds and extend from one side to the other.

The acrylic material **66** and the containers **70** are applied in sequence with the acrylic material being allowed to become tacky before the containers **70** are added such that the container with insulating material **71** would adhere to the lower fiber glass mat. The spacing between each container in which the rod is positioned is then filled with an acrylic binder, such as pumice, to form a solid mass of material **68** which when hardened around the rod and adhering to the lower mat will form a stud-like structure extending across the entire width or the height of the panel when it is removed from the mold. An additional acrylic layer **73** is then applied across the containers.

As the mold was set up, a suitable extruded keyway member made of a plastic material and extending the full width of the mold at either end of the same and as indicated at **40** or **41** is positioned in the mold with a slot **72** therein facing outwardly. Where an interior or exterior partition is to be added, an additional keyway member, such as is indicated at **41**, will be positioned in the mold above the fiber glass mat **60** if the partition is to extend from the upper surface of the panel, or the keyway member may be reversed and the slot **72** therein will be facing downwardly toward the bottom of the mold **32** in which case the fiber glass mat **60** will abut the same so as not to cover the slot therein.

With the mat **60** and the keyway members **40** and **41**, positioned within the initial mold and the container **70** of the insulating material **71** added thereto together with the acrylic and binder **68** surrounding the steel rod therein, an upper layer of fiber glass material **80** is added to the mold after an additional surfacing **73** of acrylic material is positioned on the exposed surface of the containers. The upper fiber glass mat **80** will adhere to the containers and the keyway members and define the opposite or top surface of the panel member in the mold.

The acrylic material will saturate the fiber glass mat **80** to adhere to the insulating material and the areas of the rods so that a composite panel will be constructed in which an upper and lower or inner and outer layer **60**, **80** of fiber glass will be exposed in the completed panel which will be separated by containers **70** positioned therebetween and with insulating material therein together with a series of spaced steel rods **65** surrounded by an acrylic filler **68** which will provide the stud structure within the panel for strength of the same. The keyway members positioned in the panel at the ends of the same or intermediate the extent of the same will provide the means for connecting the panels to one another and will facilitate ready and simplified assembly of the panels through the use of key members.

Where an outside wall or an interior wall is to have an opening therein such as a window or door opening, window and door frames will be positioned in the mold, such as is indicated in Figure 1.9c, which frames will have a depth dimension equal to the desired depth or thickness of the panel such that the frame will project to either side thereof.

Similarly, where a keyway member, such as keyway **41** is to be positioned facing the bottom of the mold, it will be inserted along with the frame members or in the same manner as the frame members before the initial fiber glass layer **60** is positioned in the mold. Where such a keyway member is to be positioned facing upwardly toward the top of the resulting panel, the keyway member **41** will be added to the mold above the lower fiber glass mat, at the same time the rods are positioned therein.

Where the mold construction is to be for a ceiling or roof panel, the lower fiber glass mat is positioned in the mold such as is indicated in Figure 1.9d and then the stringers or rafters **50** are added above the layer such that the acrylic material will cause the same to adhere to the lower layer and form a part of the panel. The ceiling and roof panel construction differs from the wall panels which utilize the steel rods with the acrylic material surrounding the same in that they use the rafter with the parallel spacing of approximately 16 inches within the panel. The overall ceiling and roof panel will thus be formed of an inner and outer layer of fiber glass material with reinforcing wooden stringers distributed along the extent of the same and between the container of insulating material.

The upper surface of the panel has a flange surface formed in one edge of the same, and a recessed surface at the other side edge. These are formed by inserting removable block members in the mold at the time of construction. The opposite edges of the ceiling panels have recessed surfaces to facilitate mounting of ceiling roof panels on the wall panels.

The ceiling panels have insulation between the rafters, and the roof panels are constructed without insulation between the rafters. The roof panels are constructed in the same type of mold form as the ceiling panels, except that an initial layer of roofing material, such as pea rock, is placed on the bottom of the mold before the first layer of fiber glass mat is installed such that the acrylic material when applied to the surface of the fiber glass mat on either side thereof, will cause the rock to adhere to the lower mat which becomes the exposed roof surface.

Honeycomb Cellulosic Structure

A process described by *W.W. McCoy; U.S. Patent 3,837,989; September 24, 1974; and W.W. McCoy and J.E. Wagner; U.S. Patent 3,664,076; May 23, 1972* involves a form of prefabricated structure such as commercial or industrial buildings, houses or enclosures employing an external skin bonded to an intermediate structural and insulating filler with an inner surface skin similarly bonded to the filler material.

The external skins are preferably resin bonded glass fiber and the intermediate filler composes a random honeycomb structure made of cellulosic material, such as individual pieces of paper of random size each rigidified and bonded to adjacent pieces by a resinous binder which is compatible with the binder of the skins. The structure is shown in continuous panels and actually three-dimensional structure configurations as well as discrete structural elements which may be used to replace comparable structural elements normally made of wood. A method for the continuous manufacture of structures employing the concept of this process either in the factory or on site is shown in the patent.

The same basic structure may have various configurations. There may be a number of panel sections employing the same structural arrangement but different filler density whereby the load bearing strength of the panel may be controlled and its insulating properties varied as well. Structural strength can also be varied by changing the dimensions and concentrations of resin/fiber glass of either or both external skin surfaces.

OTHER PROCESSES

Box Beam Wall Construction

A process described by *J. Palmer; U.S. Patent 3,641,724; February 15, 1972* involves modular preassembled wall sections wherein a box beam arrangement is formed directly within the wall section which provides a great deal of strength to the wall unit. The wall unit is a complete wall structure having the necessary insulation and reflective materials directly provided therein which materials are so arranged and utilized as to provide proper spacing for the installation of electrical wiring and the like at the job site.

Referring to Figure 1.10a, a wall module generally designated **10** includes a wall framework section which provides a plurality of upright studs **11**, **12**, **13** and **14**, attached at one end to a lower baseplate **15** and at the other end to an upper header **16**.

FIGURE 1.10: BOX BEAM WALL CONSTRUCTION

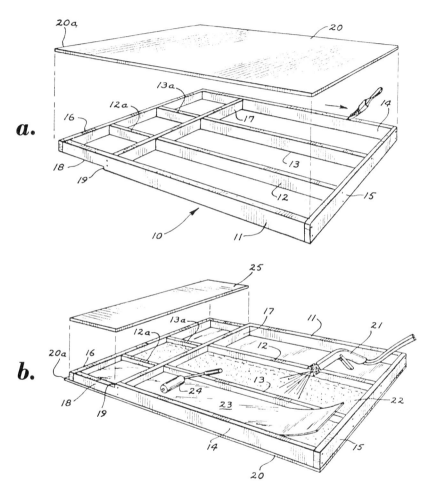

(a) Perspective view illustrating the framing and application of an exterior wall section

(b) Perspective view taken from the opposite side of wall structure illustrating the insulating and the reflective-surface techniques and box beam assembly techniques

Source: U.S. Patent 3,641,724

As illustrated in Figure 1.10a the outer stud members **11** and **14** are full length while the interior stud members **12**, **13** are substantially shorter to provide an intermediate header member **17** extending between the outer studs **11** and **14** and attached thereto with shorter stud sections designated **12a**, **13a** in aligned relation to the inner studs **12**, **13**, arranged on the opposite side of the intermediate header **17**. The outer studs **11** and **14** are notched at the upper end such as at **18** to provide a shoulder **19** and the sizes of the short stud sections **12a**, **13a** above the intermediate header **17** are all in corresponding relation to this notched member to provide the interfitting of an additional member.

An exterior cover member designated **20** is provided which is of a predetermined dimension such that the upper end **20a** thereof may be slightly above the upper end of the upper header **16** to provide a top plate partial housing section. This exterior cover member **20** is also offset slightly on one of the outer studs **11** or **14** such that an overlap is provided. This exterior member **20** is rigidly connected to the framework provided through the utilization of glue and nailing. This gluing and nailing process provides a rigid structure having a tremendous advantage over a simply nailed member.

To complete the framework structure, a discussion of Figure 1.10b is now made wherein the framework as illustrated in Figure 1.10a is turned over to present a plurality of cavities bounded respectively by the structural members. The process of providing insulation and reflective materials into these cavities includes as illustrated, a foaming process using foaming urethanes which may be applied through a mixing gun **21** directly into the cavities and thereby provide filling thereof to a desired depth with complete adhesion to the members forming the cavities. These cavities are filled to approximately two-thirds of the depth of the cavity and this foaming in place urethane material **22** will not only provide an insulating effect, but will greatly increase the rigidity of the entire structure.

After foaming and while the foaming material is still in a liquid or semiliquid state, a reflective material **23** is applied as in the form shown by a roller **24** being applied against the outer surface. This particular reflective material **23** provides in its most preferred form, both a vapor barrier source and a reflective source such that heat will be reflected back into the building and thereby provide a guard against heat loss. By bonding this material **23** directly to the foam, it is obvious that a superior situation is provided as to those arrangements wherein the reflective material is simply nailed to the individual structural elements.

As further illustrated in Figure 1.10b, the last element of the box beam is afforded and is properly attached into position after completion of the insulating and reflective installation. This element is designated **25** and includes a member to extend and fit properly into the aforementioned notched area **18** and abut with the shoulder **19**. This box beam element again is attached into its position through proper nailing and gluing procedures and at this point, the entire box beam element is formed and includes the intermediate header **17** and upper header member **16** as the flanges of the beam with the exterior surface member **20** and the interior member **25** serving as the webs of the box beam structure.

It should be noted that this box beam extends entirely along any one of the individual sections and therefore affords a proper and sufficiently strong header for heading above window sections, other openings and for the installation of truss supports or truss rafters at any position along a wall section rather than locating them with particular reference to a vertical stud or providing solid beam headers over window openings with the trusses positioned on the solid headers. It should be noted in Figure 1.10b that as the foam material **22** and the reflective material **23** assume approximately two-thirds of the depth of each of the cavities that an additional area is afforded on the inner side of the wall to permit the proper placement of wiring.

The wall structure as illustrated in Figure 1.10b can be used to represent the utilization of the concept when openings must be provided for windows or the like. In this particular case a box beam area extends entirely across the wall module **10** with a window opening being provided therein through the utilization of a window header below the window opening with vertical studs located thereunder.

Facade Panels

According to a process described by *L. V. Jochman; U.S. Patent 4,019,296; April 26, 1977; assigned to The Dow Chemical Company* buildings are clad with facade panels by applying a generally moisture-impermeable layer to the exterior wall, applying a layer of a closed-cell water-impermeable insulating foam to the water-impermeable layer and suspending facade cladding panels in space relationship to the foamed insulation. It is not necessary to seal or otherwise waterproof the joints between the facade cladding panels.

In Figure 1.11a there is schematically depicted a building or structure in accordance with the process generally designated by the reference numeral **10**. The building **10** comprises a base **11**, two walls **12** and **13** and a remaining two walls not shown. Disposed above the two walls **12** and **13** and the two walls not shown is a roof **14**. A doorway **15** is formed in the wall **12**, a plurality of facade cladding panels **17** are disposed on the exposed surfaces of the walls **12** and **13** and the two walls not shown.

FIGURE 1.11: FACADE CLAD BUILDINGS

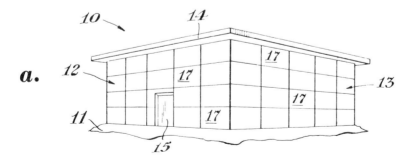

(continued)

FIGURE 1.11: (continued)

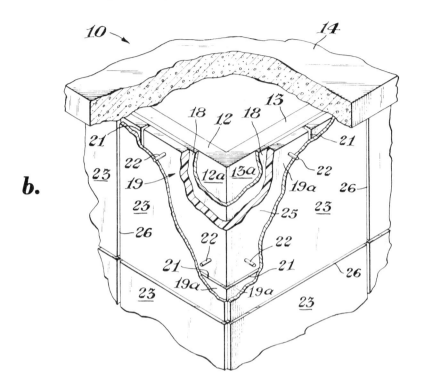

b.

Source: U.S. Patent 4,019,296

In Figure 1.11b there is depicted a fractional schematic cutaway view of the building **10** showing the uppermost juncture of the first wall **12**, the second wall **13** and the roof **14**. The walls **12** and **13** define exterior wall surfaces **12a** and **13a** respectively. Disposed on surface **12a** of the wall **12** is a water-impermeable membrane **18** which extends also over the surface **13a** and to the surfaces of the walls of the building not shown.

Immediately adjacent the membrane **18** and remote from the surface **12a** of the wall **12** is a layer of thermal insulation generally designated by the reference numeral **19**. The layer **19** comprises a plurality of panels **19a** which define adjacent edges of a plurality of water-permeable channels **21**. The panels **19a** forming the layer **19** are disposed in edge-to-edge relationship. The channels **21** provide communication between space remote from the membrane **18** and the surface of the membrane adjacent the panels **19a**.

A plurality of support means or studs **22** extend through layer **19** and membrane **18** and are generally rigidly affixed to the walls of the structure such as walls **12** and **13**. The studs **22** extend outwardly from the walls beyond the layer **19**. A plurality of facade cladding panels **23** are disposed in fixed, spaced, generally parallel relationship with the layer 19. The layer 19 and the panels 23 define therebetween a space 25. The studs 22 are connected to the panels 23 by any convenient means commonly employed to attach facade panels. The edges of the panels 23 define a plurality of passageways 26. The passageways are water permeable.

A wide variety of materials may be employed in the fabrication of structures. The walls may be of any material suitable for the fabrication of walls, such as brick, concrete blocks, cast concrete slabs, wood including plywood, hardboard, chipboard or like materials suitable for the particular application. The water-impermeable membrane may comprise or consist of a wide variety of water-impermeable materials including asphaltic and bituminous compositions alone or in combination with fibrous reinforcing materials such as roofing felt employing organic or inorganic fibers. In certain instances, the water-impermeable membrane can be formed of synthetic thermoplastic resinous film or sheet such as polyethylene, polyvinylchloride and the like.

The thermal insulating layer suitable for the process beneficially is a closed cellular material which is substantially water impermeable, that is, it does not absorb or hold substantial quantities of water within its cells. Particularly beneficial and advantageous cellular plastic foams of the closed-cell configuration include styrene polymer foams, styrene-acrylonitrile copolymer foams, styrene-methylmethacrylate copolymer foams and styrene-maleic anhydride copolymer foams preferably containing less than about 30 wt % maleic anhydride copolymerized therein, polyvinyl chloride foams, polyethylene foams and other water-impermeable materials available in cellular form which are well known to the art.

Foamed glass is particularly advantageous if the facade cladding panels are spaced apart in such a manner that substantial exposure of the foam to sunlight will occur. Polyurethane foams are also usable in applications where excessive moisture is not encountered. When organic foams are employed and substantial exposure to sunlight will occur, it is usually desirable to coat the exposed surface of the foam with a material resistant to sunlight. A variety of materials may be employed as a sunlight-resistant coating including organic latex compositions containing a high proportion of inorganic pigment.

A thin layer of mortar may be employed as a protective layer and is convenient when the foam surface is of the so-called cut-cell variety, that is, closed-cell foam wherein the exposed surface has been formed by slicing through the body of the foam. The exposed surface consists primarily of partial or open cells into which the mortar may flow, harden and provide a mechanical lock between the foam and the hardened mortar. The foam or thermal insulating layer is attached to the wall over the water-impermeable layer by an appropriate adhesive or by the cladding panel supports.

The precise means of attachment will vary depending upon the selection of the water-impermeable layer. If the water-impermeable layer is affixed to the wall over its entire surface such as a hardened polyurethane composition which was initially painted or otherwise spread onto the surface of the wall before hardening and the membrane adheres well to the wall, the foam insulation may be adhesively bonded to the water-impermeable membrane. In the event that the water-barrier layer is a material such as polyethylene and is held to the wall by stapling the foam desirably is connected to the wall by nails, bolts, studs, screws or like mechanical connectors. If maximum water resistance of the water-impermeable membrane is required, it is generally desirable to apply a sealant to the mechanical fasteners at the locations where they pass through the water-impermeable membrane.

In applying the foamed insulation, it is not necessary that each foam insulating membrane be in edge-to-edge sealing engagement and narrow spaces such as may result from manufacturing tolerances may be left between adjacent foam insulating sheets or panels without causing a significant loss of insulating value.

Fire Screening Glazing Panel

F. Jacquemin; R. Terneu and J.-P. Voiturier; U.S. Patents 3,974,316; August 10, 1976; and 3,997,700; December 14, 1976; both assigned to Glaverbel-Mecaniver, Belgium describe a fire screening glazing panel comprising a first structural ply formed by a vitreous sheet and at least one other structural ply. The panel is characterized in that there is provided between the plies at least one layer composed of or incorporating a material which when sufficiently heated becomes converted to form a thermally insulating barrier or barriers, and a protective stratum located between the first structural ply and a heat convertible layer, such protective stratum being composed so as to inhibit interaction between the barrier forming material and the first ply.

Vitrocrystalline material is formed by subjecting a glass to a thermal treatment so as to induce the formation of one or more crystalline phases therein. Preferably, the barrier forming material comprises a hydrated metal salt. Examples of metal salts which can be used in hydrated form are as follows.

Aluminates	Sodium or potassium aluminate
Plumbates	Sodium or potassium plumbate
Stannates	Sodium or potassium stannate
Alums	Sodium aluminum sulfate or potassium aluminum sulfate
Borates	Sodium borate
Phosphates	Sodium orthophosphates, potassium orthophosphates and aluminum phosphate

Hydrated alkali metal silicates, e.g., sodium silicate, are also suitable for use in the heat-convertible layer. Such substances have very good properties for the purpose in view. They are in many cases capable of forming transparent layers which adhere well to a protective stratum. On being sufficiently heated, the combined water boils and the layer(s) foams, so that the hydrated metal salt is converted into an opaque solid porous or cellular form in which it is highly thermally insulating and remains adherent to the protective stratum.

This feature is particularly important, since even if all the structural plies of the panel are cracked or broken by thermal shock, the panel may retain its effectiveness as a barrier against heat and fumes since the fragments of the plies may remain in position bonded together by the converted metal salt.

Panels according to the process may be used to form or to form parts of fire-proof doors or partitions in buildings, and for various other purposes.

Example: A fire screening glazing panel as shown in Figure 1.12 was made which comprised two sheets **1, 2** of soda-lime glass each 4 mm thick and each bearing on one face a protective stratum **3** of anhydrous aluminum phosphate 500 A thick. A heat convertible layer **4** of hydrated sodium silicate 4 mm thick was sandwiched between the protective strata **3**. The panel was transparent.

FIGURE 1.12: TRANSPARENT FIREPROOF PANEL

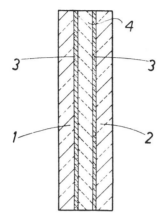

Source: U.S. Patent 3,974,316

The protective strata **3** of anhdyrous aluminum phosphate were deposited onto the glass sheets by the following method. A solution in alcohol containing one mol of anhydrous aluminum trichloride and one mol of anhydrous phosphoric acid was prepared. Because of the reaction between the aluminum trichloride and phosphoric acid, a solution of one mol of anhydrous aluminum phosphate was formed. This solution was applied to the upper faces of the sheets of glass which had previously been laid horizontally, and allowed to spread out to cover the whole area of the sheet faces, and then dried.

The treated sheets were placed in a furnace heated to 400°C, and when removed, the sheets were found to have an anhydrous aluminum phosphate protective stratum strongly adherent to one face. The sheets were then cooled and one of them was laid horizontally with its protected face upward. A 4 mm thick layer of hydrated sodium silicate was then laid onto that protected face of the glass sheet. The sodium silicate was applied as an aqueous solution having the following properties:

Proportion by weight	SiO_2/Na_2O = 3.4
Viscosity	200 centipoises
Density	37° to 40° Baume′

This layer was dried for 12 hours at 30°C in an atmosphere having a relative humidity of 35%. The face of the other sheet of glass bearing a protective stratum of anhydrous aluminum phosphate was then applied to the heat convertible layer thus formed in order to obtain the glazing panel shown in Figure 1.12.

When such a fire screening glazing panel is exposed to fire, the heat convertible layer, in this case of hydrated sodium silicate, is dehydrated and becomes opaque. The layer also thickens and gives rise to a porous body which forms an efficacious barrier against fire. During the course of this conversion by heat, the bound water is driven off, and this helps to limit the increase in the temperature of the panel.

When comparing a panel according to this example with another panel which lacks the protective strata but is otherwise similar, it is found that the panel not formed according to the process will deteriorate during the course of time. In particular, even in the absence of fire, a panel without a protective stratum tends to become opaque. A panel according to the example on the other hand does not appreciably deteriorate in this way, even during prolonged periods, and can remain as transparent as when it was made. It is the presence of protective strata, in this case of anhydrous aluminum phosphate, between the heat convertible layer, here of hydrated sodium silicate, and the adjacent glass sheets which enables the panel to maintain its transparency.

As a variation, a panel of similar construction was made using vitrocrystalline sheets 1, 2 in place of the glass sheets, and incorporating a hydrated potassium silicate layer in place of that of hydrated sodium silicate. Again, anhydrous aluminum phosphate protective strata were used. It was found that this variant panel had substantially the same properties and the same advantages as the panel described above.

In another variation, the glass sheets 1, 2 of the panel first described were subjected to a chemical tempering treatment before the protective strata were applied. This tempering treatment was a process involving the exchange of sodium ions from surface layers of the glass for potassium ions from a bath of molten potassium nitrate maintained at 470°C. This panel had the same advantages as the others described above, and in addition had a greater resistance to breakage due to thermal shock during the buildup of a fire than did the others.

Styrene-Maleic Anhydride Random Copolymers as Foam Core

H.G. Parish; A.J. Palfrey and E.R. Moore; U.S. Patent 3,637,459; January 25, 1972; assigned to The Dow Chemical Company describe a structural panel, the structural panel having a foamed core, the foamed core having first and second major surfaces and edge portions, the structural panel having at least a first skin member of substantially higher density than the core, and the skin member being rigidly affixed to the first face of the core.

The major improvement comprises making the core of a copolymer of styrene and maleic anhydride containing chemically combined therein from about 85 to 65 wt % styrene and from 15 to 35 wt % maleic anhydride, the core having a flexural strength of from about 60 to 110 psi, the polymer having a solution viscosity of from about 3 to 12 cp (viscosity of a 10 wt % solution of polymer in methyl ethyl ketone at 25°C), and beneficially a viscosity of from about 3 to 9 cp.

The process requires the use of the so-called uniform copolymers, i.e., random polymers of styrene and maleic anhydride. One method of making such a uniform polymer is described in U.S. Patent 2,769,804.

Panels in accordance with the process are readily prepared employing conventional fabricating techniques. Usually, a suitable adhesive is applied to a face of the core, a skin member or to both and the surfaces to be joined placed in contact and the adhesive cured. Such adhesives may be applied by any suitable means, depending upon the form of the adhesive. If a liquid adhesive is employed, brushing, rolling, dipping, spraying and the like are generally satisfactory. Solid film adhesives such as hot melt adhesives may be applied as a dry preformed film or extruded as a hot melt. Curing or solidification of the adhesive, depending upon its nature, may be readily accomplished by aging at room temperature, dielectric heating, heating in a platen press or any other technique convenient to the particular adhesive being employed.

Sulfur and Gypsum Wallboard

M.A. Schwartz and T.O. Llewellyn; U.S. Patent 3,929,947; December 30, 1975; assigned to the U.S. Secretary of the Interior describe a process for manufacturing wallboard and the like comprising a basic composition of sulfur and gypsum where sulfur is foamed by the vaporization of the water contained in the gypsum, thereby producing a stable foam that may be compressed under pressure between sheets of wallboard covering material.

The basic procedure of the process comprises melting a batch of sulfur at a temperature range of approximately 130° to 155°C. Viscosity increasing agents, such as P_2S_5 and styrene, may be added and permitted to react with the molten sulfur. Also, flame retardants such as dicyclopentadiene and the like may also be added to the reacting mixture. Inert materials such as talc or clay may also be added and thoroughly mixed with the molten sulfur mixture. Gypsum, $CaSO_4 \cdot 2H_2O$, is added last to the hot sulfur mixture with rapid mixing or agitation of the entire composition. At the temperature of the molten mixture, the gypsum loses approximately 75% of its water and transforms to the hemihydrate form, i.e., $CaSO_4 \cdot \frac{1}{2}H_2O$.

The water is rapidly vaporized into a gaseous state which produces foaming of the entire melt. The resulting stable foam produced is then molded under pressure to form the wallboard panel structure. The foamed sulfur achieves almost instantaneous maximum strength on cooling from the melt fabrication temperature of approximately 130° to 155°C. By contrast, prior art gypsum composition wallboards require much longer periods of time to set and cure in order to achieve their maximum strength.

The composition utilized in the process is basically all sulfur with only a very small percentage of gypsum added to achieve foaming of the sulfur melt. The amount of gypsum utilized is not critical, the only criterion being that of sufficient quantity so as to achieve foaming of substantially the entire sulfur melt. The viscosity increasing agents, inert materials and flame retardants may also be included in varying small quantities to achieve the desired cumulative result of their known qualities.

An example of a typical composition which can be employed in the process is as follows:

	Parts by Weight
Sulfur	200
P_2S_5	6
Styrene	6
Talc	15
Gypsum	5.5

Because of the small percentage of gypsum needed to achieve the required foaming of the sulfur and the extreme efficiency of the foaming reaction, it is not necessary to use pure gypsum. In fact, for economic considerations, waste gypsum is quite adequate for achieving the desired results.

The process can be practiced in either a batch manner or in a continuous manner. Referring to Figures 1.13a and 1.13b, there is depicted an example of the type of apparatus which can be utilized to form individual wallboard panels by the process in a batch manner. A mold assembly **1** includes a lower box section **3** containing the mold cavity and a top press section **5** which telescopically fits within the cavity of lower section **3**. Section 3 includes a flat bottom **7** and a plurality of vertical wall sections **9**, **11**, **13** and **15**. In use, a wallboard covering sheet **17**, of paper or other material, is laid in the mold cavity of lower section **3** on top of flat bottom **7**.

FIGURE 1.13: WALLBOARD MANUFACTURING PROCESS

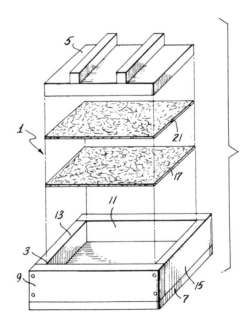

a.

(continued)

FIGURE 1.13: (continued)

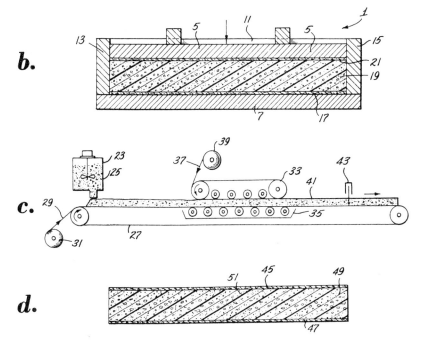

(a) Exploded perspective view of one type of molding apparatus
(b) Front elevational view, taken in cross section, of the mold of Figure 1.13a in its assembled condition
(c) Schematic representation of another type of apparatus
(d) Cross-sectional view of wallboard panel

Source: U.S. Patent 3,929,947

The stable foam composition **19**, prepared as previously indicated above, is then placed into the mold cavity on top of sheet **17** and a second sheet of wallboard covering **21** is placed over the foam. By applying downward pressure with section **5** against the upper sheet covering **21**, the foam **19** is caused to be compressed into the desired final panel thickness and forms a laminate with covering sheets **17** and **21** which are securely adhered to the opposite sides of the stable foam body **19**.

Figure 1.13c depicts a system for practicing the process where the wallboard panels are produced in a continuous manner. A hopper **23** maintains a supply of the stable foam **25** which is fed onto a conveyor **27** which is initially provided with a moving layer of wallboard covering material **29** which is fed from supply roll **31**. The foam **23** fed onto covering sheet **29** is then passed between an upper pressure platen **33** and a lower pressure platen **35**. Platens **33** and **35** may be in the form of rollers, conveyors or other suitable pressure applying means well-known in the prior art. A suitable system is that shown in U.S. Patent 1,890,674.

A second sheet of covering material 37 is fed from a second supply roll 39 and serves to cover the upper portion of stable foam 23 just prior to the pressure treatment between platens 33 and 35. Compressed to the desired thickness and laminated between the coverings 29 and 37, the assembled wallboard structure 41 is then conveyed to a cutter means 43 which divides the continuously formed wallboard into sections of predetermined sizes.

The basic structure of the improved wallboard panel produced by the process is shown in Figure 1.13d. Outer sheet coverings 45 and 47 are securely adhered to the foamed inner core composition 49. As noted, core 49 is characterized by a multitude of voids or pores 51 produced by the rapid vaporization of the water contained in the gypsum composition. As seen in Figure 1.13d, pores 51 are largest in the middle portion of the core 49 and gradually diminish in size towards the outer surfaces where the core 49 is laminated to covering sheets 45 and 47. This variation in density is due to the higher degrees of pressure experienced at the surfaces of the composition during the pressure application of the covering sheets when the foam is being compressed.

During the manufacture of the wallboard, the application of pressure serves to securely attach the outer covering 45 and 47 to the stable foam composition 49 while the latter is still in an adhesive state. As such, the dense outer surfaces produced provide a high degree of smoothness and secure acceptance of nails and other fastening means as desired in a wallboard panel structure, while the high internal porosity provides low density and high degrees of thermal and acoustical insulation.

Resilient-Edge Wallboard

J.L. Scott, F.H. Zajonc and J.H. Crumbaugh; U.S. Patent 3,998,015; Dec. 21, 1976; assigned to United States Gypsum Company describe a predecorated wall panel for use in forming walls in which the joints between the panels are not to be finished. The panel includes a rigid core of green hydrated gypsum with a layer of resilient deformable material along at least one edge being adhesively affixed thereto and a decorative sheet covering the front surface and side edge surface of the panel with a decorative sheet being under tension in the vicinity of one edge surface thereby partially compressing the deformable material and enabling two or more panels to be positioned with a core edge surface adjacent to each other thereby presenting an attractive appearance.

A wall 10 is partially illustrated in Figures 1.14a and 1.14b, where conventional studs 12, here shown to be "C" channels, secure adjacent panels 13 in an upright position, attachment being made either by conventional adhesive, or by fasteners, not shown, having a decorative head matched to the decorative coloring of the panel.

The panels 13 comprise a conventional core 14 and a decorative sheet or envelope 16, which may be vinyl or other suitable, and conventional, material. As shown particularly in Figures 1.14c and 1.14d, the core comprises a body of rehydrated gypsum over which a paper cover sheet 18 may be conventionally adhered. The sheet may be all one piece, or comprise a front and back sheet as shown. The core is defined by a front surface 20 and spaced apart side edge surfaces 22 and 24.

FIGURE 1.14: RESILIENT-EDGE WALLBOARD

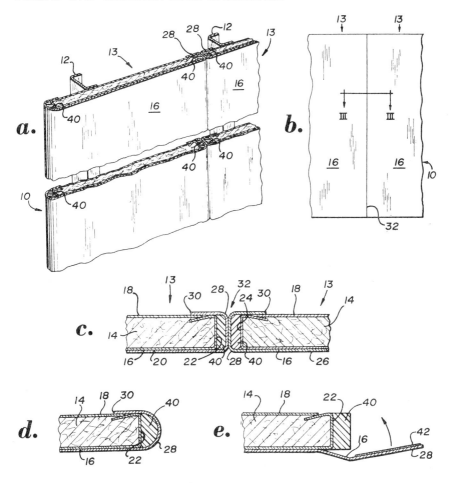

(a) Fragmentary perspective view of a portion of a wall assembly
(b) Fragmentary front elevational view of the wall assembly shown in Figure 1.14a
(c) Sectional view taken along the line III–III of Figure 1.14b
(d) Fragmentary sectional view similar to Figure 1.14c
(e) Fragmentary sectional view schematically illustrating the method of assembly of the panel shown in Figure 1.14d

Source: U.S. Patent 3,998,015

The sheet **16** in turn has an exposed front face **26** and edge portions **28** defining the side edge portions of the panel. Extreme edges **30** of the sheet **16** are preferably secured to the back of the panel by any suitable, conventional adhesive. When the panels are assembled, adjacent edge portions **28** define a joint **32**, which when constructed according to the process, is free from gaps (Figure 1.14b).

A layer of resilient, deformable material **40** is positioned under the sheet **16** along at least a portion of the edge surfaces **22** and **24**. The material is preferably secured to the core by a conventional adhesive, and partially compressed by the sheet **16** which is wrapped over it. The partial compression is best achieved by pulling the sheet **16** under tension, as described below, prior to securing the extreme edges **30**.

Material **40** is preferably a compressible foamed plastic or elastomer. In addition, it may also be formed from nonfoamed elastomers, or even a felt or corrugated paper. In any event, the panel edge, prior to assembly with other panels into a wall, must have a residual compressibility or resiliency permitting the edge to be compressed at least about 10 mils, in a direction perpendicular to edge surface **22** and **24**, when a pressure of at least about 0.01 lb per linear inch is applied. Such further compressibility permits two panels to be abutted readily, edge-to-edge, so that the sheets **16** thereof are in contiguous and intimate contact along the entire edge portions of the panels, leaving no gaps between them.

The thickness of the material **40** as measured from the edge surface **22** or **24** to the sheet **16**, prior to assembly in a wall (Figure 1.14d) is preferably between about 0.02 inch and about 0.5 inch. The actual selection will depend upon the degree of expected deviation in planar edges caused by manufacturing tolerances and other factors. Deviations smaller than 20 mils are not generally noticeable, and a preferable thickness is approximately 0.03 inch.

In accordance with another aspect of the process, it has been found necessary that the material **40** be given a partial compression by the cover sheet. Otherwise, a head will form in sheet **16** at edge portion **28**. This compression is readily achieved, Figure 1.14e, by securing the material **40**, which can have any desired cross-sectional shape, to the edge surface **22** by an adhesive, and then wrapping sheet **16** around material **40** to partially compress it into generally a half-moon shape. With the sheet **16** thus under tension, it is secured by adhesive **42** to the back of the panel.

An example was prepared as follows: A squared strip of polyester foam 0.5 inch thick and 0.5 inch wide having a density of 2 lb/ft^3 was obtained with a pressure-sensitive adhesive on the back side. This was adhered to the edge surface of a half-inch wide, conventional vinyl-enveloped gypsum board, and the sheet of vinyl film or sheet attached to the front surface of the board was pulled over the foam about 0.25 to 0.3 inch. The film was secured to the back of the panel. The resulting panel edge had a residual compressibility of 0.010 inch for an applied force of 0.019 lb per linear inch. Neither gapping nor beading occurred when the panel was tested in a wall assembly.

ROOFING, BUILDING BLOCKS
AND OTHER PROCESSES

ROOF CONSTRUCTION

Built-Up Roof Construction

According to a process described by *M.A. Hyde, P.D. Chalmers and W.B. Mackie; U.S. Patent 3,763,614; October 9, 1973; assigned to The Dow Chemical Co.* a roof is prepared wherein the water barrier layer is placed on a metal roof deck and waterproof thermal insulation is placed over the water barrier layer. A noncombustible fire resistant insulating layer is disposed beneath the water resistant layers and above the metal roof deck. The water barrier layer is not subjected to the extremes of temperature that are encountered when the water barrier layer is the outermost element of the roof structure and the structure is resistant to fire originating above or below the metal deck.

In Figure 2.1 there is illustrated a schematic, isometric representation of a roof structure designated by the reference numeral 10. The roof structure comprises in cooperative combination a metal roof deck 11. The roof deck has an upper surface 12 and a lower surface 13. The roof deck has supporting means 11a. A noncombustible thermally insulating layer 14 is affixed directly to the upper surface of the deck.

Affixed to the insulating layer is a water or moisture barrier layer 15. Beneficially, the water impermeable membrane may comprise a plurality of alternating layers of felt and a bituminous material or asphalt. A thermal insulating layer 16 having a lower surface 17 and an upper surface 18 is adhered to the surface of the barrier layer remote from the roof deck surface 12. The thermal insulating layer 16 is of closed cell configuration and is water resistant and water impermeable. A protective layer 19 is disposed on the surface 18 of the thermal insulating layer. A plurality of spaces or fissures 20 is defined by the layer 16.

Beneficially, in the fabrication of the roof, thermal insulating panels such as planks or sheets of cellular polystyrene or other cellular material are positioned adjacent each other in edge to edge relationship and no attempt has been made to seal the cracks or fissures therebetween. Indeed, in some installations employing incompletely cured or stabilized synthetic resinous foams, shrinkage of the foam occurs wherein the foam cracks in random patterns similar to mud cracking and mortar on the surface thereof ruptures in a similar pattern. Such cracking does not appear to cause loss of serviceability or desirability of the roof structure.

FIGURE 2.1: ROOF CONSTRUCTION

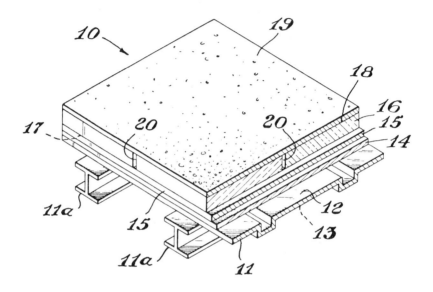

Source: U.S. Patent 3,763,614

Roof structures in accordance with the process do not appear subject to damage by freezing of water in the minor spaces between adjacent foam insulating elements. The foam insulating elements appear to have sufficient resilience to resist rupturing by the expansion of freezing water in crevices. Furthermore, in installations on a heated building the temperature adjacent the water resistant membrane usually does not reach freezing temperatures. In buildings having a roof applied in accordance with the process, little or no tendency is observed for moisture to condense on the inner surface of the roof deck.

Generally for most applications the thermal conductivity of the lower or noncombustible insulating layer will be from about 100 to about 1,000% of the upper or closed cell insulating layer, and most beneficially from about 100 to 500%. Thus, the normal temperature cycling of the membrane layer will be relatively small compared to the cycling of the ambient external temperature while reasonable protection from internal fire is obtained; i.e., a time lag of many minutes is obtained between initiation of the fire and melting of the moisture barrier layer and/or the closed cell insulating layer.

By way of further illustration, a plurality of roof panels are evaluated for fire resistance by evaluating the panels for life in a test furnace. The furnace is a hollow, square, firebrick structure having an upper opening 12 inches square. All samples are prepared using a 14 inch square of 22 gauge (about 0.025") steel roof deck sheet having 1½" deep by 1" rectangular ribs formed therein by adhering on a sheet. The ribs are removed adjacent the edges thereof so that the ribs fit into the upper opening of the furnace. A plurality of roof panels are prepared, each having the general structure shown in Figure 2.1.

In each sample the moisture barrier layer is three layers of 15 lb per square roofing felt, each layer of felt being bonded to adjacent layers (including both insulating layers) by asphalt. The closed cell insulating layer is 1" thick polystyrene foam having a density of about 2 pcf. The upper surface of the polystyrene foam is covered with pebbles to provide a coating weight of 0.1 lb. A propane gas burner is positioned within the furnace; the burner is upwardly facing and is about 4" below the ribs of the steel roof deck sheet. A thermocouple is positioned immediately below the steel deck sheet and temperatures indicated during evaluation of the samples at given times are as follows:

Time, min	Temperature, °F
0	Room temp
2	1100
5	1200
10	1350
15	1400
25	1475
30	1500

Five samples are prepared employing the following materials as the layer **14** of Figure 2.1, together with the time in minutes to sample failure. Sample failure is considered to be ignition of the combustible asphalt or polystyrene foam or collapse of the polystyrene foam.

Sample	Material of Layer 14	Time to Failure, min	Remarks
1	½" thick gypsum board containing about 12% glass fibers and vermiculite	–	Did not fail in 30 minutes
2	½" thick gypsum board	17	
3	1" wood fiberboard	20	Fire and smoke
4	1" composite board of mineral fiber and expanded volcanic glass commercially available as Fescoboard	13	Fire and smoke
5	Layer 14 omitted	2	Foam collapsed; asphalt bubbled

Laminated Polystyrene Board

A process described by *R.W. Sterrett and F.J. Jacob; U.S. Patent 3,677,874; July 18, 1972; assigned to W.R. Grace & Co.* concerns a laminated foamed thermoplastic resin insulation board suitable for use in built-up roofing. The insulating board is provided by heat-laminating, to a core of foamed thermoplastic resin such as foamed polystyrene, a surface film of foamed thermoplastic resin such as foamed polystyrene of lower density than the foamed resin core. The film can be embossed to provide a decorative finish as well as other desirable effects.

Both components of this board are chemically identical but differ in weight, thickness and physical form. For example, a 1" thick 1 lb/ft³ polystyrene board can be laminated with a 0.020" thick 6 to 7 lb/ft³ polystyrene film. The resulting laminated board exhibits quite different physical properties than the unlaminated board. It has a harder, more uniform surface, and a higher surface heat capacity. The higher heat capacity of the laminate makes thermal adhesion of subsequent material (such as saturated felt roofing sheets) more practical than with the unlaminated board. The laminated surface is also more receptive to adhesive bonding of membranes of any type.

The uniform laminated surface can also be embossed. Embossing produces attractive finishes. It also produces an improvement when heat is used to bond other materials to the embossed surface. Any heat bonding system used requires a controlled application of heat. The heat must be adequate to either activate a thermosetting adhesive, to soften a thermoplastic adhesive or to fuse the foamed polystyrene if no other adhesive system is used. However, if too much heat is applied the polystyrene surface will collapse. If the surface is embossed the range of heat necessary to produce satisfactory adhesion is greatly broadened.

It is thought that this improvement results from the fact that the portion of the surface depressed (by the embossing operation) receives heat at a much lower rate and is also not in direct contact with the material being applied. The thin channel of air separating it from the laminate delays the passage of heat to the depressed surface. If the heat applied is adequate to collapse the portion of the surface which is in thermal contact, the depressed portion (which receives heat at a lower rate) will provide a firm support limiting the collapse to the depth of the embossing.

This characteristic is particularly important in preparing built-up roofing where a hot bituminous coated roofing felt is applied to the polystyrene board. Hot pitch or asphalt is normally mopped on the uncoated roofing felt after the coated side has been put in contact with polystyrene. Such an operation has many variations such as weather conditions, temperature of the materials, rate of mopping, etc. The broad range of acceptable heat exposure obtained with embossing reduces job problems and insures good adhesion.

Polyisocyanurate Foam Core

H.G. Nadeau; U.S. Patent 3,814,659; June 4, 1974; assigned to The Upjohn Co. describes thermal barrier laminates having a polyisocyanurate foam core, outer facing sheets and one or more inner layers of heat resistant material extending through the core serving to separate the core into two or more distinct layers. This structure leads to greatly enhanced resistance to heat penetration and total degradation of the thermal barrier upon exposure to fire. Incorporation of the thermal barrier laminate into a built-up roof not only provides thermal insulation but forms a fire retardant barrier adequate to meet existing construction codes for such structures. The thermal barrier laminates can also be employed in a variety of ways in which existing barriers are employed, e.g., building panels, fire doors, pipe insulation and the like.

Example 1: A series of sheets of surface dimensions 2' x 3' and thicknesses varying from ¾" to 2", was cut from a polyisocyanurate foam bun (52" wide, 26" high, 120" long) of nominal density 2.0 pcf which had been prepared using a commercial foam bunstock machine equipped with high speed mechanical mixing head and dispenser. The foam was prepared from the following formulation (all parts by weight) using essentially the procedure described in Example 1 of British Patent 1,212,663:

Polymethylene polyphenyl isocyanate (equivalent weight = 133)	134
Polyepoxide	
Epoxy novolac resin (equivalent weight = 181)	2
Glycidyl ether of tetrabromobisphenol A (equivalent weight = 378)	14
Polyester (chlorendic acid based, equivalent weight = 161)	30
Trichlorofluoromethane	25
Dimethylaminoethylphenol	8
Organosilicone surfactant	2

The series of sheets was then treated as follows before being subjected to the Bureau of Mines Flame Penetration Test as described by Mitchell et al, Bureau of Mines Report of Investigations 6366, 1964.

(1) Two sheets, one of thickness 1.5" and one of thickness 2", were each covered on their two major surfaces with aluminum foil of thickness 0.0015", the foil being bonded to the foam surface in all cases by spreading onto the side of the foam to be contacted with the foil, polyurethane adhesive in a layer approximately $\frac{1}{64}$" thick (20 g/ft^2 application rate).

(2) Two sheets, each of thickness 0.75", were bonded using the above adhesive in approximately $\frac{1}{64}$" thickness of layer to either side of an aluminum foil sheet of thickness 0.0045", the sheet having the same dimensions as the major surface of each of the foam sheets. There was thus obtained a sandwich having the aluminum foil sheet symmetrically disposed between the two foam sheets. The outer major surface of each foam sheet was then bonded to aluminum foil sheets of 0.0015" thickness using the above adhesive in approximately the same thickness as used for the center sheet. The resulting structure was a laminate according to the process with an inner foil sheet separating two foam layers and outer facing foil sheets covering the exterior of the foam layers. The total thickness of foam in the laminate was 1.5".

(3) A second laminate in accordance with the process was prepared in exactly the same manner but using two foam sheets each of thickness of 1" to produce a resulting laminate having a total thickness of foam of 2".

The results obtained in the Flame Penetration test, which measures the time required under standardized conditions for a flame of temperature 2040°F to penetrate from one face to the other of the test sheet, were as follows:

	Test Sample	Burn-Through Time, hr
(1)	1.5" total thickness with foil outer skins	5.25
	2" total thickness with foil outer skins	10.25
(2)	1.5" total thickness with foil center and outer skins	15.7
(3)	2" total thickness with foil center and outer skins	19.2

It will be seen from the above results that the center foil laminates of the process showed a markedly superior behavior to exactly comparable laminates without the center foil sheet. In the case of the 1.5" thickness of foam, the center foil laminate of the process had a burn-through time approximately 3 times as great as the corresponding laminate without the center foil. In the case of the 2" thickness foam, the center foil laminate of the process had a burn-through time approximately 2 times as great as the corresponding laminate without the center foil. The result is noteworthy in that the melt temperature of the aluminum foil is only 1200°F, far below the temperature of the flame used in the test. Hence the increased resistance to burn-through shown by the laminates of the process cannot be attributed to heat resistance supplied by the center foil itself.

Example 2: Using the polyisocyanurate foam and the laminating procedure described in Example 1, there was prepared a center foil laminate using two sheets of polyisocyanurate foam, each of thickness of 1" and dimensions of 4.5' x 5', with a center sheet of aluminum foil of thickness 0.0045" and the same overall dimensions as the foam sheets and exterior facing sheets of aluminum foil of thickness 0.0015" and the same overall dimensions as the foam sheets.

Each of the outer aluminum sheets and the inner aluminum sheet were bonded to the respective sheet surfaces using a polyurethane adhesive applied as a coating of approximately $1/64$" (20 g/ft² application rate) in thickness. The foam laminate so prepared was then incorporated into a section of a test built-up roof by coating a section (4.5' x 5') of 18 gauge fluted (5" span) steel roof deck with steep asphalt at 400°F at a rate of 18 lb/100 ft² and placing the foam laminate on the coated deck. The upper surface of the foam laminate was mopped with steep asphalt (400°F) at a rate of 15 lb/100 ft² and 2 plies of Philips Corey fiberrock asbestos cap roofing sheet type 15 were applied with a coating of asphalt at 15 lb/100 ft² applied between the layers of sheet. The upper surface of the roofing sheet was flood coated with steep asphalt at 400°F.

The test section of built-up roof so prepared was then submitted to the Factory Mutual Construction Materials Calorimeter test; see, N.J. Thompson et al, *National Fire Protection Association Quarterly,* January 1959, page 187. The sample was rated acceptable as Class 1 roof construction material, having an average Btu/min/ft² release of 231 compared with the maximum figure of 270 permitted in order to pass the test.

The above test demonstrates the utility and acceptability of the center foil laminate as a component of built-up roofing. A test sample of built-up roof prepared exactly as described above but omitting the center sheet in the laminate was subjected to exactly the same test under precisely similar conditions. The test sample showed complete failure after only 10 minutes exposure (normal duration of the test is 30 minutes) and was clearly highly inferior to the above test sample. The rate of heat release in this case was in excess of 500 Btu/min/ft².

Sheet Metal Structure

F.E. Carroll; U.S. Patent 3,962,841; June 15, 1976; assigned to Decks, Incorporated describes an insulated roof structure and method which utilizes a sheet metal structural shape and provides superior fire protection and insulation properties. The roof structure is generally a poured gypsum or other poured concrete-like roof deck system where gypsum form board is laid on a sheet metal structural shape subpurlin or purlin structure. A foamed synthetic organic polymer board having holes to permit moisture from the poured concrete to penetrate to the gypsum form board for drying is placed adjacent and above the form board. Reinforcing wire mesh, the poured concrete and a standard weatherproof barrier is then applied resulting in a unitized structure affording high strength, high insulation properties, fire resistance and design versatility.

The sheet metal structural shape used in this process provides excellent structural characteristics while reducing weight and providing a structural shape which can be readily fabricated from sheet metal. It is highly desirable to fabricate structural shapes from sheet metal to minimize energy requirements in production and to conserve steel. Prior attempts to utilize sheet metal shapes in poured roof construction have not been satisfactory.

Some prior attempts have utilized sheet metal $\perp$ shapes as substitutes for bulb tees in roof deck construction. These sheet metal $\perp$ shapes while providing sufficient strength in the composite assembled poured roof do not have satisfactory strength characteristics themselves and in the erection, bend over or roll when walked upon by the erectors. This results in a very dangerous situation for the workers. The sheet metal structural shapes of this process provide desirable strength characteristics themselves and sufficient strength characteristics to be walked upon during erection without dangerous bending or rolling.

Referring to Figures 2.2a and 2.2e, the sheet metal shape of this process is symmetrical about a vertical bisecting plane. The shape has a central vertical web **23** from which two legs **22** project downwardly for equal lengths at an angle, shown in Figure 2.2e as a, of about 45° to about 75° to the horizontal. Each leg has a substantially horizontal flange **21** projecting outwardly at its lower extremity. The upper edge of the web has a structurally stiffening member such as a flange or a triangle.

The horizontal flange may vary in length suitable to hold the desired form board or other decking material. It has been found from about ½" to about 1" is suitable. The vertical depth of the legs may be varied to suit the strength requirements of the desired span. About 1⅛" to about 4" is satisfactory when using the shapes as subpurlins and about 4" to about 10" is satisfactory when using the shapes for purlins. The angle of the legs with the horizontal are suitably about 45° to about 75°. When used as purlins, this angle is preferably about 60° to about 75°. When used as subpurlins, this angle is preferably about 50° to about 60°, about 55° being especially preferred.

The web is important to supply vertical strength and also to prevent bending or rolling of the shapes when they are walked upon by erection workers. Regardless of the depth of the legs, a suitable dimension for the web is about ⅜" to ⅝", about ½" being preferred. As pointed out above various forms may be utilized as stiffeners on the upper edge of the web. A preferred shape of stiffener is an inverted isosceles triangle as shown in Figure 2.2a having sides **24** and base **25**.

It is preferred that the sides be about ³⁄₁₆" to about ½", preferably about ¼" when the shape is used as a subpurlin and about ⅜" to about ¾", preferably ½" when the shape is used as a purlin. It is preferred the base be about ⁵⁄₁₆" to about ½", preferably about ⅜" when the shape is used as a subpurlin and about ½" to about 1¼", preferably about ¾" when the shape is used as a purlin. The stiffener at the upper end of web **23** may also be in the form of a horizontal flange shown as **26** in Figure 2.2b, a box shape shown as **27** in Figure 2.2c, or a circular shape as **28** in Figure 2.2d. It is desirable that the shape permit the poured concrete to flow both under and over the stiffener to prevent vertical displacement or uplift.

FIGURE 2.2: INSULATED DECKING STRUCTURE

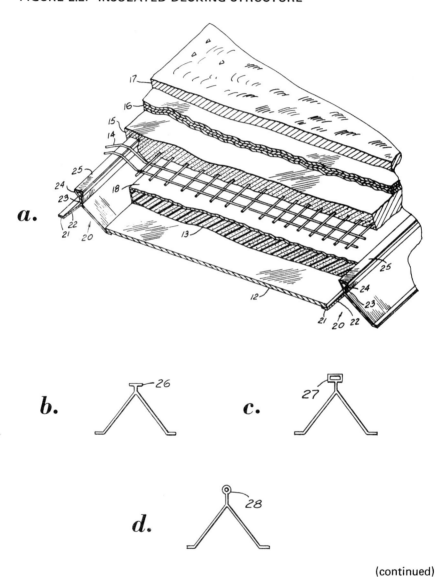

a.

b.   *c.*

d.

(continued)

FIGURE 2.2: (continued)

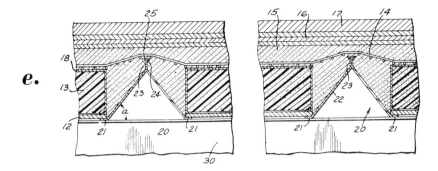

    (a)   Perspective cutaway view of an insulated roofing
            structure
(b)(c)(d)   Different configurations of the sheet metal structural
            shapes which may be used
    (e)   Sectional view of an insulated roofing structure

Source: U.S. Patent 3,962,841

The sheet metal sections may be fabricated by well-known roll forming techniques from sheet steel from about 20 gauge to about 14 gauge. Engineering data for exemplary sheet metal thicknesses and leg depths are as follows, given for the sheet metal shape itself prior to incorporation into a composite structure which would greatly increase the strength characteristics.

Sheet metal shape having suitable gauge and depth for use as subpurlin:

> 18 gauge
> 75° leg angle to horizontal
> 0.974 lb/ft
> 1.25" vertical depth of diagonal legs
> 0.1411 moment of inertia
> 0.175 section modulus
> 5.68' span at steel working stress of 48,000 psi

Sheet metal shape having suitable gauge and depth for use as purlin:

> 16 gauge
> 75° leg angle to horizontal
> 4.767 lb/ft
> 9.0" vertical depth of diagonal legs
> 14.23 moment of inertia
> 3.01 section modulus
> 19.8' span at steel working stress of 48,000 psi

These sheet metal shapes are particularly advantageously utilized in poured and precast roof deck construction. As shown in Figure 2.2a, sheet metal shape **20** holds form board **12** on flanges **21**. Sheet insulating material **16** is placed on top of the form board and is approximately the same width as the form board providing space between the sides of the insulating material and legs **22** for the poured concrete to flow into. After the concrete is poured it is seen that the concrete stiffens the sheet metal shape against spreading. Further, the fact that the concrete is adjacent the legs of the sheet metal shape increases the fire resistance of the sheet metal shape. The insulating material **13** is advantageously of a thickness such that its top surface is about even with the bottom of web **23**, or at least within the depth of the web.

The Λ configuration on the inside of the structure resulting from the use of sheet metal shapes of this process provides space for wiring, plumbing, lighting and the like and when so utilized the opening may be covered with any suitable opaque or translucent covering. Roof level solar energizers will employ auxiliary components which, too, may be housed in the Λ configuration. The Λ configuration on the inside of the structure also serves as a noise baffle to reduce noise levels.

Referring to Figure 2.2a, subpurlins **22** may be supported by any suitable structural members such as open web joists and I beams, such as shown in Figure 2.2e as **30**, spaced at proper intervals making a suitable roof support member system. Any roof support member system suitable for support of the poured roof is satisfactory. Gypsum form board, shown as **12**, having a desired thickness of perforated synthetic organic polymer foam shown as **13** in contact with the upper side of the gypsum form board is supported by the subpurlins.

The form board and foam may be utilized in prepared panels with the form board and foam laminated or may be built-up on the job site. The synthetic organic polymer foam has spaces vertically providing communication between the volume above the polymer foam to the upper surface of the gypsum form board. The spaces through the foam may be perforations of any shape providing sufficient drying area. Perforated polymer foam boards are available commercially. Such boards have previously been used for insulation over metal roof decks to enable the drying of lightweight concrete poured over the foam board.

The roof structure provides properties which are being called for by newer building regulations. The first such property is fire ratings which, following suitable ASTM testing, result in 2 hour fire ratings for the roof structure. The second important property is thermal insulation combined with the satisfactory fire rating. Present energy conservation considerations result in a "U" value of 0.10 and less being desirable.

Calculations show that roof structures of this process utilizing the sheet metal shape as a purlin and using polystyrene and gypsum concrete result in U values of 0.06 and less. When the sheet metal shape is utilized as a subpurlin with ½" gypsum form board, 1½" polystyrene foam board and 2" gypsum concrete the U value is 0.10. Thus, an inexpensive deck is provided having both a 2 hour fire rating for Class 1 fire rated construction and insulation properties resulting in U values of 0.10 and less. Further, a range of desired insulating properties may be achieved by varying the thickness of the synthetic polymer foam.

Lattice Support Straps

R.J. Alderman; U.S. Patent 3,969,863; July 20, 1976 describes a roof system wherein a lattice of support straps is positioned between the rafters and the purlins, thick insulation material is placed on the lattice of support straps between the purlins, and the hard roofing material is attached to the purlins over the insulation material. In addition, relatively narrow strips of insulation material are applied to the upper surfaces of the purlins, between the purlins and the hard roofing material, to minimize the transfer of heat from the hard roofing material to the purlins.

The lattice of support straps rests on the rafters, and some of the support straps can be connected to the purlins, as may be desirable in the particular roof structure. The insulation material is applied to the roof structure by mounting reels of insulation material on support frames and moving the support frames along the length of the purlins and unreeling or paying out the insulation material from the reels as the frames are moved. One or more layers of insulation material can be applied between the purlins, as may be desired.

Figure 2.3a shows a partially completed roof structure **10** which includes a plurality of rafters **11** which are positioned parallel to one another and equally spaced along the length of the building. The rafters are usually inclined and peaked at the centerline of the building. A plurality of purlins **12** extend along the length of the building, across the lengths of the rafters. Each purlin usually extends between adjacent ones of the rafters, and the purlins are mounted on the rafters.

The purlins are approximately Z-shaped (Figure 2.3e) and include central web **13**, upper flange **14** extending in one direction from the web, and lower flange **15** extending in the opposite direction from the web. The upper and lower flanges have their edges **16** and **17** bent further back toward the central web to form rims or minor flanges. The configuration of the purlin is such that relatively thin light material can be used to fabricate the purlin and the purlin retains enough strength to form adequate support in the roof structure. Each purlin is connected at its ends to adjacent ones of the rafters, and the purlins are parallel to one another and each purlin extends in a horizontal attitude.

As is shown in Figure 2.3a, a lattice **19** of metal straps is positioned over the rafters. The lattice includes insulation support straps **20** which extend across the rafters and are located between and parallel to the purlins. Cross straps **21** extend across and beneath the support straps. As is illustrated in Figure 2.3e, the cross straps are connected to the purlins by means of clips **22**. Each clip (Figure 2.3d) is U-shaped and includes base leg **24** and side legs **25** and **26**. The side legs are longer than the base leg, and slots **27** and **28** are formed in the side legs and are angled upwardly in the side legs from the base leg.

The slots are of a width approximately equal to the thickness of the material which forms the purlin, and the angles which the slots make with the base leg are approximately equal to the angle of the rim **16** or **17** of each purlin with respect to its upper or lower flange. The clips are mounted on the purlins by inserting the angled slots **27** and **28** of each clip over the rim of the purlin, as shown in Figure 2.3e. The straps are extended beneath the purlins and into the clips, over the base leg and between the side legs of the clips. The dimensions of each clip are such that the straps **21** must bend slightly to pass around the lower surface of each purlin and then extend upwardly over the lower or base leg **24** of each clip, causing substantial frictional engagement between the straps and the purlins and the clips.

FIGURE 2.3: ROOF SYSTEM

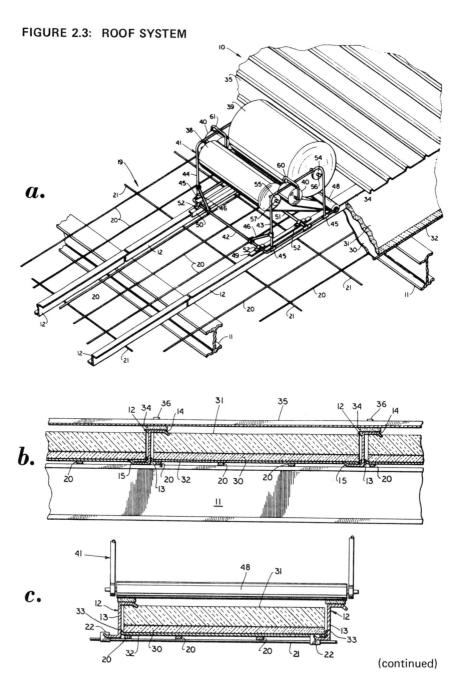

(continued)

FIGURE 2.3: (continued)

d.   *e.*

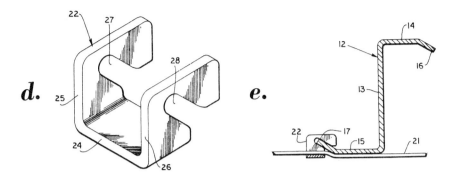

(a)   Perspective view of a partially completed roof structure and the reel of
       insulation material and the reel support framework for applying the
       insulation material to the roof structure
(b)   Side elevational view of a portion of the completed roof structure
(c)   End view of a portion of the roof structure, showing a portion of the
       reel support framework
(d)   Perspective view of the clip that supports the support straps from the
       purlins
(e)   End view of a purlin showing the manner in which the clips mount the
       support straps from the purlins

Source: U.S. Patent 3,969,863

As illustrated in Figure 2.3b, insulation material is placed on the lattice of straps.
The insulation material can comprise one or more strips or bats of material, and
the thickness of the insulation material can vary. In the embodiment illustrated
herein, the insulation material comprises two layers of strips of material, including a lower layer **30** and an upper layer **31**. The lower layer includes a layer of
vapor impermeable substance such as a vinyl sheet **32** applied to the lower surface of the lower layer which is positioned to contact the lattice **19** of straps.
The vinyl sheet is wider than the layer of insulation material and the side edges
33 of the vinyl sheet protrude beyond the insulation material and bend upwardly
and seal against the central webs **13** of the purlins.

The lower layer is thinner than the upper layer, and both layers are of a width
sufficient to reach substantially between the central webs of adjacent purlins **12**.
Additional strips **34** of insulation material are placed on the upper flanges **14** of
the purlins. Hard roofing material **35** is placed over the purlins and the strips
of insulation material **34** and connected thereto by conventional means, such as
by rivets **36**. The strips **34** of insulation material function as conduction heat
insulators between the hard roofing material and the purlins, and the lower and
upper layers of insulation material function as convection and radiation insulators
between the hard roofing material and the elements below.

As is illustrated in Figure 2.3a, the lower and upper strips of insulation material
as well as the purlin insulation material **34** are provided in reels **38**, **39** and **40**,
and a reel support framework **41** is provided for mounting the reels of insulation

material on adjacent ones of the purlins. The framework comprises a U-shaped base 42, inverted U-shaped sides 43 and 44 connected to the U-shaped base by means of clamps 45, side braces 46, roller 48, guides 49 and 50, and strut 51.

The sides are pivotal with respect to the U-shaped base by means of the clamp, and the side braces function to releasably support the sides in an upright attitude or to allow the sides to be folded over and collapsed in an attitude parallel to the U-shaped base for storage and transportation. The guides are connected to the U-shaped base and to the strut by means of U-shaped mounting brackets 52, and the brackets allow the guides to be moved toward or away from each other in situations where the spacing between purlins 12 is not uniform. Guides 49 and 50 are substantially L-shaped in cross-section, with each guide including a downwardly extending leg arranged to move between adjacent ones of the purlins and a laterally extending leg arranged to slide on the top surface of the purlins.

The inverted U-shaped sides 43 and 44 each include bearings 54 and 55, and reel support rods 56 and 57 are arranged to extend through the bearings. Roller 48 is freely rotatable and is of a length sufficient to span over the upper surfaces of adjacent ones of the purlins. The width of the strips of insulation material is approximately equal to the distance between the central webs 13 of the adjacent purlins, and the free ends of the reels 38 and 39 extend downwardly from the reels beneath the roller. The roller functions to urge the lower and upper strips of insulation material between the adjacent purlins and down into the space defined between the purlins and over the lattice 19 of straps.

Reel brackets 60 and 61 are mounted on the upper portion of the inverted U-shaped sides. The reel brackets are also of inverted U-shaped configuration with the ends of their side legs extending inwardly toward each other. The reels of purlin insulation are mounted in the brackets, and the free ends of the strips of insulation from the reels are also fed downwardly from the brackets beneath the roller. The reels are located approximately above the purlins so that the reels will pay out their insulation onto the upper surfaces of the purlin.

When the rafters and purlins have been placed in the roof structure and the roof is ready to receive its insulation and hard roofing material, a lattice of straps 20 and 21 is formed, by extending the support straps 20 across the rafters 11 and by extending the cross straps 21 beneath the support straps. The U-shaped clips 22 are inserted from beneath the cross straps and slipped over the rims 17 of the purlins (Figure 2.3e). The straps are placed under tension, and the clips lift the cross straps up into abutment with the lower surfaces of the purlins.

A plurality of reel support frameworks 41 are mounted on adjacent ones of the purlins. The relatively thin lower layers of insulation material 30 are supplied in the forms of reels 38, and these reels are mounted on the framework 40 by extending the support rods 57 through the reels and then extending the support rods through the bearings 55. The relatively thick upper layer 31 of insulation material is supplied in reels 39, and these reels are also mounted on the frameworks 41. The reels of purlin insulation material 40 are hung in their brackets on each framework.

As the workmen form the roof structure, each framework 41 is pushed out on the purlins, by sliding the frameworks on their guides on the purlins, and the workmen apply the hard roofing material 35 behind the frameworks. As the frameworks are moved along the purlins, the reels **38, 39** and **40** of insulation material pay out into the roof structure. The roller of each framework urges the relatively wide strips from the reels **38** and **39** down between the purlins onto the lattice, and the narrow strips from reels **40** at the sides of the framework are urged by the roller onto the upper flanges **14** of the purlins. Since the relatively thick insulation material is present in reel **39**, this reel must be replaced on the framework **41** more frequently than the other reels of insulation material. This reel is located closer to the hard roofing material in the normal procedure of applying the material to the roof structure, so that the reel of insulation material that must be replaced most frequently is also the reel which is most convenient to replace.

Insulation Panels Installed After Roof Panels

J.D. Studzinski; U.S. Patent 3,662,509; May 16, 1972; assigned to Illini Building Systems, Inc. describes an improved roof structure especially suited for metal buildings and which significantly increases the labor efficiency for assembling a roof by enabling the insulation panels to be installed after the roof panels are installed. Thus, on days when the weather is clear and favorable for working outside, the metal roof panels can be installed, and the installation of the insulation panels can be delayed till a day when the weather is unfavorable for working outside. The roof structure also provides for retaining the insulation panels adjacent the underneath surface of the overlying roof panels so that there are no air spaces whereby moisture can condense on the underneath surfaces of the roof panels.

In accordance with the process, an elongated channel-shaped cap member constructed of either metal or plastic is mounted on the upper flange of each purlin. Each cap member includes outwardly projecting and longitudinally extending coplanar flanges which provide support surfaces spaced under the roof panels by a predetermined distance. A strip of insulation material is attached to the upper surface of each cam member and is compressed when the metal roof panels are secured to the purlins by either screws or rivets.

After the roof panels are installed, a plurality of elastically flexible insulation panels are positioned between the purlins, and each insulation panel is flexed while opposite edge portions are inserted onto the surfaces provided by the flanges of the adjacent cap members. Preferably, each insulation panel comprises a pad of fiberglass having one side covered by a sheet of decorative vinyl material which also serves as a vapor barrier. In accordance with another example of the process, the support surfaces for the insulation panels are provided by strips of insulation material which are positioned adjacent the sides of each purlin and are supported by a longitudinally extending trim member attached to the lower surface of the purlin. Complete structural and operational details of the mode of construction are provided.

Prefabricated Roof Unit

T. Hirai; U.S. Patent 3,969,850; July 20, 1976; assigned to KK Hirai Giken, Japan describes a metal roof unit which includes a backing, a heat insulating material, and a surface metal plate. The unit can be produced by roll forming, press forming or extrusion forming on a mass-production basis in factories and can be used for roofing without special skill by expert roofers. In order to unite the metal roof units there are provided joint members which can be produced of light metal such as aluminum by roll forming, press forming or extrusion forming. The joint members are provided with a series of slits from which nails are driven through notches provided at each end of the metal roof units into a rafter which is provided under the metal roof units, and the slits and notches have sufficient allowance in order to make the joint members and metal roof units movable by heat expansion due to sun heat.

After production in factories, such roof units are adjusted to predetermined sizes on the ground, and then used for roofing. Only a simple work of joining and nailing is required for roofing using these units, and even the layman can lay roofs using these units. Since the roofing operation using the metal roof units of this process is so simple, the time required for roofing can be greatly shortened.

MASONRY BUILDING BLOCKS

Single Cavity Block

M.R. Warren; U.S. Patent 4,016,693, April 12, 1977; assigned to Warren Insulated Bloc, Inc. describes a masonry block which comprises a pair of spaced opposed side walls forming only a single cavity therebetween. The side walls define the outer side peripheral portions of the block and have mutually facing inner surfaces. At least two opposed web walls extend between the side walls. The web walls have inner surfaces extending between the inner surfaces of the side walls. The web walls include channel-forming structures situated intermediate the ends of the inner web surfaces for forming channels that extend in a top-to-bottom direction.

An insulative plate structure of thermally insulative material is disposed in the cavity with the ends thereof being mounted in the channels. The insulative plate structure is situated intermediate the side walls to divide the cavity into a pair of air cells separating the insulative plate structure from the side walls. The block is assembled by inserting the insulative plate structure into the cavity by forcing opposite ends of the plate structure into the channels. The blocks can be laid-up to form a wall and then bonded together, exclusive of mortar joints by applying a ⅛" thick layer of reinforced bonding resin to two sides of the wall.

Building blocks **10** according to the process are depicted in Figures 2.4b, 2.4c and in 2.4a forming a wall **W**. The block includes a pair of opposed side walls **12** forming inner and outer portions of the wall, and a pair of opposed web walls **14**. Each block is molded of masonry material, such as concrete, to form a hollow rectangular structure having a pair of opposed side walls and a pair of opposed web or tie walls **14** extending between the ends of the side walls. The side and web walls define a cavity **15**. The side walls and the web walls are of substantially the same thickness and define the outermost periphery of the block. The web walls have mutually facing inner surfaces **16** which extend between mutually facing inner surfaces **18** of the walls **12**.

The cavity is defined by both of the side walls. That is, there is only one cavity disposed between and defined by both side walls. One or more additional web walls can be provided internally of the outermost web walls to partition the sole cavity into cavity portions. Each web wall includes an internal channel **20** formed approximately midway between the ends of the web wall surfaces. As depicted in Figure 2.4c, the channels are in mutually facing relation, each extending in a top-to-bottom direction of the block. These channels may be formed during the molding process, or may be subsequently machined into the block. The channels are preferably fashioned so as to converge inwardly from top-to-bottom, as depicted in Figure 2.4c.

FIGURE 2.4: INSULATED MASONRY BLOCK

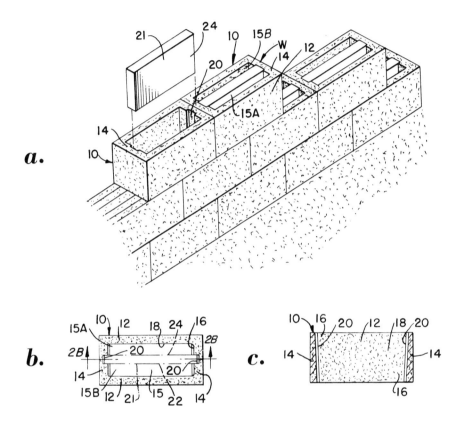

(a)   Insulated masonry blocks in a stacked condition forming a wall
(b)   Top plan view of the block depicted in Figure 2.4a
(c)   Longitudinal sectional view taken along line **2B—2B** of Figure 2.4b

Source: U.S. Patent 4,016,693

Mounted within the block **10** is a plate or panel **21** formed of thermally insulative material. The insulative plate is positioned such that outer faces **22, 24** thereof extend between the web wall surfaces **16** in directions parallel to the side wall surfaces **18**. In this fashion the plate divides the cavity **15** into a pair of air cells **15A,B** with each air cell **15A,B** being bordered by: a portion of each web wall inner surface **16**, one of the insulative plate faces **22, 24**, and the side wall inner surface **18** which is in mutually facing relation to that inner plate face.

Thus, the arrangement is such that a masonry block **10** of substantially standard design is provided with a pair of air cells separated by a layer of thermally insulative material for retarding heat and vapor transfer. The type of insulative plate material to be employed, as well as its thickness, depends upon the degree of resistance to thermal conductivity that is desired. Possible types of insulative material include polystyrene, urethane, Styrofoam, and fiberglass.

In use, the concrete block is precast in a conventional manner. Suitable channels **20** can be formed during molding or machined into the web walls **14** subsequent to the molding operation. The insulative plate **21** is manually press-fitted into the channels. The convergent nature of the channels causes the plate to be slightly compressed into a snug fit within the block. Insertion of the insulation plates can be accomplished at the block fabrication facility, or can be performed later at a building site.

The blocks are used in a conventional manner in the erection of a structure as shown in Figure 2.4a. That is, the blocks are laid-up in staggered or nonstaggered relation, with the side walls **12** forming the interior and exterior wall surfaces of the structure. The insulative plates can be inserted before or after a row of blocks are laid-up by merely pressing the plate into the channels. The convergent nature of the channels serves to firmly hold the plate in place.

Bonding of the blocks together can be performed by conventional mortar application between blocks or by the application of a bonding resin to the exterior of the structure. One conventional type of such resin is sold as Bloc-Bond. This resin has a portland cement base and is reinforced with fiberglass. Application of the resin is made in a $\frac{1}{8}$" layer onto the exterior and interior surface of a wall of free-standing blocks. By thus eliminating the use of mortar the amount of noninsulated wall area is significantly reduced.

The process is highly advantageous in that it can be utilized in conjuction with blocks of standard design. Thus, the insulation is adapted to the block, rather than the block being adapted to the insulation. No significant reduction in size of the block is required to accommodate the insulative plates. Also, the plates can be inserted into the blocks before a wall is laid-up, if desired.

Moreover, since the block involves only a pair of side walls with only one cavity disposed therebetween, the ratio of air volume to concrete volume is maximized to retain a high insulative factor. In certain prior art proposals there are employed a plurality of staggered cavities between the side walls, the added presence of masonry between the staggered cavities serving to reduce the air/concrete ratio.

Furthermore, the block of the process provides plural air gaps or cells **15A,B** to retard thermal conductivity. It is expected that a significantly higher insulative factor is achieved when the insulative plate is disposed in spaced relation from both of the side walls, as opposed to being situated flush against one or both of the side walls.

Importantly, no portion of the insulative plate of the block extends from one side wall to the other in flush engagement with a web wall, thus avoiding an enlargement of the effective heat transfer path. Since the insulative plate **21** extends across only one major dimension of the cavity **15** (i.e., parallel to the side walls) and since only one cavity is disposed between the side walls, only a minimal amount of insulative material is required per block.

Block Design

According to a process described by *M. Kosuge; U.S. Patent 4,004,385; Jan. 25, 1977* lightweight building structures are constructed by assembling a plurality of building blocks in contiguous end to end relationship. The blocks have spaced apart and parallel opposite longitudinal walls and opposite transverse members integrally connecting the walls so as to define a substantially rectangular inner space. The tops of the transverse members of at least a portion of the blocks are at a lower level than the tops of the longitudinal walls so that horizontally extending reinforcing bars may be positioned therein.

Vertically extending reinforcing bars which may be surrounded by reinforcing loops are also positioned within the inner spaces defined by the contiguous blocks. Mortar is charged to the inner spaces defined by contiguous blocks to unite the blocks and reinforcing members. V-shaped vertical edges are provided on the longitudinal walls of a block to provide stronger joints between blocks. Blocks having longitudinal walls of different lengths and flat vertical edges are used to provide L-shaped, T-shaped and cross-shaped corners. In certain examples of the building structures, sound or heat insulating materials are positioned within the inner rectangular spaces to provide sound and heat insulating properties.

Insulating and Waterproofing System

A.A. Hala; U.S. Patent 3,772,840; November 20, 1973 describes a combined insulating and waterproofing system. The method comprises substantially rectangular shaped blocks of expanded polymeric plastic materials having insulative characteristics. The blocks are disposed in predetermined adjacent proximate relationship with one another and positioned between the inner and outer wythes of a wall construction with the horizontal edges of the blocks being angularly inclined in a downward direction as viewed from the inner wythe with respect to the outer wythe.

Referring to Figure 2.5a, there is shown a combination insulating and waterproofing system **10** constructed pursuant to the principles of the process. The system comprises blocks **12** preferably fabricated from an expanded polymer plastic material, such as polystyrene or polyurethane. The blocks are preferably of rigid construction but capable of deformation or tearing upon the application of a predetermined amount of pressure. The insulating blocks are disposed between the inner wythe **14** and the outer wythe **16** of a wall structure.

As shown in Figure 2.5a, the wall structure is constructed on a masonry foundation indicated by the reference numeral **18**; however, the foundation may be fabricated from steel or any other suitable material. The inner wythe is constructed from concrete blocks **20**, but may be constructed from brick or other suitable masonry or building material. The wall structure itself may employ rows of wall-ties **22** built in with the outer wythe or built-in reinforcing rod assemblies **24** having tie members **26**.

FIGURE 2.5: INSULATING AND WATERPROOFING CONSTRUCTION

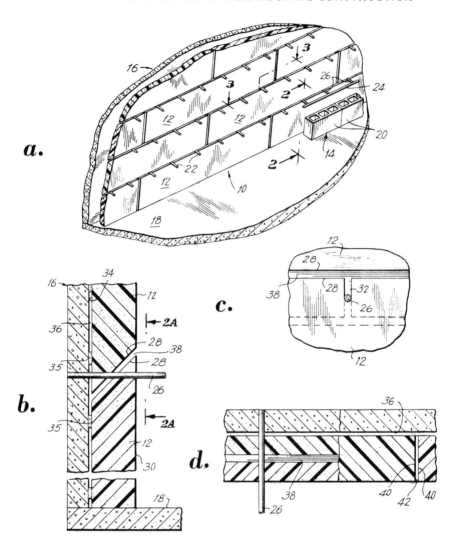

(a)   Perspective view of a combined insulating and waterproofing system
(b)   Sectioned elevational view taken on the line **2—2** of Figure 2.5a
(c)   Sectional view taken on the line **2A—2A** of Figure 2.5b
(d)   Sectional view taken on the line **3—3** of Figure 2.5a

Source: U.S. Patent 3,772,840

The insulating blocks **12** are fabricated having angularly inclined horizontal edges **28** with the upper and lower horizontal edges being disposed in substantially parallel relationship, whereby the vertical cross section of the blocks is in the form of an irregular parallelogram. As seen in Figure 2.5b, it is possible to form the block with only a singular angularly inclined horizontal edge for use as bottom or top members.

In the application of the insulating blocks or sheets 12 with respect to the wall-tie embedded outer wythe 16, which is constructed first, the block is placed flush upon the base 18 and the upper horizontal edge is then forced upon the wall-ties 22 by applying pressure upon the front face or surface 30 of the block. This force causes the deformation of the horizontal edge 28 and forms a vertical tear or slit 32 (Figure 2.5c), however, the block is easily placed in position with respect to the outer wythe. In this regard, it is to be noted that although the rear surface 34 of the block is in substantially flush or abutting relationship with the inner surface of the outer wythe, it is actually spaced therefrom due to mortar 35 or other debris which adheres to the wythe 16 during construction. Thus, there is formed a vertical channel or space 36 between the wythe and the blocks.

The blocks are forced upon the wall-ties or tie members 26 in horizontal rows or courses, usually starting from the bottom up. However, the courses may be commenced from the top or from an intermediate position without adversely affecting the completed system.

Referring to Figures 2.5a through 2.5d, it will be seen that the application of two courses of the blocks results in the upper horizontal inclined edge of the lower block and the lower horizontal edge of the upper block being disposed in juxtaposed position and forming a small horizontal angular channel or joint 38 which is directed downwardly from the inner wythe 14 to the outer wythe. The channel is formed due to the fact that in the application of progressive courses of the blocks it is virtually impossible to place the angular edges of adjacent blocks in abutting fluid-tight relationship, even though the edges are parallel.

The vertical edges 40 of horizontal blocks in the same course are also disposed in slightly spaced relationship even when attempting to place them in side-by-side abutting relationship and thus form a vertical joint 42. However, the system is constructed in a manner such that the vertical joints in one course are preferably laterally offset with respect to the vertical joints in the next vertically adjacent course.

The blocks or sheets of insulating material are preferably fabricated having heights of 16" and lengths of from 4 to 9 feet. The length is much more a matter of design than the height since it is common practice in masonry construction to place the wall-ties in the outer wythe with a vertical spacing of 16". However, if the vertical wall-tie spacing is more or less than 16" the height of the blocks can be varied accordingly.

After the system 10 has been installed, the outer and inner wythes of the wall structure completed, the structure is both waterproof and insulated. The structure is insulated by virtue of the insulating material and waterproof by virtue of the construction in the following manner. If any water should permeate between the inner and outer wythes, respectively, it will tend to flow downwardly. If the water flows down upon the rear surface of the blocks, it will flow down through channel 36 and be prohibited from permeating the blocks and also the inner wythe **14**. If water flows downwardly upon the inner surface **30** of a block **12**, it will then flow into a horizontal downwardly inclined joint **38** from whence it will flow into channel **36** and be prevented from reaching and permeating the inner wythe. Since the vertical joints of adjacent courses of the blocks are laterally offset with respect to one another, water flowing within a vertical joint **42** will reach a horizontal joint and be directed downwardly into vertical channel **36** and also be prevented from reaching and permeating the inner wythe.

It will thus be apparent that any water or vapor which subsequently moisturized will always be directed away from the inner wythe by virtue of the construction of the insulating blocks. The method of installing the system is simplicity itself in that no tools of any kind are required since the system is assembled merely by the pressure applied to the blocks by a workman.

Foaming Technique

A process described by *M. Kosuge; U.S. Patent 4,018,018; April 19, 1977* relates to an architectural block containing holes which are filled with a heat-insulator and moistureproof material. This material is merely placed in the chamber portion of the block and then allowed to foam thereby taking on its heat insulator and moistureproof characteristics. Thus, it is unnecessary to apply additional layers to the exterior of the blocks.

Another important feature of the process is the formation of grooves on one edge of each partition of the blocks. These grooves allow the insertion of a pipe through the block so that the end of the pipe can be positioned within each chamber. Thus, the process includes the structure necessary to provide a very simple technique for filling the chambers with the lightweight insulator and moistureproof material.

As shown in Figures 2.6a through 2.6d, a principal block **1** is so formed that the space between two lateral walls **2, 3** is divided by a plurality of partition walls **4** having the same height as the lateral walls to define compartments or chambers **5** passing through from the top to the bottom of the block. Both lateral sides of the block have projected end portions **2a, 2b** and **3a, 3b** of the lateral walls, the projected end portions forming grooves or recesses **6** and **7**. The top surfaces of the partitions are provided with a linear groove **8** and a pipe **9** fixed in the groove. The pipe has a length of substantially the same length as the thickness of the partition.

As shown in Figure 2.6a, the vertical grooves **6, 7** on both the lateral sides are penetrated by reinforcement steel bars **13** and filled with mortar. Furthermore, they are preferably penetrated with prolonged pipes **9'**. Numerals **11** and **12** in Figures 2.6e and 2.6f denote accessory blocks. A block **11** in Figure 2.6e has a structure similar to that of the principal block but is partitioned with a wall having no groove. One side of chamber **16** has a cover plate **15** on the bottom. A block **12** in Figure 2.6f has a structure similar to that of the principal block but is partitioned by a wall **17** without a groove. Holes **18** pass through the block.

Among the blocks as described above, the principal block and accessory block **11** are necessary for the construction of an architectural structure, but the accessory block **12** in Figure 2.6f is occasionally employed in a convenient combination according to the structure. Figures 2.6g and 2.6h show the portions for forming the walls and floor of structure **10** constructed with these blocks. Figure 2.6g is a plan view of the foundation structure of combined blocks, while Figure 2.6h is the longitudinal view thereof, which foundation structure is employable both for wall and floor. As shown in Figure 2.6g, the structure is so built that the chamber **5** may be positioned vertical or jointed horizontally as shown in Figure 2.6b to construct the base for the fundamental structure, and then, as shown in Figure 2.6h, blocks **12** are placed over and under the principal block **1** so that the chamber **18** of block **12** may pass through on the both sides of the chamber **5** of block **1**.

FIGURE 2.6: ARCHITECTURAL BLOCK

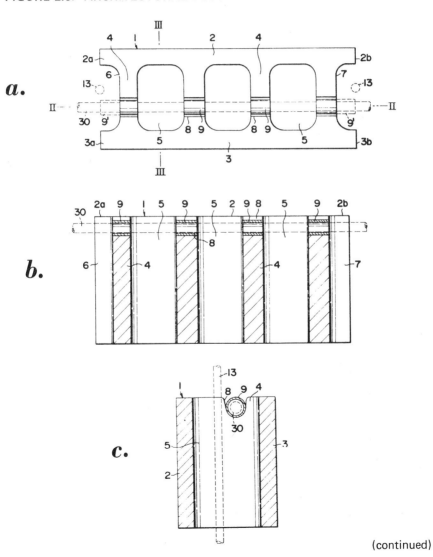

(continued)

FIGURE 2.6: (continued)

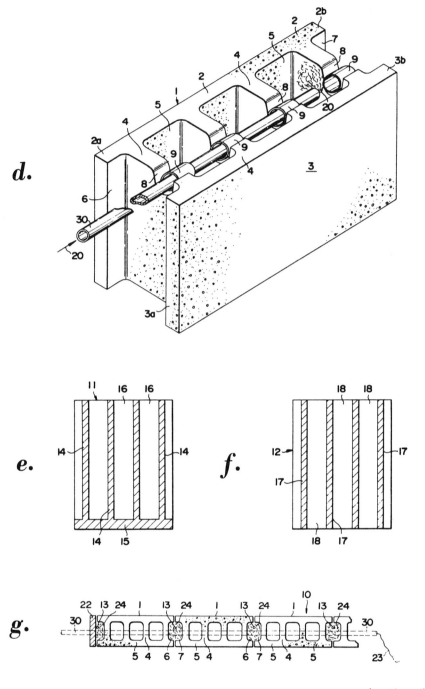

(continued)

FIGURE 2.6: (continued)

h.

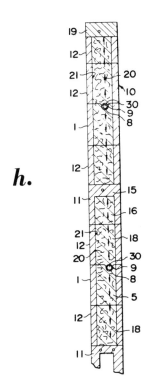

(a) Block design
(b) Fragmentary elevation along line II—II in Figure 2.6a
(c) Cross-sectional view along line III—III of Figure 2.6a
(d) Perspective view
(e) Cross-sectional view of an accessory block having a cover plate
(f) Cross-sectional view of an accessory block having holes
(g) Partial plan of a structure (wall or floor) composed of principal
      blocks jointed in series
(h) Partial longitudinal cross-sectional view of the structure con-
      structed of the blocks

Source: U.S. Patent 4,018,018

When further the blocks **11** having cover plates **15** are overlaid jointly, the chambers of three blocks are covered with this cover plate to define a chamber **21**. This chamber may also be covered with a block plate **19** in place of block **11** having the cover plate. When a chamber **21** has been defined in this manner, the injection pipes **30** are inserted, as shown in Figures 2.6a through 2.6d, into different pipes **9** disposed along linear lines in different partitions **4** of block **1**.

The composition **20** comprising a blowing-agent-containing synthetic resin mixed with glass fiber, asbestos and other inorganic lightweight insulating and moistureproof materials is injected from the injection pipe **30** by an air stream. After one chamber is filled, the injection pipe is withdrawn and inserted into another chamber to inject the composition. For the purpose of injection, an injection pipe may be placed directly in a groove **8** whether or not a pipe **9** is used, but this pipe may preferably be employed because the grooved walls may be broken and the smooth insertion of the injection pipe could be interrupted.

After the whole cavity of the block is filled with the composition, the blowing agent in the composition generates foam owing to the natural drying, and the expanded mass containing the insulating and moistureproof materials fills the chambers of each block. A piece of string is attached to the end of the injection pipe and kept outside the block. The string is pulled in after the complete withdrawal of the injection pipe and is pulled out after the insertion of the injection pipe. Thus, the expanded mass is bored to form an orifice, which serves to aerate the expanded foam and to prevent it from hardening. The outmost orifice is usually closed with a stopper.

When the effectiveness of the expanded mass has been reduced after use for a long period, the injection pipe may be pulled through to let the fresh composition set up the expanded mass. The vertical grooves **6, 7** are penetrated by a reinforcing bar **13** and filled with mortar **24**, thus being reinforced. An oblong cover plate **22** is mounted to the end of the reinforcement portion. Since the hollows in the blocks are filled with heat insulator and moistureproof material, the structures constructed with the blocks maintain coolness in summer and warmth in the winter within the interior.

Foam Layer

J.C. Barnhardt, Jr.; U.S. Patent 4,002,002; January 11, 1977 describes an insulating building block which is a composite of masonry material and a foam material. The insulating building block is obtained by applying a layer of foam material to a longitudinally extending side face of each cavity within the hollow block. By the process, a block manufacturer is enabled to produce directly at his facility a complete, unitized product having improved energy-saving features over those of the prior art.

With the composite construction, a building wall is properly insulated as soon as the building blocks have been laid. Furthermore, the layer of foam material does not hinder the use of the cavities of the blocks as finger access openings for workmen handling the blocks during construction. An efficient and economical utilization is made of the foam material to provide a high degree of thermal insulation with a relatively small amount of foam.

Keystone Shaped Insulation Insert

D.A. Selby; U.S. Patent 3,817,013; June 18, 1974 describes a composite building block comprising outer and inner concrete members affixed by interlocking shape to an internal layer of insulating material.

Figure 2.7a shows a wall **5** built up of a series of building blocks **6**, set in courses **7**, preferably with the blocks of one course being offset with the blocks of adjacent courses in known manner. Each of the blocks is composed of an outer concrete member **8** and an inner concrete member **9** as shown in Figures 2.7b and 2.7c and separated from each other by an inner insulation insert **10**.

FIGURE 2.7: INSULATED CONCRETE BLOCK

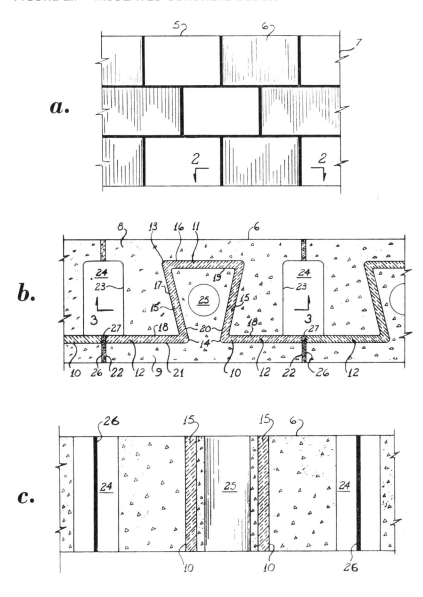

(a) Partial front elevation of a section of wall incorporating the building blocks
(b) Enlarged sectional plan view of a building block in one course, taken on the line 2–2 of Figure 2.7a
(c) Sectional elevational view taken on the line 3–3 of Figure 2.7b

Source: U.S. Patent 3,817,013

The three members 8, 9 and 10 forming the blocks 6 are cast together by placing the insulation insert 10 in the mold of a conventional concrete block machine
and filling the mold in known manner or, the insulation insert could be formed
in place between precast inner and outer concrete members 8 and 9. Alternately,
the concrete members 8 and 9 could be formed by extrusion on both sides of
the insulation insert, and the blocks cut to size by sawing.

The insulation insert is keystone shaped, having one member 11 spaced inwardly
of and parallel with an outer surface of the block, and having a pair of members
12 spaced inwardly of and parallel with an opposite outer surface of the block,
with the outer ends 13 of the member 11 being joined to the inner ends 14 of
the members 12 by the members 15 which are set at an angle relative to the
planes of the members 11 and 12.

The outer, or first, concrete member 8 has its inner surfaces 16, 17 and 18
molded to fit the adjacent surfaces of the members 11, 12 and 13 of the insulation insert, while the inner surfaces 19, 20 and 21 of the inner, or second, concrete member 9 are molded to fit the adjacent surfaces of the members 11, 12
and 13 of the insulation insert. The members 12 of the insulation insert extend
outwardly into line with the end surfaces 22 of the inner concrete member to
align with the members 12 of adjacent blocks.

The end surfaces 23 of the outer concrete members are recessed inwardly to reduce the weight of the blocks and to form an air space 24 between adjacent
blocks. Thus, it will be seen that the outwardly facing surfaces of the members
8 and 9 together with the ends 23 form a lateral outer periphery and the insert
extends from a first point on this outer periphery to a second point on this
outer periphery.

Each of the inner concrete members of the blocks is provided with an internal
cavity 25 formed in that portion of the member which is bounded by the members 11 and 15 of the insulation insert. As seen in Figure 2.7b, the blocks are
mortar joined by the mortar 26. The resulting gap between adjacent ends of
the members 12 of the insulation inserts are filled by a compressed insulation
filler 27 to ensure continuous insulation throughout the wall 5.

In the construction of the wall, a strip of insulation may be laid in the horizontal mortar joint parallel to the wall face and behind the mortar which is
adjacent to the front and back faces of the block. This insulation strip would
be contiguous with the insulation inserts in the blocks above and below it.
Special architectural finishes could be applied to either or both of the faces of
the concrete members. Where desired, lightweight concrete could be used to
decrease the weight of the insulated blocks and the insulated blocks could also
be combined into fabricated panels for use as room dividers or partitions.

Preformed Insert

R.W. Whittey; U.S. Patent 3,885,363; May 27, 1975; assigned to Korfil, Inc.
describes a preformed masonry building block having a preformed insulating insert disposed in the cavity. The inserts are of molded expandable polystyrene
or the like and are characterized by cut-away portions for finger access to marginal walls of the block cavities and by elongated vertical openings in the inserts
for enhanced cross-sectional compressibility. The cross-sectional dimensions of

the blocks vary substantially requiring such insert compressibility. In one form, an insert has a four-sided shape with an elongated slot in each side, in another an integral flange across the bottom of the insert closes a through internal vertical passageway. In a third form, right angularly arranged slots extend downwardly from a top portion of the insert and terminate short of the bottom and side walls. A further form of the insert includes side walls which are serrated viewed in cross section and a pair of vertical openings separated by a flange and closed at the bottom.

A substantial degree of nonuniformity in cavity dimensions is accommodated with the inserts of the process. Interference with manipulation of the blocks during block manufacture and subsequent construction is wholly avoided with the use of finger access openings for workmen handling the blocks and assembled inserts, and efficient insulation and fire retardation, together with retardation of sound and moisture transmission, are accomplished with the use of the inserts in a wall or other structure formed with the insulated blocks. Assembly of blocks and inserts can be accomplished with a high degree of ease and convenience and at economic advantage with the features of lateral or cross-sectional insert compressibility.

Insulating Filler

A process described by *P.R. Grants; U.S. Patent 3,704,562; December 5, 1972; assigned to I.F.S. Incorporation* generally relates to hollow building blocks and more particularly insulating fillers or inserts for such blocks and an insulating strip associated with the inserts to provide a continuous insulating barrier or wall formed by a plurality of courses of hollow building blocks.

Figure 2.8a illustrates a building wall 10 constructed of a plurality of vertically superimposed courses of hollow building blocks 12 joined together by use of a conventional mortar 14 forming a joint. The lowermost course of the blocks is supported on a footing 16 with an interior floor 18 connected thereto in a conventional manner and an exterior grade 20 also being provided with the wall structure being conventional with the building blocks 12 and 14 including an inner wall and an outer wall separated by vertical webs 22 which extend for approximately one-half of the height of the inner and outer walls. Thus, a regular hollow building block is provided except that the webs are depressed and are approximately one-half the height of the building block although this relationship may vary where desired. Such building blocks are provided and they include corner blocks and standard procedures may be employed in installing the blocks so that reinforcement is provided if necessary; air vents are provided if necessary.

The insert or filler is generally designated by the numeral 24 and includes a body of foam plastic material 26 such as foam urethane or the like which is noncombustible and has very good insulating properties. One longitudinal edge of the generally rectangular block is provided with a plurality of notches 28 therein which are spaced apart a distance to receive the webs or the blocks therebetween.

As illustrated in Figure 2.8c, four notches are provided with the notches being generally one-half of the vertical dimension of the insert 26 so that when the notches receive the webs, the space between the webs will be filled by the material between the notches and the remainder of the vertical space between the walls of the building blocks will be filled by the remainder of the insert **26** thereby providing a vertical insert of insulating material which occupies the vertical and horizontal dimensions of the block except for the space occupied by the webs. The thickness of the block **26** is substantially less than the space between the inner and outer walls of the building block as illustrated in Figure 2.8a.

FIGURE 2.8: INSULATING FILLER FOR HOLLOW BUILDING BLOCK

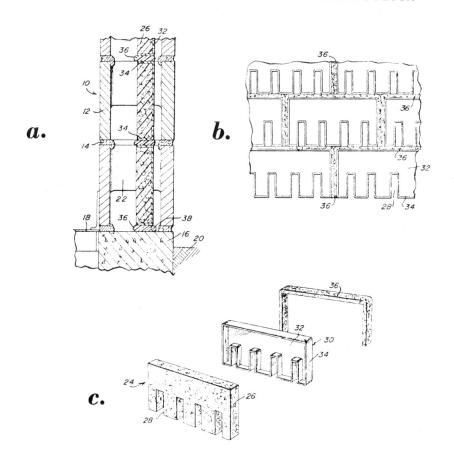

    (a)   Sectional view of a hollow wall formed by a plurality of hollow building blocks with the insulating filler and insulating strip incorporated

    (b)   Side elevational view of an assembly of the insert filler and insulating strips showing the continuity of the insulating barrier formed thereby when assembled

    (c)   Exploded perspective view illustrating the insulating filler, the enclosure and the insulating strip

Source: U.S. Patent 3,704,562

An enclosure **30** is provided for the insulating block and includes a facing member **32** and a peripheral edge member **34** which completely encloses one face of the insulating block and the periphery of the insulating block including the periphery of each of the notches **28**. The enclosure may be constructed of preformed plastic sheeting or treated paper which will provide a vapor barrier for the insulating block or insert **26** and also provide additional mechanical strength to prevent damage to the insert during assembly with the building blocks.

The enclosure may be used to shape the foam plastic when it is being formed and the enclosures may be connected together by a perforated or prescored line for separation of the inserts. This would enable a plurality of the inserts to be formed by using the enclosures as a form with the enclosed inserts then being separable from each other after hardening so that they may be assembled with the building blocks **12**.

An insulating strip **36** is constructed of a compressible insulative material which is engaged with the continuous top peripheral edge and the two side peripheral edges of the enclosure so that the insulative strips will compressibly engage each other where they pass along the side edges **34** and will engage the bottom edge of an overlying and staggered course of inserts **24** as illustrated in Figure 2.8b so that a continuous insulating barrier will be formed by the inserts **24** and the enclosures when combined with the insulating strips. Where the lowermost filler or insert engages the footing **16**, the insulating strip may be provided with a continuous bead of mastic **38** to assure better waterproofing of the joint.

Styrenated Chlorinated Polyester Resin

According to a process described by *J.L. Head; U.S. Patent 3,925,090; Dec. 9, 1975* a cellular cement composition is obtained from reactants consisting essentially of portland cement, water, a styrenated chlorinated polyester resin, aluminum flakes, and an alkali.

The compositions are obtained by mixing vigorously and reacting in the following order these reactants: (a) from about 22 to about 28% by weight of water and from about 65 to about 70% by weight of portland cement, (b) from about 5 to about 10% by weight of a styrenated chlorinated polyester precursor mix, (c) from about 0.04 to about 0.07% by weight of fine aluminum flakes, and (d) from about 0.10 to about 0.15% by weight of an alkali.

The end product is a low density, soundproof, heat-impermeable cement which is useful for the manufacture of structural units, including building blocks, beams, slabs, pipe, and the like, as well as walls, floors, and exterior surfaces. Cellular cement is also used in small decorative items such as statuary and bird baths, as well as in bath tubs and burial vaults.

The following example is typical of the manner in which the cellular cement is made: 32 gallons of water (266.6 lb, 24.3% by weight) of water and eight 94 pound bags of portland cement, type I (752 lb, 68.6% by weight) were mixed vigorously for 2 minutes at ambient temperature. Then 75 lb (6.9% by weight) of a styrenated chlorinated polyester precursor mix consisting of 65 parts of chlorinated polyester resin (Hetron 197), 35 parts of styrene, 0.5 part of cobalt naphthenate, and 1.0 part of methyl ethyl ketone peroxide was mixed vigorously with the foregoing water-cement mixture for 5 minutes. Then, 0.7 lb of fine

aluminum flakes (300 to 400 mesh; 0.06% by weight) was added to the foregoing and vigorous mixing was continued for about 30 seconds. Finally, 1.4 lb
of sodium hydroxide (0.12% by weight) was added. Vigorous stirring was continued until incipient gelation was achieved, that is, until the mixture had plastic
strength such that discrete bubbles remain separate and continue to grow in
size. This stage was usually reached within about 30 seconds of vigorous stirring after the addition of the sodium hydroxide. The mixture was then poured
into an appropriate construction form and allowed to set overnight. The forms
were then removed.

Radiant Energy Reflectors

H.S. Morton; U.S. Patent 3,654,740; April 11, 1972 describes a concrete construction block having protruding flanged arms on one side, the flanged arms
being angled so as to hold insulating radiant energy reflectors. The reflectors
are constructed of a resilient material and are held between the flanged arms under tension. The arms are so angled with respect to the block face that the resilient radiant energy reflectors form alternating concave and convex surfaces
along the interior wall surface. The angle of protrusion of the arms also permits an overlap of the adjacent reflecting surfaces to insure maximum efficiency.
A corner block having a protruding arm adapted to receive a concave or convex
reflecting surface provides proper mating with divisible segments of the basic
construction block so no break in the continuity of the reflecting surfaces exists
at the corners of the building.

OTHER CONSTRUCTION TECHNIQUES

Reinforced Epoxy-Concrete Structures

A process described by *R.G. Reineman; U.S. Patent 3,922,413; November 25,
1975* provides lightweight, high strength, reinforced concrete constructions
which are formed of a plurality of substantially coextensive, parallel, spaced
apart membranes. These membranes consist of alternate, integrally bonded
layers of fiber-reinforced epoxy resin and epoxy-resin-containing concrete interconnected by continuous longitudinal and lateral transverse webs of epoxy-resin-
containing concrete that form a unitary structure having a plurality of enclosed
cavities which are each filled with a hollow-form core. The construction can be
in the form of a flat sheet or panel of any desired size, or it can have a curvi-
linear configuration.

Referring to Figure 2.9, the numeral 10 designates a lightweight, high strength,
construction comprised of a pair of substantially coextensive, parallel membranes
12 and 14 of alternate integrally bonded layers of fiber-reinforced epoxy resin
and epoxy-resin-containing concrete interconnected by longitudinal transverse
web 16 and lateral transverse webs 18 of epoxy-resin-containing concrete. The
construction is bounded by longitudinal transverse members 20 and end transverse members 22, which form a unitary structure defining a plurality of enclosed
cavities filled with a hollow-form core material 24, the hollow-form core material
filling cavity 24a being deleted to illustrate the structural features of construction 10.

FIGURE 2.9: LIGHTWEIGHT REINFORCED CONCRETE CONSTRUCTION

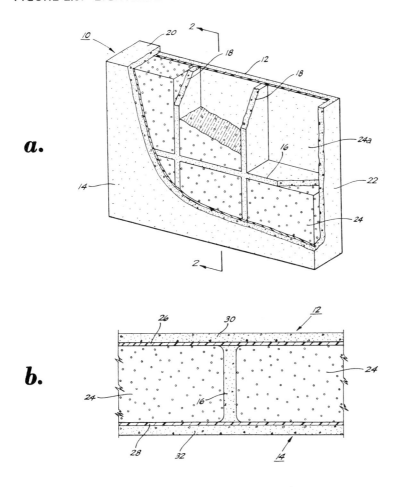

(a)   Partially cut-away isometric view of construction
(b)   Partial cross-sectional view taken along the line 2–2 of Figure 2.9a

Source: U.S. Patent 3,922,413

As more particularly illustrated in Figure 2.9b, parallel membranes **12** and **14** are formed of inner layers **26** and **28**, respectively, of fiber-reinforced epoxy resin and outer layers **30** and **32**, respectively, of epoxy–resin-containing concrete.

To manufacture a construction in accordance with the example illustrated in Figures 2.9a and 2.9b, the layer **32** of uncured epoxy–resin-containing cement is applied to the mold and overlayed with layer **28** of epoxy-resin-saturated fibers. Next, blocks of hollow-form core materials **24** are placed on epoxy resin layer **28** and suitably spaced to provide the desired configuration of transverse webs. Separate units of the core material can be placed in the mold, or the core material can be placed in the mold in the form of one or more large slabs, in which case the grooves for the webs are formed by cutting out or routing the desired pattern of grooves in the core material.

After the hollow-form core material is in place and properly positioned or cut, the grooves between adjacent pieces of core material and around the periphery of the core material are filled with uncured epoxy-resin-containing cement. Layer 26 of epoxy-resin-saturated fibers is applied directly to the surface of the core material and the wet epoxy-resin-containing cement in the web and peripheral grooves. The construction is completed by applying the layer 30 of epoxy-resin-containing cement.

This final layer of uncured cementitious material can be finished in any conventional manner to provide the desired finish, such as by trowelling, floating, rubber floating, brooming, and the like, or it can be left unfinished. It is sometimes preferable to apply a light coat of epoxy resin over this final layer to act as a sealer and to inhibit crazing of the cementitious material. After the wet cement mixture and epoxy resin have set sufficiently that the structure has sufficient strength to be handled, the construction can be removed from the mold. Also, where desired for decorative purposes, a light finish coat of stucco or plaster, not shown, can be applied to the exterior surfaces of the construction.

The products of this process are relatively lightweight, high strength, rigid, precisely formed members exhibiting low heat and sound conductivities and possessing excellent fire, weather, and vermin resistances. Because of their high flexural strengths and impact resistance, these constructions have high earthquake and shock loading properties.

Outer Wall Insulating System

A process described by *M.S. Todorovic; U.S. Patent 3,906,693; September 23, 1975* constitutes an outer wall insulation system which includes a plurality of blocks secured to the outer wall of a residential or industrial structure. Positioned between the blocks is any desired type of insulation, for example, fiberglass, or rockwool. The insulation is held in place by a plurality of elongated members which are secured to the outer surface of the blocks. The elongated members possess a pair of transverse flanges which are adapted to facilitate the securement of the siding at its desired location. After the elongated members have been utilized to secure the insulation in place, a conventional siding material, such as aluminum siding, is placed over both the insulation and the elongated members which hold the insulation.

Wooden Wall Log

V.M. Vizziello; U.S. Patent 3,992,838; November 23, 1976; assigned to New England Log Homes, Inc. describes a building construction element in the form of an insulated wood log with flat planed upper and lower surfaces. A longitudinally extending groove is cut into each surface to a depth about half the thickness of the log, the grooves being on opposite sides of the vertical medial plane of the log, and each groove is filled with a foamed plastic mass having thermal insulating properties substantially equivalent to those of polyurethane plastics.

Referring to Figure 2.10a and 2.10b, a natural wood log **10** is normally debarked, then sawed and planed to form flat parallel upper and lower faces **11**, **12**, each provided with a medially disposed longitudinal groove or kerf **13**, **14**. A wider and deeper groove **15**, **16** is cut into each planed face, adjacent and parallel to the kerfs, the grooves being on opposite sides of the vertical plane through the kerfs and each groove extending approximately half-way through the log. As orders of magnitude, the wall logs shown in Figures 2.10a and 2.10b usually average 9" in diameter with variations between 7" minimum and 11" maximum, the kerfs being ½" wide and the grooves being 1" wide. The logs being about 6" thick, each groove has a depth of about 3". These stated dimensions are by way of example, only, and may be varied as circumstances dictate or suggest.

Each of the grooves is filled with a suitable foamed plastic **17**, **18**, preferably foamed in situ to ensure good adhesion and complete filling of the space, and the stability and rigidity of the combination may be further aided by the provision of ties in the form of headless metal spikes or pins **20**, driven horizontally into the sides of the log at suitable intervals and passing through the plastic at a level of slightly less than half the depth of the groove. Long screws could be substituted for the spikes with the same stabilizing effect but at a higher cost. Proper vertical alignment of the logs, and basic sealing of the horizontal cracks between them, is effected by the provision of splines **21**, designed to fit in the kerfs of adjacent logs as shown in Figure 2.10b.

FIGURE 2.10: INSULATED WALL LOG

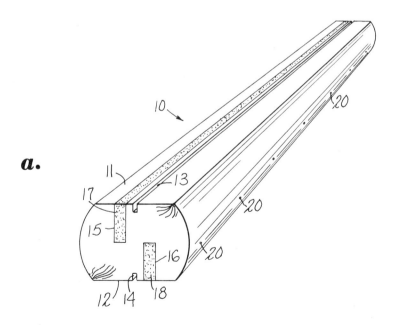

a.

(continued)

FIGURE 2.10: (continued)

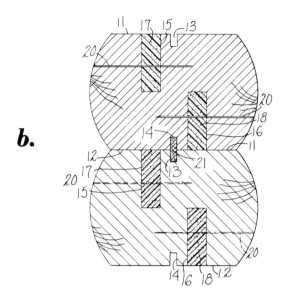

_b._

(a)   Perspective view of a log with insulation-filled grooves
(b)   Vertical transverse section through two logs, one superimposed
         on the other

Source: U.S. Patent 3,992,838

In each of the logs shown the inner ends of the filled grooves are at least 1"
apart so that the integrity of the log is not impaired and the foamed plastic,
particularly urethane base plastics, is rigid, durable, strongly adhesive, fire resistant and characterized by excellent thermal insulating qualities. (R = 9.09 in a
typical urethane foam.) Wood, though somewhat insulating, is generally a better
conductor of heat than urethane; by providing plastic foam inlays, disposed as
shown, the heat path through the log is attenuated, made tortuous and thus
elongated, resulting in substantially improved insulation. Wherever the horizontal thickness of the log is less than maximum, a plastic barrier to horizontal
heat transfer is encountered. In each log shown the maximum horizontal thickness is substantially greater than the vertical thickness, so that a wall built as in
Figure 2.10b has excellent thermal insulating qualities.

Urethane base plastics are particularly suitable because of their insulating, bonding and fire resistant qualities, but other materials such as polyethylene, ABS,
polypropylene and polystyrene are adapted for foaming and could be used.

Air-Supported Multiwall Structure

J.E. Choate and J.R. Hall; U.S. Patent 4,021,972; May 10, 1977 describe a structure which includes a closed envelope formed from a flexible, generally air and
moisture retaining, impervious, wall, skin, sheet or web preferably of plastic.

This main or primary sheet is contoured to form a dome, namely an inverted concaved member anchored to an appropriate footing or foundation and provided with an air impeller, blower or compressor, taking by suction air from the exterior and discharging into the interior of the envelope to maintain the envelope inflated. In one form of the process refrigeration units maintain the air in the interior at below freezing temperatures and artificial snow or ice is disposed on the interior ground surface, in the event the structure houses a ski slope or ice skating rink.

According to the process, a pervious flexible, secondary wall, skin, sheet or web, which is preferably in the form of an open mesh plastic net, is secured to the inside portion of the main sheet by means of flexible ribs. Thus, with the main sheet, this secondary sheet defines a plurality of juxtaposed, insulation containing, compartments or cavities which are filled with pop-corn-like, form retaining, lightweight, expanded plastic particles or aggregates of insulation.

In some cases, the compartments are vertically disposed, being separated by the parallel plastic ribs. In such an arrangement, the secondary wall or sheet is overlapped adjacent an upper portion of the structure so that a nozzle of an air gun can be inserted to discharge, in situ, the air impelled particles into the various cavities, successively. In another example these particles are discharged from the outside ground level, up through built-in, flexible plastic tubes leading to the upper area of the cavity, the tubes being prefabricated with and secured along the inner surface of the secondary wall. In either event quite large and well insulated structures are provided.

The process thus provides an enclosure, which is flexible, lightweight and inflatable, the enclosure being capable of ready assembly on the site with the insulation being capable of ready assembly on the site with the insulation being quickly and simply installed, in situ, after inflation of the enclosure.

End-Joined Cans for Home Structure

A.E. Moore; U.S. Patent 3,982,362; September 28, 1976 describes a low-cost building, building section or wall, usable as a house, mobile home, vehicle, or part of these or other structures, having numerous parallel rows of end-joined used or new cans. Each can preferably consists of a tube and can-cover elements at ends of the tube and containing thermal insulation, which may be only dead air in the relatively small hollow can space but preferalby includes loose small portions of low-cost insulating material, such as sawdust, vermiculite, cotton linters, dry sand or dust, ashes, cinders, ground bark, rice or other seed hulls.

Abutting pairs of can cover elements are tightly held together by connecting means which may be bonding material (epoxy putty, solder or the like) or bands of slightly stretchable adhesive tape which encompass and adhere to portions of juxtaposed pairs of the end-joined cans. A layer of mesh sheaths one side of each group of the can rows, and the other side is sheathed with wall material which may be mesh or insulating solid panels. Rod-like elements (bolts, rivets, pieces of wire or the like) go through a layer of mesh, between adjacent can rows and through the other layer of wall material, clamping them in assembled relation; and stucco impregnates and coats the mesh. The cans may be of the type that contain liquid drink, paint, or coffee, or may be glass jars. Optionally, used cans, each of which has only one end cover and its other end open, may be utilized.

Spandrel Units

H.E. McKelvey; U.S. Patent 3,999,345; December 28, 1976; assigned to Shatter-proof Glass Corporation describes a building element which is primarily adapted for use as a spandrel in the construction of the exterior walls of buildings. A spandrel is an exterior wall panel which usually fills the space from the bottom of the window sill to the top of the window below. The process provides insulated spandrels including two spaced parallel sheets or plates of glass sealed together around their peripheral edges to provide a dead air space therebetween and a substantially rigid sheet or layer of a cellular plastic insulating material disposed between the glass sheets and substantially filling the space.

Figure 2.11a illustrates diagramatically a portion of the front wall of a building including spandrels **10** which are mounted between vertically spaced windows **11**. Figure 2.11b shows the spandrels and windows supported by horizontal mullions **12** spaced from one another by vertical mullions **13** which are suitably attached to the floors or construction beams of the building as indicated at **14**. The mullions are preferably in the form of hollow I-beams to provide channel portions **15** in which the spandrels and windows are mounted.

As shown in Figures 2.11b and 2.11c, the spandrel consists of an insulated unit comprised of inner and outer spaced parallel sheets or plates of glass **16** and **17** respectively, preferably tempered, which are held in properly spaced relation by a substantially rectangular hollow spacing member **18** arranged between the glass sheets at their edges. A substantially U-shaped metal frame **19** surrounds the edges of the glass sheets and comprises a base portion **20** disposed opposite the peripheral edges of the glass sheets and bridging the spacing member. The base portion of the frame is provided with inwardly directed flanges **21** and **22** which overlie the outer marginal faces of the inner and outer glass sheets respectively and which are adapted to exert a clamping pressure to urge the glass sheets against the spacing member.

FIGURE 2.11: SPANDREL UNIT CONSTRUCTION

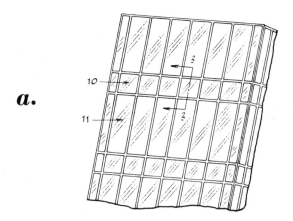

(continued)

FIGURE 2.11: (continued)

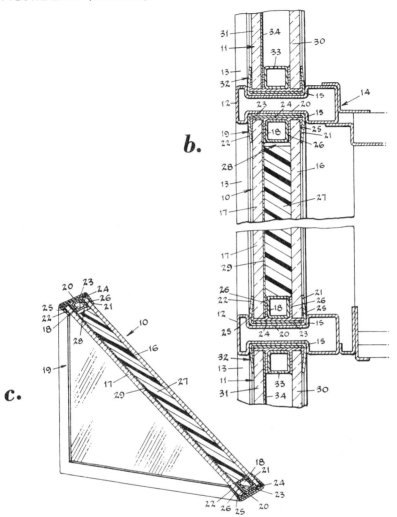

b.

c.

(a) Fragmentary view of a portion of the exterior wall of a building showing the relative arrangement of windows and spandrels
(b) Vertical transverse section taken substantially on line 2–2 of Figure 2.11a
(c) Persepctive sectional view of the insulated spandrel unit

Source: U.S. Patent 3,999,345

An elongated strip of sealing material **23**, such as asbestos tape or paper, is used to line the inner side of the base portion **20** of the mounting frame **19** and disposed inwardly of the sealing strip is a layer of a suitable adhesive material **24**, such as, for example, polyisobutylene, which also overlaps the outer marginal portions of the glass sheets, as indicated at **25**. A thin layer **26** of the adhesive material can also be applied between the spacing member and the glass sheets.

According to the process, there is disposed between the two glass sheets **16** and **17** a substantially rigid sheet or layer **27** of insulating material consisting of an expanded or foamed cellular plastic material. The cellular plastic materials used include expanded polystyrene plastic material composed of an agglomerate of expanded particles of polystyrene or polyurethane foams in a substantially rigid state so that they are self-supporting.

The space between the inner and outer sheets of glass may be of any desired width but is preferably in the range of from ¼" to 2". The layer of cellular plastic material is of a corresponding thickness so that it fills the space between the glass sheets and preferably contacts the inner surfaces thereof without, however, exerting sufficient pressure as would deform the glass sheets. The vertical and horizontal dimensions of the sheet of plastic material are slightly less than those of the space between the glass sheets to provide a small clearance between the peripheral edges of the sheet of plastic material and the base portion **20** of the mounting frame **19**, as indicated at **28**, to allow for thermal expansion.

By way of example, a typical spandrel unit consists of two sheets of ¼" tempered glass with a ½" dead air space between that is essentially filled with a sheet of expanded polystyrene having a density of 1.0 pcf, mean test temperature of 40°F and K factor of 0.24. By approved testing procedures, such a unit has been found to have a winter U-factor of 0.334 Btu/hr/ft²/°F based on an outside wind velocity of 15 mph and an inside wind velocity of 0, and a summer U-factor of 0.325 Btu/hr/ft²/°F based on an outside wind velocity of 7½ mph and an inside wind velocity of 0. While expanded polystyrene is ordinarily white in color, it may be painted or dyed black or a dark grey as desired.

The process also comprehends the provision of a very thin coating of a reflective metal **29**, such as, for example, gold or chromium which may be applied to the inner surface of the outer glass sheet **17** in any desired manner such as by the sputter-coating process. While it has been found that this coating does not have any appreciable effect on the insulating qualities of the spandrel unit it does enhance the appearance.

The spandrel unit is preferably used in combination with double-glazed windows **11** which, as shown in Figure 2.11b, are comprised of the spaced inner and outer glass sheets **30** and **31** mounted in a frame **32**, similar to mounting frame **19** of the spandrel units **10** and maintained in spaced relation by a spacer **33**. The windows are also mounted in the horizontal and vertical mullions **12** and **13**. By providing the inner surfaces of the outer glass sheets **31** of the windows with a thin coating of a reflective metal similar to that on the inner surfaces of the outer glass sheets of the spandrel units, the exterior wall of the building will have an overall mirror-like appearance.

Double Pane Window Containing Heat-Absorbing Fluid

R.T. Furner; U.S. Patent 4,024,726; May 24, 1977; assigned to Enercon West describes a system which substantially reduces the solar heat gain of building interiors through building windows by absorbing from the sunlight heat or infrared radiation before such heat can reach the building interior. It thus enables a substantial cost reduction in the initial installation of building air conditioning systems and it further significantly reduces the running costs of the air conditioning system since less energy is required.

Generally speaking, the process contemplates the provision of a double-pane window in which the two panes are spaced apart to define a window space. This space is filled with a heat-absorbing fluid, preferably clear water which is an ideal infrared radiation absorbing medium, so that part or all of the heat of the incoming sunlight will be absorbed by the fluid. The fluid is thereby correspondingly heated. A heat exchanger coupled to a refrigeration system is in contact with the fluid and cools it, that is, withdraws therefrom the heat energy absorbed from the sunlight. Consequently, the solar heat gain of building interiors is greatly reduced. The system further includes means for activating and deactivating the refrigeration system in response to the presence or absence, respectively, of sunlight entering the window so as to not unnecessarily cool the fluid at night or on cloudy days.

The coolant for the heat exchange supplied by the refrigeration system can be obtained from the central air conditioning unit for the building. In accordance with another aspect of the process, however, the coolant is obtained from a special absorption refrigeration system which is powered by sunlight, that is, by the heat energy obtained from sunlight. A heat absorption plate is installed adjacent, e.g., above the window in question on the building exterior so that it is exposed to sunlight whenever sunlight penetrates the window.

The heat absorbed by the plate is then used to power the absorption refrigeration system and cool the fluid in the window space without consuming costly energy (as, for example, consumed by the central air conditioning unit for the building). This construction has the advantage that the refrigeration action is automatically discontinued when the exposure to sunlight of the window ceases. Furthermore, as the infrared radiation of the sun increases, that is, as the sunlight becomes hotter due to atmospheric conditions, the angle of the sun, etc., the cooling rate of the refrigeration system increases correspondingly because of the greater heat absorption by the absorption plate. Thus, the system is self-regulating.

Stick-On Insulators

A.T. Franklin; U.S. Patent 3,729,879; May 1, 1973 describes an insulation assembly for a building structure which includes a device for quickly and easily securing the insulation to wall joists so as to eliminate the job of stapling insulation to the joists. The insulator consists of soft woolly insulation material such as glass fiber placed between paper or equivalent sheets which are glued together along their side edges. These side edges on one outer side have a pressure sensitive adhesive which is covered by a removable protective strip of paper that can be peeled off to expose the adhesive for placements against the wall joists.

Reflectorized Flexible Wall Covering

O.A. Becker; U.S. Patents 3,917,471; November 4, 1975; and 3,811,239; May 21, 1974 describes flexible composite elements useful as heat and cold insulating structures of improved heat and cold insulating power. The process involves preventing the transfer of thermal energy by heat radiation by means of highly reflective, metallized reflector foils and radiation chambers associated with them, while substantially limiting thermal conduction. They reflect the heat rays impinging on them to the extent of about 90% and radiate off only about 10% of the energy taken up. By flexible, insulating spacers, consisting preferably of soft, elastic plastic foam blankets which are perforated or slit and then expanded, radiation chambers for the reflector sheets are formed. By the perforating or slitting of the insulating agent, the chamber walls are reduced to thin foam ribs. The chambers are small and tightly sealed-off so that no convection of the air can take place.

Figure 2.12a shows in vertical cross-sectional view two reflector sheets or foils **1** and **2** consisting, for instance, of plastic sheets both of the surfaces of which have been vapor-coated in a vacuum with a highly reflective layer of aluminum. A perforated or slit and expanded plastic foam cover **3** is arranged between the two reflector sheets as spacer means forming insulating small radiation chambers **4**. The reflector sheets which preferably consist of plastic material coated by vaporization with aluminum on both sides, are connected and welded at their edges in air- and vapor-impervious manner to form an envelope or casing.

The inner air is displaced and replaced by predried air through valves **5** and **6**. For the protection of the envelope formed by the sheets and to prevent deposition of dust and precipitation of water of condensation, it is surrounded on all sides by a wide mesh, grid-shaped fabric **7** and **8**. Instead of a fabric, there can also be used, for instance, a flexible plastic foam sheet with or without perforations, and when not perforated, with strongly profiled surfaces, for instance, with protruding ribs extending in opposite directions on the two surfaces.

This composite element is closed off and sealed on all sides from the outside by further similar or, for instance, decoratively printed plastic sheets or foils **9** and **10** the edges of which are also connected and sealed to form a hermetic envelope. In this connection the inner surfaces of the sheets can also be coated by vapor deposition with highly reflective aluminum. The grid-shaped fabrics thereby form radiation chambers containing quiescent air. At the same time they assure the spacing between the reflector sheets. All parts of the composite element are flexible. In this way changes in air pressure and temperature can be accommodated by corresponding change in volume. If the second envelope formed by the sheets also consists of reflector sheets coated with aluminum by vapor deposition, a third replaceable decorative envelope can be provided.

Depending on the purpose which this composite element is to serve, for instance, as insulating covering for the walls of rooms, it may be advisable to connect the individual composite elements with a main pipeline or a hose line which in its turn is connected with an air-drying system. In this way, in case of change of volume in the envelope **1** and **2**, air can escape via the drying system and, conversely, in case of cooling, dried air can be supplied again via the drying system to the envelope.

The air-drying system can consist, for instance, in known manner of a housing with insertable screens to receive air-drying agents, for instance, calcium chloride. Air cooling units of known type can also be provided for the removal of water vapor either alone or in addition. Valves can be associated with the composite element and the air-drying system. Thus, for instance, an upper valve can open upon an increase of the pressure in the composite element and a lower valve in case of a decrease of the pressure can allow additional air to enter via the air-drying system.

FIGURE 2.12:   FLEXIBLE INSULATING COMPOSITE

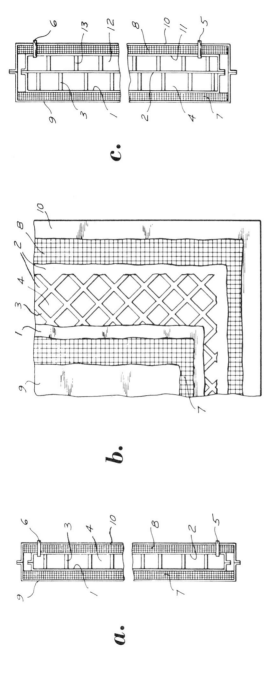

(a)   Composite element in cross-sectional view
(b)   Top view of the composite element of Figure 2.12a, but with layers partially removed
(c)   Another composite element in cross-sectional view

Source: U.S. Patent 3,917,471

Figure 2.12b shows the sequence of the layers of the composite element, commencing with the outer, for instance, decorative, protective sheet **9** and terminating with the opposite protective sheet **10**. The top view of the foam cover **3** shows the expanded slits or the radiation chambers **4** formed thereby.

Figure 2.12c is identical to Figure 2.12a, except for the insertion of a foam cover **13** forming further radiation chambers **12** and another reflector sheet **11** which is hermetically sealed to the reflector sheet **1** along the entire edge. There can be provided as required any desired number of insulating groups consisting of reflectors sheets **1** and **2** and the spacer means forming radiation chambers **4** and **12** to increase the insulating effect. The thin reflector sheets or foils and the fine surface coatings with vapor deposited aluminum are of no special importance with respect to the heat conduction, despite the fact that the edges of the sheets or foils are joined to each other.

The same applies also to the outer sheets **9** and **10**. The very soft foam layers **3** and **13** which are reduced to thin ribs, also conduct the heat to only a very slight and inconsiderable extent. Convection of the quiescent dry air enclosed in the chambers is also practically excluded for energy transfer. The depth of the chambers is preferably only about 10 mm. The wide-mesh, grid-shaped outer fabrics **7** and **8** also composed of insulating material, transfer the heat from the outer sheet to the inner sheet only to a slight extent. If the foam covers or layers and the fabric are also vapor-coated with aluminum then the insulating effect is further increased by reflection. The grid-shaped fabric can also be made of other material, for instance, of rubber or cords or the like. The products of the process are especially suitable for tents, wall and ceiling coverings, and floor coverings.

Porous Envelope Containing Particulate Material

J.T. Hughes and J.A. MacWilliam; U.S. Patent 3,869,334; March 4, 1975; assigned to Micropore Insulation Limited, England describes a panel of thermal insulating material which comprises a porous envelope containing a dry particulate insulating material, the particles of the insulating material being bonded to each other by pressure.

In one example of the process a microporous thermal insulating material is prepared by mixing microporous silica aerogel and titanium oxide as an opacifier. The amount of opacifier is preferably about one-tenth to twice the weight of the silica aerogel. A fibrous material can also be included in the mix, for example, if the insulating material is to be used at elevated temperatures, ceramic fibers can be used. For lower temperature applications, other fibrous materials can be used. Other microporous particulate materials which can be used instead of silica aerogel include finely divided microporous metal oxides such as alumina, zirconia and titania of similar particle size and structure as silica aerogel.

When an intimate mixture of silica aerogel or other microporous particulate material, opacifier and optionally the ceramic or other fibers has been prepared, the insulating material is placed in a porous envelope in the form of a bag of fiber glass cloth of low permeability. To fill the bag with the insulating material, the bag is attached to a discharge nozzle at the end of a feed pipe for the insulating material. The bag is contained within a chamber connected to a suction device and the pressure within the chamber is reduced to obtain an induced feed of the insulating material into the bag. When sufficient insulating material has been fed into the bag, the application of suction is discontinued, the bag is removed from the discharge nozzle and the open mouth of the bag is closed as by sewing or by application of a resinous sealant.

The bag containing the insulating material is then placed on the bottom die of a press and the upper die of the press is then lowered whereby the bag of insulating material is formed into a semirigid panel. The two dies of the press are so formed as to permit the escape of air through the pores of the bag. There is a build-up of reaction forces within the microporous insulating material which creates a tension strain in the fiber glass.

On application of the pressure to the bag of insulating material which causes air to escape from the bag through the pores of the fiber glass cloth, the particles of microporous silica aerogel become bonded to each other and the fiber glass bag, at least part of the bonding between the bag and the particles of silica aerogel resulting from penetration of the pores of the fiber glass bag by the silica aerogel particles. The silica aerogel particles thus become effectively mechanically interlocked with the material forming the envelope.

The tension strain created within the fiber glass cloth remains after release of the applied pressure and this induced tension strain provides additional rigidity for the insulating panel structure. If the fiber glass cloth is composed of a smooth yarn, the yarn is preferably pretreated by deposits thereon of a ceramic oxide or a chemical "starch" to increase the degree of bonding between the yarn and the insulating material.

One important application of a panel construction is in thermal storage heaters in which, because of the high insulating efficiency of the insulating panel, the thickness of the panel which is used can be substantially reduced as compared with previous insulating panels so that the size of the cabinet or like structure of the storage heater can be reduced giving a considerable saving in size and materials.

Carbonized Natural Grains in Core Material

E.I. Robinsky, J. Timusk and V.R. Riley; U.S. Patent 3,830,903; August 20, 1974 describe lightweight material in the form of expanded stabilized natural grains or cereals such as puffed wheat or puffed rice which are treated by a stabilizing means and used either alone or combined with other materials. The treatment of the grains, after expansion by stabilizing the same, such as by carbonization, enhances their properties as a building material by improving mechanical properties, by reducing the softening effect of water, by increasing resistance to biological attack, etc. The carbonized expanded natural grains may be used as an insulating material in discrete particles, or bonded together into boards or other bodies or as a lightweight insulating aggregate for use in concrete or in spray-on applications.

A method of expanding the grains is well-known and is the same as that presently employed to produce breakfast cereals such as is described in U.S. Patent 3,505,076. Moist grains are placed in a closed vessel in which heat and pressure are applied. The pressure is then suddenly released, and instantaneous formation of steam within the kernel causes expansion. Another known method of expanding the grain is by extrusion.

Example 1:   Treatment of Expanded Natural Grains — The expanded grains are stabilized by being heated at a suitable temperature for a sufficient duration of time so that the grains are carbonized. This has the effect of stabilizing the grains as well as increasing their resistance to biological attack. As an alternative, unexpanded cereal grains may be heated to the same temperature range and this results in both the expansion and carbonization of the grains.

Example 2:   Stabilization of Natural Grains and Their Use with Urethane Foam — This example demonstrates a method of stabilizing the expanded grains. Before being expanded or puffed, unexpanded grains were soaked in a chemical solution of one ounce of phenylmercuric acetate per gallon of water. The saturated grains were then passed through a dryer to reduce the moisture content to the amount required for expansion and were then expanded by conventional puffing procedures.

The residue of phenylmercuric acetate remaining in the expanded grains acts as a bacteriostat to inhibit the growth of living organisms which could cause degradation of the grains. The expanded grains were then mixed in a pan mixer with an equal weight of a liquid chemical which produces an expanded urethane foam. The particular chemical formulation used was 16F-1402 [General Latex and Chemicals (Canada) Ltd.]. The mixture of expanded grains and liquid urethane was then placed in a rectangular mold and covered. Within 10 minutes, the liquid urethane expanded to engulf the beads and fill all interstitial voids between the beads. This created a rectangular insulating board of a density of 1.8 pcf.

Example 3:   Stabilization of Natural Grains and Their Use with Asphalt — This example demonstrates a method of stabilizing the expanded grains. 0.33 ft^3 of expanded grains were sprayed with a chemical solution of one ounce of phenylmercuric acetate per gallon of water to inhibit the growth of bacteria. The grains were manually mixed while being sprayed. The expanded grains were then passed through a dryer at 180°F until all the evaporable moisture had been driven off, and were then mixed with a roofing asphalt heated to 400°F in the amount of 4.5 lb of asphalt for each cubic foot of beads. The type of asphalt used was 140°/150°F softening point and was supplied by Imperial Oil Ltd., Canada. The mixing was carried out in a slow speed rotary mixer (30 rpm). The asphalt/expanded grain mixture was then poured into a rectangular mold. After cooling, a rectangular insulating board was produced.

Example 4:   Insulating Board — This example demonstrates a method of stabilizing the expanded grains. Expanded grains were heated at 600°F for 15 minutes in a closed container. This had the effect of producing expanded carbon beads. The beads were then mixed with a roofing asphalt heated to 400°F in the amount of 4 lb of asphalt for each cubic foot of beads. The type of asphalt used was 140°/150°F softening point. The mixing was carried out in a slow speed rotary mixer (30 rpm). The asphalt/expanded carbon bead mixture was then poured into a rectangular mold. After cooling, a rectangular insulating board was produced. The asphalt coating on the carbon beads strengthens the beads and renders them impervious to moisture.

Example 5:   Stabilization of Natural Grains — This example demonstrates a method of stabilizing the expanded grains. Expanded grains were heated at 600°F for 15 minutes in a closed container. This had the effect of producing expanded carbon beads. These beads were then soaked in a solution of colloidal silica for 1 hour. The particular solution used was Ludox colloidal silica type HS. The carbon beads were then passed through a dryer to remove all evaporable water. The residual coating of silica remaining on the carbon beads strengthened the beads and increased their resistance to moisture absorption.

Example 6: Building Panel Sandwich Core Material — A building panel sandwich core material is prepared by mixing the carbonized expanded natural grains of Example 1 with a Portland cement paste with or without small amounts of fillers such as sand or asbestos fibers. Preferable proportions are as follows: 80% carbonized expanded natural grains to 20% Portland cement paste by volume. The Portland cement paste preferably includes about 5% by weight of asbestos fibers or about 10% by weight of granular fillers such as sand to reduce shrinkage, and preferably has a water-cement ratio of about 0.25 by weight. The mixture of carbonized expanded natural grains and paste may be sprayed or cast and cured by conventional methods as for ordinary concrete. It has been found that the resulting composite has a density of approximately 15 pcf and a compressive strength of approximately 50 psi. This material provides an excellent core material for sandwich building panels because of the following reasons:

(1) It has a substantial shear modulus or stiffness, and this allows the efficient transfer of shearing stresses between the sandwich faces.

(2) It provides an acoustic and thermal insulation barrier.

(3) It will bond directly to the sandwich faces due to the setting of the cement provided that a rough surface or mechanical fasteners are present.

Corrugated Boxboard Based Fiber

D.L. Ruff and N.U. Siddiqui; U.S. Patent 3,701,672; October 31, 1972; assigned to Grefco, Inc. describe a building product which is comprised of from 20 to 95% expanded perlite, preferably 65 to 85% by weight; 5 to 80%, preferably 10 to 35% by weight of optimally refined corrugated boxboard based fiber and 0 to 15%, preferably 4 to 6% by weight of asphalt.

The corrugated boxboard fiber employed in the process can be new corrugated boxboard stock, if desired. However, it is a particular benefit of the process that the source can be waste corrugated boxes. Some of these waste boxes are entirely composed of kraft paper; most are composed of a corrugated layer of neutral sulfite semichemical stock (NSSC) covered on each side by a layer of regular unbleached or bleached kraft paper. The actual stock used may range from 100% NSSC to 100% kraft; however, it is preferable to employ stock containing about 40 to 75%, preferably 55 to 65% by weight, of the kraft base and 25 to 60%, preferably 35 to 45% by weight of NSSC. Especially advantageous is a stock mix consisting of about 60% kraft and about 40% NSSC as the fiber component.

A typical sample of corrugated box waste pulp refined in a Valley type beater in the range of 0 to 80 minutes was tested for Schopper-Reigler freeness. Fiber index values were determined on the pulp at the end of a 60 minute refining cycle. The refined stock was fibrous and nongelatinous. The following data were obtained.

Schopper-Reigler Freeness Values (cc)

Minutes Refined in Valley Type Beater	
0	880
10	640
20	530
40	230
60	180
80	120

Fiber Classification and Fiber Index of 60 Minute Refined Corrugated Box Waste

Sample Number	Mesh Size	Wt Oven Dried, g	%	Fiber Length, mm
1	+28	0.05	1.0	0.60
2	+48	1.60	32.0	0.29
3	+100	1.15	23.0	0.15
4	+200	0.75	15.0	0.074
5	-200	1.45	29.0	0.037
Total 5 cuts	–	5.00	100.0	–

$$\text{Fiber Index (L)} = \frac{(W_1L_1) + (W_2L_2) + (W_3L_3) + (W_4L_4) + (W_5L_5)}{W_t}$$

$$= \frac{(0.05 \times 0.6) + (1.60 \times 0.29) + (1.15 \times 0.15) + (0.75 \times 0.074) + (1.45 \times 0.037)}{5.0}$$

$$= 0.154 \text{ mm}$$

Where the W's = oven dried wt/cut in grams
Where the L's = average fiber length/cut in mm
W_t = total wt of oven dried pulp in grams

In contrast, when one employs kraft paper alone, refined as in the prior art until gelatinous and nonfibrous, one obtains Schopper-Reigler freeness values of less than 100 cc and a fiber index of 0.058 to 0.061 mm. In this process, one obtains valuable products in the Schopper-Reigler freeness range of 120 to 530 cc especially in the range of 120 to 230 cc and one must adhere rigidly to the identity and ranges of components in order to achieve the superior products of the process.

Example 1: A waste corrugated box comprising about 60% kraft and about 40% NSSC and weighing 1,000 grams air dry were cut into 4" x 4" pieces and added to 14.28 liters of water at 130°F. The 7% slurry was then pulped for 15 minutes, using conventional equipment at 1,800 rpm, and then 10.72 liters of water at 100°F were added to give the slurry a consistency of 4%. The pulping was continued for 30 minutes (total pulping time, 45 minutes) at which time the water had a temperature of 98°F. 3 liters of the 4% slurry were removed and reduced to a 2% slurry by the addition of water, in which the final slurry temperature was 77°F. The slurry was then placed in a Valley type beater. Samples were taken at the beginning of the refining process and at regular intervals and tested for Schopper-Riegler freeness values giving readings as shown in the first table above.

Example 2: A board containing 0% asphalt was prepared as follows: 31.2 grams (oven dried basis) of optimally refined waste corrugated boxboard based fiber was weighed out and slurried with 2,315 grams of tap water warmed to about 125°F. The above slurry was mixed for 5 minutes in a 4 liter beaker after which

103.8 grams of typical expanded perlite was slowly mixed with the aqueous slurry. After an additional 5 minutes of mixing, the slurry was poured into a forming box, dewatered, pressed, dried and tested. The resulting product had a density of 9.9 pcf and a MOR of 115 psi.

Tests on lab boards (all of which had the same basic formulation by weight, 23% fiber, 72% perlite and 5% asphalt emulsion solids) prepared from refined pulps reveal that maximum modulus of rupture (MOR) values are obtained using refining times of at least 40 minutes, preferably 60 minutes, using a Valley type beater, although values superior to present commercial products using different fiber stock are obtained after about only 20 minutes of refining time.

For instance, with board by weight consisting of 23% refined fiber (about 40% NSSC and about 60% kraft), 72% expanded perlite and 5% asphalt solids made to densities of 9 and 10 pcf, MOR values consistently exceeded 100 psi and typically reached 120 and above with retention of the other desirable characteristics. For instance, with the same boards there resulted internal bond values of 14 to 20 psi and compression resistance of 168 to 212 psi at 50% consolidation with extremely low water absorption.

The building product can be prepared for commercial use as an insulating board, form board, acoustical board, panel board, core board and structural board and the like in a wide range of densities, but particularly in the range 8 to 15 pcf. Throughout the range of densities just stated the products exhibit superior flexural strengths as evidenced by the modulus of rupture according to ASTM C-203. Typically, one obtains values of about 85 psi for board of 8 pcf density; values of up to 160 psi and typically at least 120 with 10 pcf density board; and values of up to and greater than 260 and typically at least 220 psi for 15 pcf density board.

In contrast, similar board using newsprint as the fiber portion and having a 10 pcf density gives results of about 65 psi. As a matter of fact, because of the relatively low MOR results that are typical of newsprint fiber perlite board, that board is limited from a practical basis essentially to a minimum density of about 10 pcf. No such restriction is imposed on the product of this process. The product exhibits high strength and thereby minimizes breakage during manufacture and handling, even at densities as low as 7 pcf.

The building product shows abrasion resistance superior to corresponding perlite newsprint board and thereby reducing the generation of dust. The board is at least comparable to newsprint board in having high water resistance and low thermal conductivity (a low K factor), and accordingly suffers no property detriments while achieving enhanced values in other categories, as described above.

METAL CASTING AND FURNACES

RISER SLEEVES AND HOT TOPS

Paper Fibers

D.A.J. Patterson and T.A. Williams; U.S. Patent 4,012,262; March 15, 1977 describe a composition for use in the manufacture of thermally insulating, refractory articles which is of plastic consistency and contains fibrous organic material and particulate refractory material, the fibrous material constituting from 1 to 20% by weight of the composition. The composition would usually include one or more other constituents, for example, a binder, a sintering agent, a lightweight filler and possibly inorganic fibrous material.

The composition is a stiff paste having a consistency similar to that of mortar used for bricklaying. Thus, when a heap of the composition is deposited on a flat surface, for example, the floor, it merely slumps without showing any tendency to flow out over the surface.

Such a composition, which may also be described as a plastic solid, is a stable mixture that is not prone to separate out after preparation. It can readily be handled and can easily be measured by weight or by volume into amounts having a consistent solids content for molding. Because of its low liquid content, it can easily be placed into a mold having a small number of vents for liquid extraction and it can be subjected to pressure in the simplest of presses in order to extract sufficient liquid to enable it to be firm enough to be removed from the mold for stove drying. The fibrous organic material of the composition preferably consists of paper fibers and/or cardboard fibers. The process is described in the following examples, in which parts are by weight.

Example 1: A pulp was prepared by mixing 10 parts of dry newspaper with 110 parts of water in a Z-blade mixer. After 60 minutes mixing, the newspaper and water had formed a fibrous pulp, to which were added, in the mixer 85 parts of crushed firebrick having particle size not exceeding 10 mesh (BSS sieve) and 5 parts of a phenol-formaldehyde novolak resin as a binder. Mixing was continued for a further 20 minutes after which time the mixture had a plastic consistency. A measured quantity of this mixture was inserted in an open-topped tile-forming mold, using a scoop, and was roughly spread within the mold cavity. The mold pressure plate was then fitted into the mold over the mixture and a pressure of about 50 psi was applied to express the greater part of the water and compact the mixture into the desired shape. The water was expressed through a number of fine holes in the mold side walls and also through the fine gaps between the mold pressure plate and the mold side walls. The pressed composition was then removed from the mold cavity and placed in a drying stove in which it was dried for a period of 3½ hours at a temperature of 150°C to remove the remaining water and cure the resin.

The resultng tile had satisfactory insulating, refractory and mechanical properties when used as part of the insulating lining for the top of an ingot mold and its dimensional accuracy and surface finish were excellent. When the tile-making process was repeated, large numbers of tiles were produced having very consistent dimensions and mechanical properties and of very consistent weight and density.

Example 2: A pulp was prepared by adding 2½ parts of dry cardboard to 45 parts of water in a hydropulper. This mixture was beaten at high speed for 10 minutes by which time it had formed a fibrous pulp. This pulp was then transferred to a Z-blade mixer where 60 parts of alumino-silicate refractory powder (-10 mesh), 30 parts of calcined alumina, 2½ parts of alumino-silicate fibers and 5 parts of phenol-formaldehyde novolak resin were added. The resultant mixture was then mixed for 20 minutes after which time it had a plastic consistency.

A second pulp was then prepared by adding 20 parts of dry paper to 200 parts of water in a hydropulper and beating this at high speed for 10 minutes to form a fibrous pulp. This pulp was then transferred to a Z-blade mixer where 72 parts of crushed firebrick (-10 mesh), 5 parts of vermiculite powder and 3 parts of ball clay were added. This second mixture was then mixed for 20 minutes after which it also had a plastic consistency.

A measured quantity of the first mixture was then inserted into an open topped tile mold and was spread evenly over the lower part of the mold. A measured quantity of the second mixture was then inserted into the mold and spread evenly over the first mixture. The proportions of the two mixtures were so adjusted as to give a layer of the first mixture of approximately one third of the thickness of the layer of the second mixture.

The mold pressure plate was then fitted into the mold cavity and the contents were pressed and stove dried in similar fashion to that described in Example 1. The resulting ingot mold tile for steelmaking had satisfactory mechanical properties and was dimensionally accurate. The pressing of the two mixtures into one integral tile form enabled a tile to be produced in which the side facing the molten steel poured into the ingot mold (the layer of first mixture) had substantially improved refractory properties and was better fitted to resist penetration by the molten steel, while the side away from the molten steel and backing onto the ingot mold wall (the thicker layer of second mixture) had substantially improved thermally insulating properties and greater mechanical strength.

Flexible Fiber Mat

E.J. Jago and K.P. Cooley; U.S. Patent 4,024,007; May 17, 1977 and R.C. Phoenix and E.J. Jago; U.S. Patent 3,876,420; April 8, 1975; both assigned to Foseco Trading AG, Switzerland describe a method of providing a lining in a cavity which comprises applying to the surface to be lined a lining material which is a self-supporting deformable fiber mat refractory composition and exerting pressure on the surface of the applied composition sufficient permanently to deform it so as to conform the composition to the contour of the surface. The process is of particular value where the cavity is an ingot mold head or a head box. Preferably either the outer surface of the fiber mat refractory composition or the walls of the cavity to be lined are provided with a coating of adhesive to join the lining material to the wall.

A deforming means of particular value is an inflatable bladder, and according therefore to a particular feature of the process there is provided a method of lining a cavity with heat insulating material which comprises inserting into the cavity a preformed flexible lining material and an inflatable bladder, inflating the bladder to urge the material against the wall and to deform the lining material to the shape of the cavity to be lined and deflating and withdrawing the bladder. The cavity may be wholly or partly lined by such a method, the bladder being surrounded by or to one side of the lining material.

The lining material, if it be of a type which needs to be caused or allowed to set to a rigid condition, may be so caused or allowed to set prior to or after deflation and withdrawal of the bladder. In some cases, the lining may be preformed to the correct shape of the cavity to be lined, and then collapsed in order to allow insertion of the lining into the cavity.

By the method it is possible easily and quickly to line cavities with heat insulating material, without the necessity of ensuring that the preformed lining shapes used correspond in shape accurately to that of the cavity to be lined. It is even possible, by the use of the method to line a cavity of square cross section using a preformed hollow cylindrical shape of lining material. A further advantage of the method is that the lining material may be pressed onto the cavity walls uniformly. Such an advantage is automatically achieved when an inflatable bladder is used. The following examples illustrate the process.

Example 1: A 1% dispersion of Kaowool aluminum silicate fiber was made in a 30% colloidal silica sol. The slurry produced was filtered onto a 60 mesh bronze cylindrical screen to produce a mat of 150 mm internal diameter, 150 millimeters long and having a wall thickness of slightly over 12 mm. Forming was accomplished using a vacuum of approximately half an atmosphere drawn on the interior of the filter screen. The cylindrical mat was removed from the screen and flattened by gentle pressure. The flattened cylinder was inserted into a cylindrical cavity of approximately 175 mm internal diameter and opened sufficiently to enable insertion of a rubber bladder so that it extended out either end of the matted fiber sleeve. Air pressure was applied to the interior of the bladder by a hand pump, causing it to expand and force the mat into close contact with the metal pipe. On deflation of the bladder, the mat remained in place.

Example 2: A 150 mm diameter, 150 mm high, 12 mm thick cylindrical mat was prepared from Fiberfrax aluminum silicate fiber, novolak resin and methanol.

The fiber content was 1% and the resin content 5%. The wet mat was raised to 70°C for 15 minutes, at which time it was found to be dry, of good shape and flexible. Following collapsing of the cylinder, it was inserted in a 175 mm cylindrical cavity, preheated to 260°C. The cavity was part of the riser system of a large cast iron casting. The bladder was inserted and inflated causing the sleeve to be pushed against the cavity wall. The bladder was removed and the sleeve allowed to harden in place, the methanol being evaporated by the residual heat in the cavity. Other similar cavities, also constituting risers for the same large casting were lined with commercial riser sleeves. Molten cast iron at 1350°C was poured into the runner system of the casting, and the casting allowed to solidify.

After casting the riser sleeves were examined. The commercial sleeves were extensively eroded and the risers showed a quantity of pipe indicating low thermal insulation and high chill. The riser sleeve of this process was substantially undamaged, and the solidified riser had a fairly flat top, indicating high thermal insulation and low chill.

Refractory Fiber

W.C. Miller; U.S. Patent 4,014,704; March 29, 1977; assigned to Johns-Manville Corporation describes a refractory composition capable of being formed into riser sleeves, hot tops, ladle liners and the like for containing molten ferrous metal, which comprises in percentage by weight:

	Percent
Refractory fiber	30–50
Granular silicon carbide	1–35
Inorganic binder	5–45
Organic binder	2–10
Refractory filler	5–35

The refractory fiber must be one which is entirely or at least predominantly alumino-silicate fiber. A principal component of the composition is refractory fibers. These refractory fibers are inorganic fibrous materials which are synthetically formed as opposed to natural mineral fibers such as asbestos. In the composition of this process, it is required that more than half of the fibrous component be synthetic alumino-silicate fibers. These are fibers formed from melts of alumina and silica or predominantly of alumina and silica with lesser amounts of added oxides such as titania, zirconia, or chromia.

Typical of such materials are commercial refractory fibers known as Cerafiber and Fiberchrome. It is preferred that the refractory fiber component be entirely composed of synthetic alumino-silicate fibers.

Fibrous Wollastonite

J.A. Bognar; U.S. Patent 3,804,701; April 16, 1974; assigned to Oglebay Norton Company describes insulating compositions and structures for use in hot topping. The composition comprises on a dry basis from about 5 to 95 weight percent fibrous Wollastonite; at least about 3 weight percent of a second fibrous material selected from the group consisting of inorganic fibrous materials (i.e., other than fibrous Wollastonite), organic fibrous materials and mixtures thereof, the inorganic fibrous material being present in an amount within the range of 0 to

20 weight percent and the organic fibrous material being present in an amount within the range of 0 to 8 weight percent; an inorganic particulate filler present in an amount within the range of 0 to 90 weight percent; and from about 2 to 8 weight percent of a binder selected from the group consisting of inorganic or organic binders.

Wollastonite is a naturally occurring calcium metasilicate. It is the only pure white mineral available commercially which is wholly fibrous. Fiber lengths average 13 to 15 times the diameter. The following examples illustrate the process.

Example 1: A uniform slurry of 52% by weight water, 1.22% of rock wool, 18.3% of a 5% aqueous dispersion of pulp stock, 2.52% of F-1 fibrous Wollastonite, 24.43% of C-6 Wollastonite powder and 1.53% of phenol-formaldehyde resin was made by stirring to produce uniform dispersment. The composition was then poured into molds, dewatered and dried. The resulting composition had a density of about 55 pounds per cubic foot. On a dry basis, this produced a composition in the following proportions:

	Weight Percent
Fibrous Wollastonite	8
Rock wool	4
Paper fiber	3
Powdered Wollastonite	80
Resin binder	5

Example 2: A lower density product was prepared in a manner similar to that described in Example 1, the ingredients being present in the proportions indicated below (dry basis):

	Weight Percent
Fibrous Wollastonite	94
Paper fiber	2.5
Glass fiber	1
Resin binder	2.5

This composition produced a structure which was much softer than that produced from the composition of Example 1, and which had a density of about 31 pounds per cubic foot. This composition was deemed to be more useful in the manufacture of gaskets than panels or boards, while the denser composition of Example 1 was deemed to be more useful in the manufacture of panels or boards. The compositions produced insulating panels exhibiting a high degree of impact resistance and an ability to undergo moderate bending without breaking.

Heat-Expandable Composition

M. Takashima; U.S. Patent 3,923,526; December 2, 1975; assigned to Aikoh Co., Ltd., Japan describes a heat-insulating board for covering the top surface of a feeder head of molten metal which contains a material which expands upon being heated, or a mixture of such materials. The heat-insulating board is used in combination with a heat-insulating sleeve or slab for heat-insulating the side surface of the feeder head, and the heat-insulating board is expanded during use when the heat-expandable material(s) expand by the action of the heat from the molten metal to be cast, whereby any gap between the heat-insulating board and the sleeve, slab or a side wall of a casting mold is closed. The composition containing heat-expandable materials of the process preferably consists of the following ingredients:

	% by Weight
Natural mineral heat-expandable material and/or chemically treated heat-expandable material	5–80
Refractory material	0–92
Easily oxidizable metal	0–30
Oxidizing agent	0–30
Fibrous material	0–40
Carbonaceous substance	0–20
Oxidation accelerating agent	0–10
Binder	3–20

As heat-expandable natural minerals, vermiculite, shale, obsidian, perlite, pitchstone, bloating clay, etc., are suitable. As chemically treated heat-expandable materials, flake-like graphite, pitch, etc., treated with an acid and/or an oxidizing agent are suitable. These heat-expandable materials are known, and have been used as an ingredient for a feeder head heat-insulating composition.

Referring to Figure 3.1a, a heat-insulating board is provided in a mold **1** in such manner that the hollow space of the upper portion of the mold is covered with the board, which mold is lined with a slab **2** at the inner surface of the upper part thereof for the purpose of heat-insulating the side surfaces of the feeder head.

FIGURE 3.1: FEEDER HEAD COVER

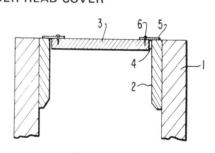

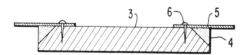

(a) Sectional view of an upper part of a mold with heat-insulating board.

(b) Enlarged section of the heat-insulating board of Figure 3.1a.

Source: U.S. Patent 3,923,526

The board consists of a plate-like molding **3** and a composition **4** attached to the outer rim of the molding **3**, which composition **4** contains a heat-expandable material or a mixture of such materials. The board is fixed to the mold by means of a thin steel plate or a steel wire **5** fixed to the board by a nail or a rivet **6**.

Into this mold, a molten metal, particularly a molten steel, is cast by the known bottom pouring method until the surface of the molten metal or molten steel is close to or is contacted with the heat-insulating board. In this heat-insulating board, the composition **4** is expanded by the action of a heat of the molten steel to fill the gap between the board and the slab **2**, whereby the atmosphere is excluded from the surface of the molten metal and the heat-insulation of the feeder head is improved. Figure 3.1b is an enlarged section of the heat-insulating board shown in Figure 3.1a. The following example illustrates the process.

Example: (1) A composition to be provided on the outer rim of a molding was formed of the following ingredients:

	% by Weight
Natural vermiculite	50
Cellulosic material derived from paper	6
Wood meal	5
Diatomaceous earth	10
Silica sand	23
Phenol-formaldehyde resin	6

(2) A plate-like molding was formed of the following ingredients:

	% by Weight
Aluminum	8
Ferrous oxide	16
Aluminum ash	38
Cellulosic material derived from paper	10
Asbestos	6
Carbonized rice husks	12
Cryolite	3
Phenol-formaldehyde resin	7

Formation Method — To mixtures of materials as set-out at (1) and (2) water was added to form the respective slurries. Using a model for dehydrating and molding, water contained in the slurry of mixture (2) was removed to form a plate-like molding. The slurry of mixture (1) was charged on the rim of the thus prepared molding, and water was removed to form a plate composed of the mixture (2) with the mixture (1) present only on the outer face of the plate. Drying the thus formed assembly, a heat-insulating board as shown in Figure 3.1b was prepared which had a size of 830 mm (length) x 830 mm (width) x 40 mm (thickness). Using the heat-insulating board thus prepared, six steel ingots each weighing 8 tons were cast by the bottom pouring method, using the construction shown in Figure 3.1a.

Steel ingot-making using the board of this process was compared with conventional steel ingot-making (six steel ingots) by using a heat-insulating board consisting of only the above described composition (2) under the same conditions, and the results obtained were as follows: The average yield of the ingots cast using the heat-insulating board according to the process was improved 1.3% as compared with the ingots cast using the conventional board. This was due to the superior heat-insulating effect of the heat-insulating board of the process which results from the close sealing of the gap between the board and the slab.

Silica-Treated Perlite and Cellulosic Composite

C.G. Sproule, Jr. and H.F. Wagner, Jr.; U.S. Patent 3,740,240; June 19, 1973; assigned to C.G. Sproule and Associates, Inc. describe an ingot mold hot top made of lightweight heat insulation board, especially adapted for use in regulating the solidification of a molten mass of metal in a mold. The board is prepared from a composition comprising perlite, cellulose filler material and binder. At least one surface of the resulting material is treated with silica, or other refractory material.

These materials are present in the following amounts: from about 60 to 80% by weight of perlite, from about 15 to 30% by weight of cellulose filler material, from about 5 to 10% by weight of a heat destructible binder, such as asphalt binder, and from about 3 to 6% by weight of clay, such as kaolin clay. A typical analysis of the inorganic material present in the composition is shown by the following table:

| | -----Percent by Weight------ | |
	Minimum	Maximum
Ferric oxide	0.05	1.5
Titanium dioxide	0.03	0.2
Magnesium oxide	0.1	0.5
Silicon dioxide	71.0	75.0
Aluminum oxide	12.5	18.0
Potassium oxide	4.0	5.0
Sodium oxide	2.9	4.0
Calcium oxide	0.5	2.0

As previously indicated, at least one surface of the resulting heat insulating material is treated with silica in an amount between about 2% by weight and 50% by weight, and preferably between about 5% by weight and 12% by weight.

The dried heat-insulating board, i.e., board which is not green is impregnated with silica to a depth up to ¼ inch employing any suitable procedure. For example, according to one procedure the board can be dipped into a silica solution and the depth of surface treatment regulated by the depth of the solution, concentration of silica in the solution, and the like. In accordance with another procedure, the board can be sprayed with a silica solution.

The silica solution can be simply a dispersion of silica in liquid medium such as water, such as Ludox (a colloidal silica—submicroscopic particles of silica in colloidal suspension in water), or such as Psyton. If desired, conventional wetting agents such as Alkanol WXN, can be added to the silica solution for further control of the surface treatment. By thus regulating the location of the surface treatment and the depth of impregnation the thermal efficiency of the board can be modified for varying applications. As this amount of silica impregnation is increased, the thermal efficiency of the board decreases. It has also been found that there is a definite relationship between the mass of the sinkhead and the depth of refractory surface treatment necessary to maintain the integrity of the heat-insulating board. For example, it has been found that boards of the process perform properly when used with a sinkhead of about 200 pounds if they have a surface tratment of about $1/32$ inch in depth. On the other hand, it has been found that the same boards need a surface treatment of about $1/8$ inch in depth if used with a sinkhead weighing about 4,000 pounds. After the board has been treated with silica it is then dried at room temperature or slightly elevated temperatures in an oven prior to utilization.

Heat-insulating material having the composition described above provides excellent characteristics when used to control the rate of solidification of a mass of molten metal in an ingot. Due to the low thermal conductivity of this material, the heat of the melt is not dissipated rapidly through the walls of the mold and/or hot top, but is retained in the molten metal. In this manner, there is provided a reservoir of molten metal in the upper portion of a mold or hot top which feeds the pipe or shrinkage cavity as it tends to form in the solidifying ingot, thereby preventing or at least minimizing the formation of pipe in the metal ingot. The heat-insulating material of the above composition is substantially incombustible at the temperatures encountered in the production of steel ingots.

The heat-insulating board not only results in increased ingot yield and improved quality, but results in significant savings in time and labor. Hot tops can be constructed from the heat-insulating board which weigh about $1/4$ to $1/6$ of the weight of hot tops previously used. Because of their light weight and durability, hot tops formed from the heat-insulating board can be easily handled by one man without material handling equipment.

Shaped Heat-Insulating Articles

A process described by *G. Norton; U.S. Patent 4,008,109; February 15, 1977; assigned to Chemincon Incorporated* relates to shaped heat-insulating articles for forming molten-metal-contacting linings for metallurgical molds and in particular relates to preformed refractory boards, slabs or sleeves for use, for example, in lining hot tops (heads of ingot molds) and risers in foundry molds.

According to the process a metallurgical heat insulating article for forming a molten-metal-contacting lining for metallurgical molds comprises a granular and/or fibrous refractory material which has a high infrared and ultraviolet radiation opacity, an exothermic mixture of a fuel and an oxidizing agent, and a binder and has a density below 0.7 gram per cubic centimeter. A heat-insulating lining according to the process preferably is of the following composition by weight:

	Percent
Refractory material (granular and/or fibrous)	50–67.5
Fuel	12.5–30
Oxidizing agent	2–4
Fluoride	0–2
Colloidal silica sol	2–15
Organic binder	5–8.5

The preferred method of forming a lining is to form an aqueous slurry of the composition, dewatering the slurry on a porous former of the required shape, and drying the shape so formed in an oven. A dispersing agent, such as aluminum sulfate or ferric chloride, may be added to assist the formation of the slurry. The following examples illustrate the process.

Example 1:

	Percent
Aluminosilica fiber	40.0
Calcined magnesia (–100 +400 BSS mesh)	11.5
Aluminum powder (–36 +400 BSS mesh)	21.5
Barium nitrate (–60 +200 BSS mesh)	4.0
Cryolite (–100 BSS mesh)	2.0
Colloidal silica (30% solids)	15.0
Starch (pregelled, partially cationic)	3.0
Phenol-formaldehyde two-stage resin (cures at 150° to 180°C)	2.0
Ferric chloride	1.0

Example 2:

	Percent
Aluminosilica fiber	42.2
Alumina (–100 BSS mesh)	7.0
Titania (–100 BSS mesh)	7.0
Aluminum (–200 +400 BSS mesh)	21.0
Barium nitrate (–60 +300 BSS mesh)	2.0
Sodium silico-fluoride (–100 BSS mesh)	1.0
Colloidal silica (30% solids)	12.0
Starch	4.5
Resin	2.5
Aluminum sulfate	0.8

Preformed shapes were made to the composition of the above two examples by the slurry method and had densities of about 0.35 gram per cubic centimeter. It should be noted that in this example the required low density is obtained by the use of a low density granular refractory and no fibrous refractory is added. A preformed shape was made to this composition by mixing the ingredients with 15 to 20% water to give a flowable mixture. This mixture was blown (or alternatively hand-rammed) into a suitable core box and was then stripped and dried at a temperature of 180° to 190°C. The preformed shape had a density of 0.45 gram per cubic centimeter.

FURNACE INSULATION–CERAMIC FIBER BLANKETS

Compressed Blanket

A process described by *W.S. Brady; U.S. Patent 3,854,262; December 17, 1974; assigned to The Babcock & Wilcox Company* relates in general to the lining of furnaces, and more particularly to the lining of a soaking pit cover where the lining is fabricated from flexible blankets of fibrous insulating material such as Kaowool, glass fibers and the like. In the process strips of fibrous blanket material are compressed to form the sections of the walls such as in panels, where the metallic parts exerting the compression pressure on the blanket material are buried in the material and thereby protected against the high furnace temperatures to which the surface of the liner is exposed during normal operations.

A corner of a furnace as shown in Figure 3.2a includes a portion of a furnace roof 10 and a portion of one upright side wall 11 enclosing a furnace space 12. The lining of the furnace is formed of strips of fibrous material 13 which are assembled in an impaled and compressed condition in panels 14, with a plurality of panels arranged in side by side array to cover the exposed surface of a furnace wall or walls.

Advantageously, with the construction, hereinafter described, the panels 14 for any particular furnace will be assembled away from the furnace, and after assembly may then be installed on the furnace walls or roof, as desired, with either welded or bolted mounting in the furnace frame.

Each panel 14 is formed with a flat metallic base plate 15, ordinarily of steel, and will be provided with angles 16 forming the sides and ends of the panel. The strips of fibrous insulating material 13 are inserted in the panel and held in place by impaling rods, or threaded bolts 21, with the fibrous strips compressed to fit in a panel, for example in the embodiment of Figure 3.2a the panel is 3 feet by 4 feet. In this particular instance the overall dimensions of the furnace wall to be covered permit a plurality of panels of this size to complete the entire wall coverage such as roof 10.

As shown, the base plate 15 is provided with angle irons 20 which are 3 inches by 1½ inch by 10 gauge, stainless steel. These angles are perforated, as hereinafter described, and as shown each angle 20 is provided with threaded bolts 21 welded to the upright flange 22 of the angle at a position generally about ½ inch from the end of the flange. The base 23 of the angle 20 is welded to the base plate 15 so that when assembled angles 20 are approximately 6 inches apart. The threaded bolts are 7 inches long and, as shown, 8 strips of 1 inch thick and 6 inches wide fibrous insulating material 13 are impaled on the threaded bolts 21 and compressed into the 7 inch space. Each threaded bolt is provided with a washer 25 and a nut 26 (see Figure 3.2d) which compresses the eight 1 inch wide strips into a 7 inch spacing.

As shown particularly in Figures 3.2a and 3.2b the bolts 21 are alternately spaced on the upright flanges 22 of the angles 20 on approximately 10 inch spacing. Each of the angles is perforated as at 30 as shown in Figure 3.2d to accommodate the washer 25 and nut 26 of a threaded bolt 21 from the next adjacent angle 20.

With this construction, an approximately 3 inch layer of fiberized insulating material covers all of the metallic parts of the panel. It will be apparent that compressing one end portion of the strips 13 will permit the outer portion to blouse so that on the furnace side of the panel the insulating material will be continuous to protect the metallic parts of the panel.

In referring to Figure 3.2c it will be noted that each of the metallic bases 15 of each panel is provided with a structural support 33 at each end. In the embodiment shown, the supports 33 are in the form of a channel iron. One leg of the channel is welded or otherwise secured to the base **15** of the panel while the other end of the channel will be bolted to structural steel work supporting the wall. This type of construction is useable for either the roof or the side walls of the furnace.

FIGURE 3.2: FURNACE LINING

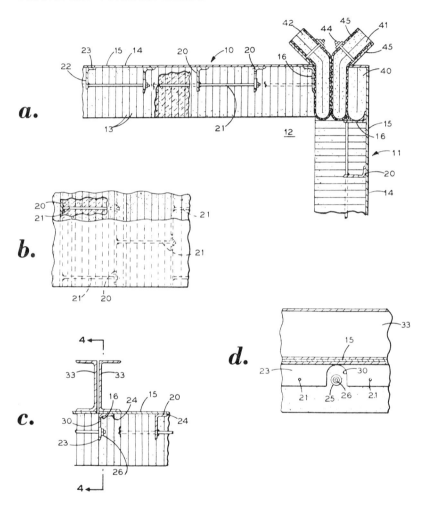

(a) Partial elevation, in section, of a furnace wall construction.
(b) Plain view, partially in section, showing a portion of the furnace wall arrangement of Figure 3.2a.
(c) Enlarged segment of a portion of the furnace wall arrangement shown in Figure 3.2a.
(d) View taken on line **4—4** of Figure 3.2c.

Source: U.S. Patent 3,854,262

Fiber Orientation

A process described by *R.A. Sauder, G.R. Kendrick and J.R. Mase; U.S. Patent 3,819,468; June 25, 1974; assigned to Sander Industries, Inc.* involves the use of a ceramic fiber mat which can be applied either directly to the interior of a high-temperature furnace or to an intermediate insulating member which, in turn, is attached to one of the furnace walls.

The ceramic fiber mat is preferably made up of strips which are cut transversely from a length of ceramic fiber blanketing which is commercially available. The strips are cut from the fiber blanket in widths that represent the linear distance from the cold face to the hot face of the insulating fiber mat. The strips which are cut from the blanket are placed on edge and laid lengthwise adjacent each other with a sufficient number of strips being employed to provide a mat of the desired width. Naturally, the thickness of the fiber blanket from which the strips are cut will determine the number of strips required to construct the mat. The strips can be fastened together by wires, or by ceramic cement or mortar which is preferably employed in the region of the cold face of the mat. The mat can be applied to the furnace wall or to an intermediate member by means of a stud welding method or by ceramic cement, mortar, or the like.

The ceramic fiber strips referred to herein are cut from a ceramic fiber blanket which is commercially available from several different manufacturers; these blankets are known as Kaowool, Fiber-Frax, Lo-Con and Cero-Felt. Most of these ceramic fiber blankets have an indicated maximum operating temperature of about 2300°F. The end or edge fiber exposure provided by the process not only provides an improved insulation up to the maximum indicated operating temperatures suggested by the manufacturers, but because devitrification and its deleterious effects are largely eliminated, also permits operation up to about 2800°F.

By arranging the fibers in an end or edgewise exposure, that is, where the fibers are oriented in planes generally perpendicular to the wall of the furnace, devitrification is not necessarily avoided but its undesirable side effects are minimized or eliminated because devitrification takes place at the ends of the fibers rather than along the lengths thereof; thus cracking and delamination are essentially avoided by the process even up to a temperature of 2800°F which is above recommended maximum temperature specifications imposed upon the fiber blankets by the manufacturers.

Another advantage which accrues from the use of the fiber blanket (or strips) in the end or edge exposure of the fibers is that the resulting mat has a certain resiliency in a direction parallel to the insulated face. Thus, where metallic fasteners are employed to attach the mat or composite block to the interior wall of the furnace or oven by burying or imbedding the fastener in the insulating member, this natural resiliency of the material will tend to keep the ends of the fastening elements completely covered at all times; this is true even if a tool is inserted in or through the fiber material to engage the metallic fastener for turning or welding purposes; after the tool has been withdrawn the natural resiliency of the fibrous material, as presently oriented, will cause the material to spring back and completely cover the outer end of the metallic fastening member.

End-On Orientation of Fibers

S.J. Shelley; U.S. Patent 3,930,916; January 6, 1976; assigned to Zirconal Processes Limited, England describes a lining element for constituting an inside surface or hot face for a furnace or oven, such lining element comprising refractory fibers oriented so that a substantial proportion of the fibers are end-on to the surface supporting the fibers.

The preferred method of making a lining element in accordance with the process is to cut into laterally extending strips a sheet of refractory fibrous material having fibers extending longitudinally parallel to the flat plane of the sheet. The strips are then turned through 90° and attached end on to a backing of unhardened refractory cement.

Referring to Figure 3.3a, the panel illustrated includes a rigid steel sheet 1, a layer 2 of Stillite Therbloc mineral wool and a layer 3 of vermiculite block. These two layers 2 and 3 are bolted to the steel sheet 1 by heat-resistant bolts 4 having heads 4' seating in recesses 5 in the layer 3.

Cemented to the layer 3 by a film 6 of refractory cements is the hot-face 7 of refractory fiber. The cement is made as follows: 300 ml of Nalfloc N 1030 silica sol is diluted with 200 ml of deionized water and to the diluted silica aquasol are added: (1) 25 grams of laponite SP synthetic clay; (2) 250 grams of sillimanite grade C 200; and (3) 1,000 grams of sillimanite grade CML 100. The powders are added individually with stirring in the order given. A thin slip is first obtained, thickening considerably on standing for 24 hours to provide the cement.

The hot-face 7 of refractory fiber is formed from strips 8 (seen end-on in Figure 3.3a) of ceramic fiber blanket (Triton Kaowool ceramic fiber blanket, one inch thickness and 8 lb cubic foot density is convenient) of the required dimensions joined edgewise to the vermiculite layer 3 through cement film 6. The strips 8 of ceramic fiber blanket are at least partially impregnated with a silica aquasol, a silica alcosol or an acid hydrolyzate of an alkyl silicate, as described in U.K. Patent 1,302,462. On heating, this makes the strips of ceramic fiber rigid and fixes the desired orientation of the fiber. Alternately, the strips 8 of ceramic fiber may be made partially or wholly rigid with an alumina gel binder.

Figures 3.3b and 3.3e illustrate the forming of the hot face 7 of refractory fiber in more detail. Figure 3.3b shows an uncut blanket 9 having fibers 9' extending longitudinally generally parallel to the plane of the blanket. Laterally extending cuts divide the blanket 9 into strips 8, as shown in Figure 3.3c. Each strip 8 is turned about its long axis through an angle of 90°, as shown in Figure 3.3d. The orientation of the fibers 9' is end-on relative to the panel. The strips 8 are now at least partially impregnated with either a silica aquasol, or a silica alcosol, or an acid hydrolyzate of an alkyl silicate.

The impregnated strips 8 are now placed on the cement film 6 applied to the vermiculite layer 3 as shown in Figure 3.3e to give the hot face 7 of the insulating panel. Alternatively, the strips 8 may be made rigid with an alumina gel binder. Impregnation of the strips 8 of ceramic fiber blanket to confer rigidity is not essential. If desired the strips 8 may, after turning through an angle of 90°, be placed on the cement film and pressure then applied to the strips 8 to make sure that good contact between the fibers and the cement is achieved over the whole surface. An essential feature of the process illustrated in this example is the turning of the ceramic fiber strips 8 through 90° to obtain the end-on orientation of the fibers 9' relative to supporting layers of the panel.

FIGURE 3.3: FURNACE LININGS

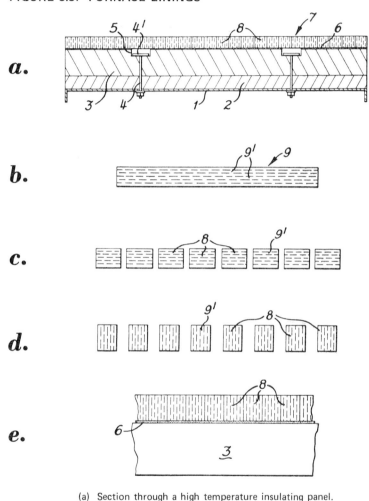

(a) Section through a high temperature insulating panel.
(b)–(e) Diagrammatic views illustrating how the hot face of
the panel of Figure 3.3a is constructed.

The modular panels shown in Figure 3.3a are desirably heated to 60° to 100°C to harden the cement and make the strips of ceramic fibers rigid thus fixing the desired orientation of the fibers. After this heat treatment, the modular units are ready for use in furnaces or oven constructions.

Other advantages arising from the use of ceramic fiber blanket in the way described in the example are that any available thickness and density of blanket can be used to obtain a wide range of thickness and density in the hot-face layer. Strips of the desired thickness are cut, then compressed to the desired density during the forming of the hot-face. This eliminates the formation of blanket off-cuts, which otherwise would be waste material.

Prefabricated Modules Using Compression Members

J.K. Balaz, R.A. Kelly and W. Schulze; U.S. Patent 3,832,815; September 3, 1974; assigned to Flinn & Dreffein Engineering Company describe a furnace lining comprising modular components fabricated of high temperature ceramic fiber blanket material, wherein the blanket material is arranged in layers in which the fibers in the layers are disposed substantially perpendicular to the hot face of the furnace lining, the layers being compressed during assembly into a resilient bundle calculated to compensate for thermal shrinkage of the blanket material. Prior compression of the blanket layers in this manner places the surfaces of the blanket material adjacent to similar surfaces of other blanket layers where mutually shared forces of compression inside the bundle prevent laminar peeling of the exposed surfaces of the felted blanket.

Compression forces inside each furnace lining module are resisted by pins threaded substantially perpendicularly through the blanket layers and disposed near the outer and cooler face of the module, that is, remote from the interior of the furnace. The pins not only serve to hold the ceramic fiber blanket layers in compression but can also be utilized to affix the modules in their assembled relation to the furnace shell, as by means of pin engaging hooks which are fastened to the furnace shell. The pins and hooks, being situated remotely from the hot face of the furnace lining, are not subjected to the deleterious effects existing within the interior of the hot furnace. Furthermore, because of their location, they do not afford a highly conductive path for heat to travel from the inside of the furnace to the outside surfaces of the furnace.

Supporting forces, transmitted to the ceramic fiber material by the pins, are thus directed substantially parallel to the direction of the greatest strength of the blanket. The layers of blanket material in adjacent modules preferably are oriented with 90° alternating rotation in the furnace lining assembly so that the preset compression forces in the modules are brought to bear against adjacent modules.

Compressional forces between modules may be advantageously augmented by forming the modules with strips of ceramic blanket of trapezoidal configuration instead of rectangular configuration, with the broader base of the trapezoidal strips facing the interior of the furnace. In this example, the modules have a substantially frusto-pyramidal configuration. As such frusto-pyramidal shaped modules are drawn into their intended position in the furnace lining, as by the aforementioned supporting hooks, they are squeezed into the configuration resembling that of a rectilinear parallelopiped.

Thus, the hot face of the furnace lining becomes laterally compressed in all directions parallel to the hot face of the lining, such compressive forces being maintained by the resilience of the ceramic fiber blanket material.

Furnaces lined with modules of the type described remain effective for months to exposure to temperatures as high as 2400°F and for much longer times without the need for repair or replacement at lower temperatures.

Prefabricated Blocks

C.O. Byrd, Jr.; U.S. Patents 4,001,996; January 11, 1977 and 3,952,470; April 27, 1976; both assigned to J.T. Thorpe Company describes an improved insulating block for lining a wall of a furnace or like equipment or for forming a wall of a furnace. An insulating blanket is folded into a plurality of folds of adjacent layers of fiber insulating materials, with a support member mounted in a fold of the insulating blanket. The support member is mounted by a suspension arm with an attachment member so that the blanket may be mounted with the wall of the furnace.

The blanket is formed with an inner surface portion exposed along an insulation surface, known as a hot face in the art, to the interior of the furnace, with a side surface portion extending outwardly at an end of the inner surface portion to a fold for receiving the support member. An inner end of the blanket inside the side surface portion extends inwardly from the fold to an interior surface of the inner surface portion opposite the hot face so that the support member is surrounded within the insulating blanket and protected from heat and corrosive substances in the furnace.

The prefabricated insulating blocks may be installed and removed easily and independently of adjacent insulating blocks lining a wall using the attachment structure of the process. For independent installation, the attachment member of the insulating block is inserted into and supported by a slide channel mounted with the wall. For independent removal of the insulating block, an access opening of the attachment member is aligned with a fastener aperture of the slide channel thereby allowing access to the fastener holding the slide member and insulating block in the mounted position for ease of installation and removal.

Alternately, plural insulating blankets are mounted and formed into composite blocks, by plural attachment means, mounted transversely with each other, to which the support members of the blankets are attached by the suspension arms, to thereby form composite blocks of a parquet-like construction for lining the furnace wall. In addition, the composite blocks may also be used to form the walls of the furnace itself.

In Figure 3.4d, the letter B designates generally the insulating block of the process for lining a wall (not shown), which may be either a side wall or a roof of a furnace or of some other high temperature equipment such as soaking pits, annealing furnaces, stress relieving units and the like. The insulating block B is preformed from folding insulating blankets, such as a blanket L, for insulating the furnace, with a support S (Figure 3.4c) mounted in certain of the folds in the folded blanket and an attachment mounting or channel M for mounting the supports and the blanket to the wall.

Considering the blanket L more in detail, such blanket is formed from a suitable commercially available needled ceramic fiber sheet, such as the type known as Cerablanket, containing alumina-silica fibers or other suitable commercially available refractory fibrous materials. It should be understood that the particular component materials of the ceramic fiber sheet used in the blankets are selected based upon the range of temperatures in the high temperature equipment in which the apparatus is to be installed.

In a first example (Figure 3.4a), the blanket L is folded into adjacent layers 10 mounted sinuously and extending inwardly and outwardly in such a manner between a first end layer 12 and a second end layer 14 at opposite ends of the attachment mounting or channel M. Adjacent ones of the layers 10 and those layers 10 adjacent the end layers 12 and 14 form inner folds 16, adjacent inner end portions 18 of the blanket L near an insulation surface 20, or hot face as termed in the art, exposed to interior conditions in the high temperature equipment. Outer folds 22 are formed between adjacent layers 10 at an opposite end adjacent outer end portions 24 at positions intermediate each of the inner folds 16.

The blanket L is supported at certain of the outer folds 22, designated 22a and 22b (Figure 3.4a) by a support beam 26, details of which are set forth in an alternate blanket example (Figure 3.4c) of the support S mounted in the folds 22. The support beam 26 is formed from a folded bar or a high temperature-resistant metal or alloy or other suitable material, although other shapes of support beams and materials may be used. The support beam 26 is mounted at a center portion 26a (Figure 3.4c) thereof within a loop 28 formed at a lower end juncture of suspension arms 30 and 32 of a suspending tab or support tab T of the attachment mounting M. The support beam 26 may be welded, such as by spot welding, and the loop 28 and the suspension arms 30 and 32 welded together for additional strength and support, if desired.

Alternately, the suspending tab T may be formed with a single suspension arm. An opening is formed in the center portion 26a of the U-shaped support beam 26, and the single suspension arm inserted to extend through such opening. The portion of the suspension arm extending through the opening is then bent to fit against one side of the support beam and secured to the support beam 26 by spot welding the suspension arm.

In the layers of the blanket L, the fibers of material normally extend longitudinally within the layer, as indicated by fibers F (Figure 3.4d). However, it has been found that a stronger and more compact insulating brick may be formed by needling adjacent layers together. In the needling process, a needle loom, such as a needle felting machine known as a fiber locker, a plurality of thin metal needles N, with plural pointed barbs 33 formed thereon are repeatedly forced through adjacent layers of the material to be fused together. As shown schematically in Figure 3.4d, the barbed needles N pierce perpendicularly into the adjacent layers and catch certain of the fibers F, changing the direction of their orientation from their normal longitudinal extension to a position where a portion F-1 of the fibers in the adjacent layers are transversely disposed to the remainder of the fibers and extend into other adjacent layers to bind the layers together into an insulating block.

In this manner, the perpendicular fibers bind the adjacent lamina or layers of the blanket together, compacting and strengthening the blanket. Further, the needling process binds the adjacent fibers together into a tougher, more homogeneous mass. An opening is formed through the outer end portions 24 of the blanket L adjacent the fold 22 receiving the support beam 26 (Figure 3.4c).

FIGURE 3.4: PREFABRICATED INSULATING BLOCKS FOR FURNACE LINING

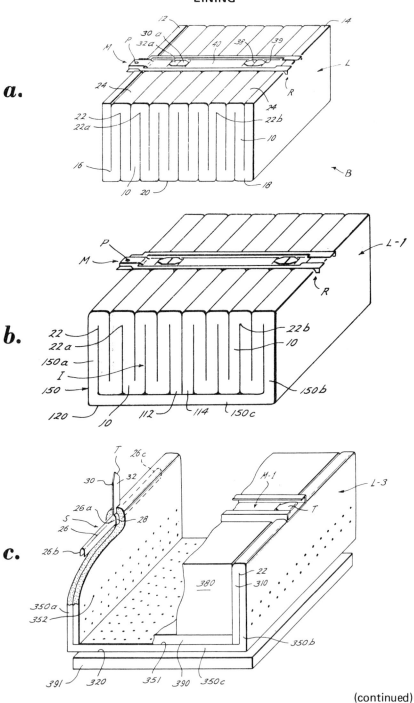

(continued)

FIGURE 3.4: (continued)

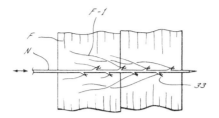

(a) Isometric view of an insulating block.
(b) Isometric view of an alternative insulating block.
(c) Isometric view of an alternate insulating block with portions broken away.
(d) Schematic cross-sectional representation of adjacent layers of fiber blankets being needled together.

Source: U.S. Patent 4,001,996

The opening so formed extends upwardly through the blanket **L** from the fold **22** for passage of the suspension arms **30** and **32** through the blanket **L**. It is to be noted that the support beam **26** is mounted to extend outwardly to ends **26b** and **26c** (Figure 3.4c) from the center portion **26a** thereof over a substantial portion of the lateral extent of the fold **22** in the blanket **L** in comparison to the width of the loop **28**.

Mounting lugs **30a** and **32a**, formed at upper ends of the suspension arms **30** and **32**, respectively, of each of the support tabs or suspending tabs **T** extend upwardly through mounting orifices **38** in a central attachment channel or a stringer channel member **40** of the attachment mounting **M**. The mounting lugs **30a** and **32a** are folded downwardly against the stringer channel member so that the block **B** may be mounted against the wall. The ends of mounting lugs **30a** and **32a** may in addition, if desired, be inserted to extend downwardly through mounting orifices **39** in the attachment mounting **M** so that sharp ends of the tabs **T** are enclosed beneath the attachment mounting **M**.

The insertion of the ends of the lugs **30a** and **32a** through the mounting orifices **39** protects the hands of installers against points or sharp surfaces at the ends and, in addition, further strengthens the connection of the supports to the attachment mounting **M**.

Additionally, each of the attachment mounting **M** has an attachment receptacle **R** formed at an end thereof and an attachment pin member **P** formed at an end opposite the attachment receptacle **R**. The attachment receptacle **R** of the apparatus receives the attachment pin **P** of an adjacent block of the apparatus, while the attachment pin **P** extends outwardly beyond the preformed insulation block **B** to provide access for welding in order to mount the block to the furnace wall. After such mounting, the pin **P** is fitted into an attachment receptacle **R** of another adjacent block **B**. A starting anchor having an attachment pin **P** mounted therewith is used at a starting location to begin installing operations. Further details concerning the attachment pin, attachment receptacle, the starting anchor and the installation of insulation blocks with a furnace wall using the attachment mounting M are set forth in U.S. Patent 3,952,470. In a second case (Figure 3.4b) a blanket L-1, formed from a single piece of suitable insulating material of the type set forth above, is folded so that adjacent layers 10 are formed which extend outwardly in a sinuous fashion from both a first end portion 112 and a second end portion 114 of the blanket L-1. The blanket L-1 is preferably mounted to the furnace wall with an attachment mounting M and supports S in a like manner to the blanket L as set forth above. The end layers 112 and 114 are preferably located at the center of the attachment mounting M.

The adjacent layers 10 and end portions 112 and 114 of the blanket L-1 form an inner portion I of the blanket L-1, which is enclosed by enclosing layer 150. The enclosing layer 150 includes side surface portions 150a and 150b extending inwardly from folds 22 formed by the side surface portions 150a and 150b and adjacent layers 10 located at the ends of the attachment mounting M. The side surface portions 150a and 150b extend to a continuous inner surface portion 150c of the enclosing layer 150. Thus, the blanket L-1 is a continuous folded member extending from the end portion 112 through layers 10, side surface portion 150a, inner surface portion 150c, side surface portion 150b, and layers 10 to end layer 114. The exterior surface 120 of the inner surface portion 150c, or hot face as termed in the art, is exposed to interior conditions in the high temperature equipment.

To obtain a stronger, more rigid, and more compact insulating block B, several or all of the adjacent layers 10 and 150 of the blanket L-1 may be needled together in the manner set forth above. The blanket L-1, formed by being folded in the manner set forth above, contains no through passages or avenues for hot gases to pass from the hot face to the cold face adjacent the attachment mounting M. Additionally, since the supports support the blanket L-1 at outer folds 22a and 22b in the manner as set forth for the first example (Figure 3.4a), the supports S are completely enclosed so that there is no path for the passage of heat and corrosive elements of the furnace atmosphere to the supports S.

In another example, (Figure 3.4c), a blanket L-3, formed from a single piece of suitable ceramic fiber insulating material, is first folded to form an inner surface portion 350c which is exposed along an interior insulation surface 320, or hot face, to interior conditions in the high temperature equipment. Side surface portions 350a and 350b of the blanket L-3 extend outwardly from each end of the inner surface portion 350c toward the wall of the furnace to a fold 22 formed therein for receiving a support S in the manner previously set forth. Inner wall member portions 310 adjacent the side surface portions 350b and 350c, respectively, extend inwardly from the fold 22 to an interior surface 351 of the inner surface portion 350c opposite the insulation surface 320.

The inner wall member portions 310 and the side surface portions 350a and 350b, respectively, are preferably needled together in the manner set forth above. This needling enables the blanket to hold itself to the support S and gives it sufficient strength to support a load. The needled blanket L-3 may be mounted to the furnace wall with a slide channel, attachment mounting M-1, and supports S in a like manner to the blanket L-2 set forth above. It should be noted that a blanket L-3 folded over the supports S in the manner just described completely encloses the supports S so that there is no direct path for furnace gases or heat to pass from the hot face to the supports S.

A large mass of bulk ceramic fiber **380**, or other lower temperature rated insulation refractory material of lower cost, is placed in an enclosure or pocket formed by surfaces **352** of the inner wall member portions **310**, the interior surface **351** of the inner surface portion **350c**, and the attachment mounting **M-1** which attaches the insulating block **B** to the wall of a furnace. This bulk material may be contained temporarily in a plastic or fiber container which will burn and be consumed when the insulating block is exposed to the heat of the furnace.

The mass of bulk ceramic fibers **380** which is compressed and retained in the plastic or fiber container also may be needled to more uniformly distribute, reorient and tie together the fibers so that when the container is removed or burned away the fibers will retain the shape of the container. However, the mass of fibers **380** need not be needled together in order to construct the block **B**.

In addition, the inner surface portion **350c** and the side surface portions **350b** and **350a** of the blanket **L-3** may be needled to the mass of bulk ceramic fibers **380**. This will prevent the fibers in the center of the insulating block **B** from falling out if the inner surface portion **350c** of the blanket **L-3** fails. It should be noted that the needled ceramic fiber insulating blanket **L-3** can support many times its own weight when the blanket is placed in tension by material placed in the above described enclosure formed by the blanket **L-3**. It should be also noted that the use of bulk fibers supported by the folded blanket **L-3** substantially lowers the cost of the insulating block **B** without impairing the ability of the insulating block **B** to withstand high temperatures.

Further, a ceramic fiberboard **390**, or alternatively a layer of relatively dense material such as 30 lb fused silica foam, may be placed in the bottom of the enclosure formed by the folded blanket **L-3** prior to insertion therein of the mass of bulk fibers **380**. The ceramic fiberboard **390** gives structural strength to the enclosure formed by the blanket **L-3**, thereby preventing the insulating block **B** from sagging.

In addition, an insulating mat **391** having a higher temperature capability than the blanket **L-3** may be mounted on the exterior surface **320** of the inner surface portion **350c** between the outer surfaces of the side portions **350a** and **350b**. The insulating mat **391** is bound to the inner surface portion **350c** by needling the inner surface portion **350c** to the insulating mat **391**. Thus, it is possible to use different strata of materials for the block **B** to provide the lowest possible conductivity and cost commensurate with the conditions to which the block **B** will be exposed.

Further, if desired, the insulating mat **391** may be mounted in an offset position with respect to the inner surface portion **350c** of the blanket **L-3**. The insulating mat **391** so mounted overlaps the gap between adjacent insulating blocks **B** when plural insulating blocks **B** are mounted adjacent each other for lining the furnace wall.

This overlapping of the gap between adjacent insulating blocks **B** by the insulating mat **391** prevents the flow of hot gases from the hot face to the cold face in the space between contacting surfaces of adjacent blocks **B**.

FURNACE INSULATION—OTHER PROCESSES

Refractory Gas Concrete Using Chrome-Alumina Slag

A process described by *K.D. Nekrasov, A.P. Tarasova, A.A. Bljusin, T.P. Avdeeva, V.A. Elin, P.A. Roizman and A.P. Denisenko; U.S. Patent 3,784,385; January 8, 1974* relates to a method of preparing a mix for producing refractory gas concrete, and to the product thus obtained, which can find application in various branches of the economy, particularly, in the metallurgical industry as a material for heat insulation and lining of heat-treating, annealing and open-hearth furnaces where a temperature of up to 1200°C is to be maintained.

The essence of the process is as follows. Chrome-alumina slag which is a waste product of the aluminothermic process of producing metallic chromium features an ability of expanding when subjected to wet heat treatment, and therefore its utilization as a filler in a mix for preparing refractory gas concrete allows a reduction of shrinkage phenomena and an increase of the temperature at which the gas concrete can be employed.

An average chemical composition of the slag used is as follows: 75 to 80 wt % Al_2O_3, 4 to 10 wt % CaO + MgO, 5 to 10 wt % Cr_2O_3, 3.5 wt % Na_2O, and 0.7 to 1.0 wt % SiO_2.

The use of the chrome-alumina slag alone as a filler in the refractory gas concrete makes possible the production of a gas concrete with a strength of 5 to 8 kg/cm²; therefore for increasing the strength characteristics of the refractory gas concrete a high-alumina refractory material is introduced into the mix, which allows an increase in the strength of the gas concrete to 12 to 20 kg/cm² and an increase of the temperature at which the gas concrete can operate. A high-alumina powder prepared from wastes of broken high-alumina articles must contain not less than 62% of aluminum oxide.

Both the high-alumina refractory material and chrome-alumina slag must be ground to such a degree of fineness, that not less than 70% of a sample should pass though a sieve with a mesh of 4,900 apertures per cm². The refractory and slag materials introduced into the gas concrete mix as fillers in the powdered state make it possible to obtain mixes with a homogeneous structure, this being a very important factor for producing cellular concretes. The refractory gas concrete produced from the mix prepared by this process has the following physico-mechanical properties: (a) up to 1200°C operation temperature; (b) not less than 12 to 20 kg/cm² ultimate compression strength after the maximum operating temperature; (c) not higher than 1% additional shrinkage at operating temperature; and (d) not less than 500 to 800 kg/m³ and over volume mass.

A gas concrete mixer is started and then charged with water preheated to 65° or 70°C, and an aqueous solution of sodium silicate and sodium hydroxide; then powdered materials are introduced into the mixer, i.e., chrome-alumina slag, high-alumina refractory, finely ground sodium silicate and nepheline slurry or ferrochrome slag. The mix having been thoroughly mixed, aluminum powder mixed with a small quantity of water is introduced into the mix, and the resulting composition is thoroughly stirred again so as to preclude the commencement of gas evolution in the mixer. On completion of the stirring the gas concrete mix is poured into metal molds preheated to 38° to 42°C, and the mix is allowed to stay in the molds at this temperature for a period of 3 to 5 hours. After preliminary hardening of the articles, the hump is cut off from them, and the shaped articles are subjected to autoclave treatment by self-curing techniques. With the help of electric heaters the temperature in the autoclave is maintained within 170° to 180°C.

The steam which evolves in the autoclave builds up a pressure which during 3 hours reaches 8 gauge atmospheres and is maintained at this level for 4 hours. Then the pressure is relieved to 0 during a period of 3 hours, and the gas concrete articles are removed from the autoclave. After that the articles are kept under shop conditions for 3 days at a temperature of 20°C to complete readiness. The following examples illustrate the compositions of the mix for refractory gas concrete.

Example 1: Composition of the mix is as follows: filler, 22 wt % chrome-alumina slag, 22 wt % high-alumina refractory; 10 wt % finely ground soluble glass; 15 wt % nepheline slurry; 0.13 wt % aluminum powder; 0.87 wt % sodium hydroxide; 12.5 wt % water; and an aqueous solution of soluble glass with a density of 1.38 to 17.5.

The refractory gas concrete produced from the mix features the following properties: 600 kg/m³ volume mass; 1200°C operation temperature; 15 kg/cm² ultimate compression strength after exposure to temperature of 1200°C; and 1.0% additional shrinkage after exposure to temperature of 1200°C.

Example 2: Composition of the mix is as follows: filler, 28 wt % chrome-alumina slag, 28 wt % high-alumina refractory; 5.8 wt % finely ground soluble glass; 8.1 wt % nepheline slurry; 0.20 wt % aluminum powder; 1.0 wt % sodium hydroxide; 13.7 wt % water; and an aqueous solution of soluble glass with a density of 1.38 to 15.2.

The refractory gas concrete produced from the mix features the following properties: 600 kg/m³ volume mass; 1200°C operation temperature; 16 kg/cm² ultimate compression strength after exposure to temperature of 1200°C; and 0.87% additional shrinkage after exposure to temperature of 1200°C.

The strength of the refractory gas concrete is such that this concrete can be used instead of lightweight refractory articles for high-temperature insulation and lining of furnaces and thermal units operating at a temperature of up to 1200°C.

Sand, Asbestos and Chamotte Composition as Space Filler

A process described by *L.B. Zilberman, S.B. Krakhmalnikov, G.A. Kudinov, D.B. Kutsykovich, J.G. Moiseev and A.M. Shneider; U.S. Patent 3,923,535; Dec. 2, 1975* relates to a heat insulating material for filling up spaces between the cooling plates and the shell of a blast furnace. The heat insulating material is prepared from solid powdery components and liquid components. The solid filler comprises: 17 to 19 wt % river sand (without shells); 8 to 10 wt % finely crushed asbestos; and 2 to 3 wt % finely crushed ferrochrome slag; with chamotte powder (a particle size up to 2 mm with up to 7% of 2 mm particles) balance.

The liquid components are as follows: 9 to 10% liquid glass as binding agent in addition to 100% by weight of the powdery components; 19 to 21% water as solvent in addition to 100 wt % of the powdery components; 0.5 to 1.0% of a foaming agent in addition to 100% by weight of the liquid components. The foaming agent may comprise petroleum sulfonic acids obtained during the processing of kerosene or diesel-oil distillate with sulfuric anhydride. Naphthene soap may also be used which comprises sodium soaps of water-insoluble organic acids extracted from the wastes of alkaline purification of kerosene, gas oil and solar-oil distillates of mineral oil.

The abovementioned components are mixed in a concrete mixer for 10 to 15 minutes to obtain a cream-like mixture. The resulting material exhibits fluidity sufficient to transport it along pipelines under a gauge pressure of 6 to 8 kgf/cm^2, e.g., by means of compressed air. Since the material exhibits rapid setting it should be used quickly after the preparation (in 3 to 5 minutes) to fill up the spaces between the shell of a blast furnace and the cooling plates.

Upon filling up the spaces the heat insulating material sets in 0.5 to 1.5 hours and does not shrink during setting due to a low content of the liquid components. The experiments have shown that the heat insulating material has a required fluidity to ensure complete filling of the spaces, and sets rapidly. The material has uniform composition, exhibits low heat conductance and high gas-proofing capacity. The blast furnace of 2,000 m^3 capacity, where the heat insulating material was used, had a temperature of about 30° to 40°C at the outer surface of the shell.

Firebrick Formulation

A.E. Booth; U.S. Patent 3,649,315; March 14, 1972; assigned to Armstrong Cork Company has found that insulating refractories of required low densities may be formed without the necessity of incorporating burnout material, such as sawdust. In accordance with the process, mineral wool, aluminum powder, hydrated lime, particles of plastic refractory clay, and alumina and/or pyrophyllite when desired, are dry-blended, mixed with water to form a moldable mix which is poured or cast into molds and vibrated.

The molded mix is allowed to expand and set after which it is dried to a moisture content of about 20 to 25%. At this point, the slab stock is readily removed from the molds. The slab stock is put on conventional conveying equipment and transferred to a kiln wherein it is fired at a temperature within the range of from about 2200° to 2850°F for anywhere from about 2 to 6 hours dependent on end use temperature requirements and mix compositions. Bricks produced in accordance with this process are of low density, good strength and uniform pore structure as compared with a ragged pore structure for firebrick produced using a burnout technique.

Closed Cell Clay Foam Brick

P.S. Sennett, J.P. Olivier and S. Ross; U.S. Patent 3,737,332; June 5, 1973; assigned to Freeport Minerals Company describe a clay foam which is characterized by a closed cell structure and the ability, in its calcined state, to float in water. The gas bubbles of the foam are substantially completely encapsulated by walls formed from clay particles. The foam is prepared by generating a foaming gas in an aqueous dispersion of clay particles to which has been added a minor amount of a fatty amine as a foaming agent. A variety of inorganic filler materials can be incorporated into the foam in order to alter or impart new properties to the foam. The foam can be calcined to further alter its properties. The foam has many uses including use as a refractory brick. The following examples illustrate the process.

Example 1: A clay slip containing the foaming agent was prepared by blending the following materials with a laboratory stirrer in the order listed: 3,000 grams water; 1,000 grams delaminated clay (Nuclay); and 10 cc (~10 grams) 1% by wt aqueous solution of lauryl amine. The resulting clay slip contained about 25% solids and had a pH of about 3.5. The slip was then transferred to a typical flotation cell and a foam prepared by bubbling air into the slip. The foam produced was removed from the flotation cell, drained of liquid, dried for several hours at 100°C and then calcined at about 1250°F for about 1 hour.

The calcined clay had a closed cell structure and floated for at least about 4 weeks in water at room temperature. Upon examination of the foam under a microscope, it was apparent that the gas bubbles were enclosed by walls made of clay particles.

Example 2: The procedure of Example 1 was repeated except that the pH of the slip was about 4.0 and the foam was calcined at 1850°F for about 1 hour. The calcined clay had properties similar to those of the clay of Example 1.

Example 3: The following materials were blended with a laboratory stirrer in the order listed: 2,170 grams water; 2,170 grams delaminated clay (Nuclay); 20 cc (~20 grams) 1% by wt aqueous solution of lauryl amine; and 2.17 grams tetrasodium pyrophosphate (TSPP).

The TSPP is added as a dispersant for the clay. The slip had a solids content of 50%. The slip was then foamed with air in a manner identical to that set forth in Example 1 after which the foam was calcined at 1850°F for 1 hour. The resulting foam was very stable, of closed cell structure with the clay particles forming the walls of the bubbles, and floated in water at room temperature for at least about 4 weeks. The calcined foam, after grinding, had a TAPPI brightness of 90.7 (not tightly packed) and 91.2 (tightly packed). The specific gravity of the ground foam was about 2.70.

Mineral Wool, Asbestos and Binder

A process described by *J.R. Roberts and H.C. Roach; U.S. Patent 3,682,667; August 8, 1972; assigned to United States Gypsum Company* relates to heat-insulating blocks of improved characteristics, and in particular blocks which under prolonged and repeated exposure to extremes of high temperature exhibit minimal change in dimension and shape. The composition made by the admixture in water slurry of finely-subdivided materials comprising an organic binder, such as starch or a water-activated cellulosic binder having in its original hydrated state a TAPPI drainage freeness of at least 10 minutes, mineral wool, clay and in a preferred form asbestos fiber and a wax-asphalt emulsion.

The mineral wool used in the process preferably is wool produced from blast furnace slag which is low in iron oxide content and is often referred to as white wool. The residual iron oxide content is less than 2 to 3%. A particular wool consists (by weight) of about 36.6% silica, 14.6% aluminum oxide, 1.1% iron oxide, 36.5% calcium oxide, and 11.3% magnesium oxide, and about 0.2% acid insolubles, and has a temperature of crystallization of about 2415°F and becomes molten at about 2450°F. Mineral wool produced from phosphate slag typically has a suitably low iron content, less than about 1%, and may be used in substantial substitution for blast furnace slag wool. Phosphate slag wool may be expected to become molten at about 2300°F and to crystallize at about 2200°F.

The water-activated cellulosic binder preferred to be used has a TAPPI freeness of at least 10 minutes. The binder is prepared by forming a slurry or furnish of cellulosic material, such as unbleached kraft pulp. The slurry is gelatinized by passing through a series of refining and gelatinizing steps to attain the requisite TAPPI freeness. This hydration is far beyond that commonly used in the paper industry such as in forming glassine paper furnish, and is more fully described in U.S. Patent 3,379,608. The TAPPI freeness is determined in accordance with the TAPPI Standard T221 os-63 and may be defined as the drainage time in seconds per gram of pulp used in the standard sheet machine at 20°C.

The asbestos preferred in the process is a short-fiber asbestos material having an average fiber length of 2.5 millimeters and an average fiber diameter of 0.8 micron. The clay utilized is a fusible ceramic clay, preferably nonswelling and having a high proportion of silica and alumina, such as fire clay, ball clay or the like. The wax used in the conventional wax-asphalt emulsion may be such as microcrystalline wax or crude or refined paraffin. A suitable emulsion would comprise about 50% asphalt and about 10% wax, by weight, and while not a critical component, usefully may serve as a sizing to stabilize the block in high-humidity storage conditions.

Blocks according to the process may be formed by a process comprising forming an aqueous slurry of starch or cellulosic binder material having a TAPPI freeness of at least 10 minutes, admixing a ceramic clay, asbestos fiber, mineral wool and a wax-asphalt emulsion with the aqueous slurry of cellulosic binder or starch to form an aqueous dispersion having a consistency of about 5 to 9%, forming a sheet by depositing and accumulating the solids of the dispersion on the screen of a board-making machine, compressing the sheet and removing excess water therefrom, cutting the sheet into blanks, drying the blanks, and cutting and trimming the dried blanks into blocks of the desired sizes and shapes.

In a typical installation, the blocks of insulating material are mounted behind the refractory brick of the furnace or kiln, matched tightly together to seal the brick from exterior portions of the installation. As the blocks first become heated to high temperature, the organic binder and the wax-asphalt emulsion, if present, burn out. However, the high temperature causes the clay to fuse and thus to serve as a binder which gives the block continued strength and basic form. The clay should be present in an amount at least equal to the organic binder, so as to serve adequately as a binder-replacement.

Siliceous Material

C.H. Noll and J.B. Andrews; U.S. Patent 3,965,020; June 22, 1976; assigned to Johns-Manville Corporation describe a siliceous thermal insulation material which can be used satisfactorily in environments in which the hot face temperature of the object to be insulated is on the order of 2000°F. The method of forming a shaped siliceous thermal insulation comprises mixing in an aqueous medium at a temperature in the range of 170° to 210°F components comprising 8 to 14 weight percent clay, 60 to 85 weight percent diatomite powder, 5 to 10 weight percent hydrated lime and 1 to 8 weight percent synthetic fiber; allowing the mixed components to stand for a period of 0.5 to 2.0 hours; molding the desired shape; and thereafter drying the shaped siliceous thermal insulation.

As an example of the process, shaped insulation materials were formed in the following manner from components in the table below:

	% by Weight	Pounds
Calcined diatomite powder	67	900
Natural diatomite powder	11	150
Hydrated lime	7	100
Bentonite clay	11	150
Synthetic fiber	2	20*
Natural organic fiber	2	25

*10 pounds glass fiber and 10 pounds polyester fiber.

These were mixed with 1,400 gallons of water for 9 minutes at about 190°F. The mixture was then pumped into a storage tank and allowed to stand for 1 hour during which time the temperature dropped from 190° to 150°F. Thereafter, the slurry at a temperature of about 150°F was piped to filter molding presses and molded into blocks and half-rounds of various sizes.

Filter press pressure and time of molding were regulated to produce blocks which following drying had a density in the range of from 18 to 26 lb/ft^3. Drying of the smaller thickness molded pieces was for a period of about 8 to 16 hours (overnight) while the larger thickness pieces were generally dried for a period of up to 40 hours. The various shaped products were thereafter shown by laboratory testing and sealed use to be eminently suitable for insulation at a hot face temperature of 2000°F.

LIQUEFIED GAS STORAGE

PRESTRESSED CONCRETE STORAGE TANK

R. Marothy; U.S. Patent 3,852,973; December 10, 1974 describes a prestressed concrete structure for the storage of liquefied gas which comprises an inner and outer concrete facing maintained in spaced relationship. Insulating material is positioned between the facings and the inner facing is saturated with moisture. When the structure is filled with liquefied gas, the tension due to thermal stresses is partially relieved and both facings act together to withstand the forces on the structure.

Referring to Figures 4.1a through 4.1c, a prestressed concrete storage structure **10** for storing liquefied gases is shown. The tank **10** has three major components: a floor **12**, a substantially cylindrical wall **14**, and a roof or dome structure **16**. Each of the major components has concrete inner facings **12a, 14a** and **16a** and concrete outer facings **12b, 14b** and **16b** respectively. The concrete facings are preferably formed of lightweight concrete weighing about 100 lb/cu ft. The facings are preferably about 6 inches thick and are maintained in spaced apart relationship with the space between the faces which is preferably about 4 feet filled with insulating material **18**.

The insulation material is preferably a rigid, insulating concrete weighing approximately 20 to 50 lb/cu ft. Zonolite, Perlite, Permalite and Mearlcrete are examples. The rigid insulation transmits shear between the facings, causing the facings to share any loading stresses placed upon the structure. The facings thus act together to insure a substantially stronger structure than was heretofore known.

Referring to Figure 4.1c there is shown in more detail an example of the process. Thus, as shown, concrete facings **12a, 14a** and **16a** are connected by dowel reinforcement **32** to the outer facings **12b, 14b** and **16b** in order to prevent separation of the facings and to further increase the shear resistance of the insulation **18** and increase the load carrying capability of this structure. The dowels **32** may advantageously be made of strands or wires which have high

136

FIGURE 4.1: PRESTRESSED CONCRETE STRUCTURE FOR THE STORAGE OF LIQUEFIED GAS

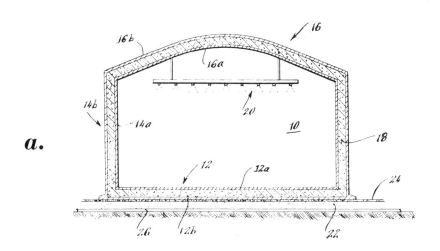

a.

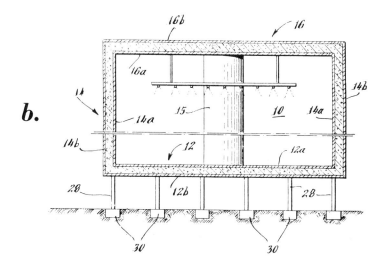

b.

(continued)

FIGURE 4.1: (continued)

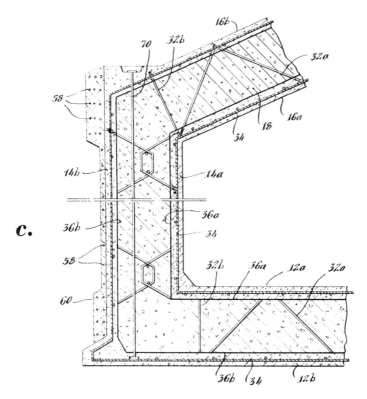

c.

(a)   Vertical cross-sectional view of tank
(b)   Partial vertical cross-sectional view of the tank with
        alternate roof and foundation design
(c)   Partially fragmentary view of a wall, floor and roof
        section of the structure of Figure 4.1a

Source: U.S. Patent 3,852,973

strength and are not embrittled at cold temperatures. The facings **12a, 12b,
14a, 14b** and **16a, 16b** may also be advantageously reinforced with regular mild
steel reinforcement **34**, in order to control shrinkage and cracking and to increase the strength of the structure. The surfaces between the insulation and
concrete facings may advantageously be sealed against moisture by, for example,
epoxy paint as shown at **36a** and **36b**.

To construct the storage structure, facing **12b** is poured on the ground with
dowels **32** anchored in place. Then sealant **36b** is applied and insulation concrete **18** poured. After drying, the upper sealant **36a** is applied and finally concrete facing **12a** is poured. The construction of the roof is similar. Unless it
is precast in a horizontal position, the construction of the wall requires a

somewhat different approach. Wall facings **14a** and **14b** may be slip formed first and insulation **18** poured between them. Alternatively, insulation **18** may be slip formed first and concrete facings **14a** and **14b** applied by a pneumatic (Gunite) process. In either event, dowels **32a, 32b** must be cast against the concrete forms and bent into final position after the insulation is placed.

The walls, floor and dome are prestressed in two directions. Vertical tendons **70** in the wall prestress the wall facing in the vertical direction, while a series of wires or tendons **58** are wound around the entire periphery of the tank to prestress the tank in the other direction. These tendons are further covered by a protective concrete coating **60**. Curved tendons placed over the dome in two perpendicular directions prestress the dome. The floor is prestressed by placing additional tendons on the wall or floor. As shown, all facings are monolithically connected with each other without the need for joints or other connectors.

Returning to Figure 4.1a, as an important feature of the process a spray system of fog nozzles **20**, preferably arranged in a circular ring, is attached to the roof of the structure. Prior to using the tank, water is emitted from the nozzles to fully saturate the concrete inner facings of the floor, wall and roof.

Since the elastic, shrinkage and creep properties of concrete at different temperatures are markedly dependent on the level of water saturation, by maintaining the inner facing in saturated conditions while the outer facing is dry, differential elastic, shrinkage and creep strains can be set up between the two facings. When the tank is filled with liquefied gas at low temperature, the strength and elastic modulus of the fully saturated concrete inner facings increase, partially relieving the tension due to thermal stresses. Furthermore, while at low temperature there is no shrinkage in the inner facing, shrinkage will occur in the outer facing further decreasing the compression of the outer facing, thus equalizing the stresses between the two facings.

Also, due to the freezing of the moisure trapped in the pores of the concrete when the low temperature liquefied gas is introduced into the tank, an impermeable vapor barrier is created. Furthermore, the necessity of sliding joints between the different parts of the structure can be eliminated by uniformly cooling down the inside of the structure prior to the introduction of the liquefied gas. This can readily be accomplished by spraying a predetermined amount of liquefied gas into the structure through nozzles **20**.

Tank **10** may be constructed directly on the ground as shown in Figure 4.1a, in which case a layer of fine sand **22** and friction relieving material **24** is placed under the floor to allow the floor to slide relative to the ground. Heating cables **26** may be included to prevent freezing of the soil underneath the structure.

Tank **10** may also be advantageously be elevated above the ground by pendulum support columns **28** as shown in Figure 4.1b, which rest on reinforced concrete footings **30** directly on the ground. These pendulum supports are designed to allow radial motion only while resisting tangential motion to insure the lateral stability of the tank. As also shown in Figure 4.1b, the structure can be constructed with a flat roof which is supported by a circular central column **15**.

INSULATION MEMBER

H.S. Smith, Jr.; U.S. Patent 3,924,039; December 2, 1975; assigned to The Dow Chemical Company describes an improved method and article for cryogenic construction. The article is particularly useful when employed with the spiral generation method of applying insulation. An insulating member is provided having an elongate cross section and a plurality of backing members also of elongate cross section adhered to a face of the backing member. Fracturing of the insulation on temperature cycling is substantially reduced.

FIGURE 4.2: CRYOGENIC INSULATION

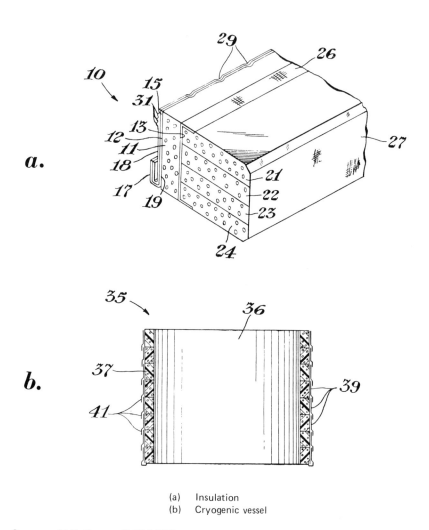

(a)   Insulation
(b)   Cryogenic vessel

Source: U.S. Patent 3,924,039

In Figure 4.2a there is shown a schematic isometric representation of one end of an elongate insulating member generally designated by the reference numeral **10**. This insulating member comprises a first or face member **11**. The facing member has an external face **12** and an internal face **13**. The member **11** has a width generally commensurate with the height of the insulating member **10** and extends entirely across the height of the member **10** as shown in Figure 4.2a.

A laminate **15** is affixed to the surface **12** of the member **11**. The laminate has a generally J-shaped configuration where one edge is folded outwardly about 180° to form a flange **17**. The laminate **15** comprises a first or outer layer **18** which beneficially is metal, such as aluminum. The laminate has a second layer **19** which advantageously is a heat sealable adhesive; for example, a copolymer of about 80% ethylene and 20% acrylic acid. Other well-known adhesives such as epoxy resins and the like are also employed with benefit.

Affixed to the internal face **13** of the facing element **11** are a plurality of internal insulating members **21**, **22**, **23** and **24** in adjacent generally contiguous relationship. The members **21**, **22**, **23** and **24** together cover the internal face **13** of the facing member **11**. A reinforcing scrim **26** is disposed at the interface between the facing member **11** and the internal elements **21**, **22**, **23** and **24**, and is folded over the adjacent internal elements **21** and **24**. Beneficially, the scrim **26** is an open weave glass fabric. Disposed on the elements **21**, **22**, **23** and **24** remote from the scrim **26** beneficially is a second scrim **27**.

Adjacent faces of the elements **21** and **22** are not joined, but are free to separate when the temperature of the inner members remote from the laminate **15** are reduced. The second scrim or fiber reinforcement **27** is affixed to the inner members **21**, **22**, **23** and **24** at a location generally parallel to and remote from the scrim **26**. The reinforcement **27** markedly reduces the tendency for crack initiation as the temperature of the face carrying the scrim **27** is reduced.

Advantageously the scrim **27** is slit at locations corresponding to the interfaces between the internal members. A plurality of transverse corrugations **29** are formed in the laminate **15**. The corrugations **29** extend toward the J-shaped edge. A plurality of generally continuous beads **31** of a heat activable adhesive are disposed on the outer surface of the laminate **15** remote from the J-shaped fold.

Figure 4.2b schematically depicts a structure prepared employing elements of Figure 4.2a. The structure is generally designated by the reference numeral **35**. The structure **35** comprises a vessel **36** such as a cryogenic vessel surrounded by a generally cylindrical insulating body **37**. This cylindrical insulating body is made up of a plurality of turns of insulation material **39** having a structure generally similar to that depicted in Figure 4.2a. Each of the turns has a flange **41** equivalent to the flange **17** which has been bent downwardly, flattened against the outer layer of the adjacent lower turn and sealed to provide a generally liquid-proof exterior surface.

Generally in preparing articles in accordance with the process, the corrugations such as the corrugations **29** extend partly across the laminate and provide for controlled buckling of the metallic laminate when the strip is wound onto a surface such as the spherical surface. The controlled buckling introduced by the corrugations provides a relatively shallow depression which occurs at regular intervals. Beneficially, the quantity of a heat activable adhesive **31** such as a strip of bitumen, ethylene-acrylic acid copolymer or ethylene-vinyl acetate co-polymer is sufficiently large that on heat sealing of the skin member to the adjacent strip, such as is depicted in Figure 4.2b, the minor depressions corresponding to the corrugations are filled and a liquid–tight outer surface is obtained.

Particularly advantageous insulating members are employed using synthetic resinous closed cell thermoplastic foams such as polystyrene foam, thermoplastic polyurethane foam, polyvinyl chloride foam, polyolefin foam such as polyethylene foam and polypropylene foam, and the like, having a closed cell structure.

CAPILLARY INSULATION

J.L. McGrew; U.S. Patent 3,755,056; August 28, 1973; assigned to Martin Marietta Corp. describes a capillary insulation comprising a cellular structure which provides a plurality of contiguous discrete cells with a capillary cover closing one side of the cells and capillary openings in the cover communicating with each cell. The cellular structure in so formed that the cell walls have an excess of material between points of interconnection. In the preferred case, this excess material is characterized by cell walls which are S-shaped in cross-sectional configuration. This is achieved by forming the cells from a plurality of strips of material, each of which is in the form of a sine wave with adjacent strips being staggered and connected at spaced points. The resultant cells have S-shaped cell walls which permit the individual cell walls to expand and contract relative to the other cells and prevent stress accumulation in the panels.

Figure 4.3a illustrates a capillary insulation assembly indicated generally by the reference numeral **10**. The capillary insulation is secured to and carried by a support wall **12** which may be the wall of a tank or any other surface which it is desired to insulate from a low temperature boiling point liquid.

The insulation assembly **10** comprises a cellular structure **14** which includes a plurality of discrete cells **16** and a closure means in the form of a capillary cover **18** extending across the cells and secured by suitable means such as an adhesive **19** to the cellular structure **14**. A suitable filling, not shown, such as rock wool or polystyrene chips, may be provided in the cells to reduce radiation and convection currents. Capillary openings, holes or pores **20** are formed in the cover **18** with each opening being associated with one of the cells **16**.

The cellular structure **14** may be fabricated of any lightweight material which is compatible with the liquid being insulated and which has a low thermal conductivity. For example, plastic impregnated Kraft paper may be used. Both materials are relatively flexible in all directions transverse to the plane of the paper. The capillary cover **18** may be made from a suitable plastic film such as one mil Mylar film.

The capillary insulation contemplates the formation of a plurality of discrete gas columns within the cells **16** with the gas columns extending between the surface of the wall **12** and the liquid. The size of the openings **20** are such that a stable capillary interface or membrane is formed at the interface of the gas columns and the liquid, with the membrane preventing liquid from penetrating

FIGURE 4.3: CELLULAR INSULATION

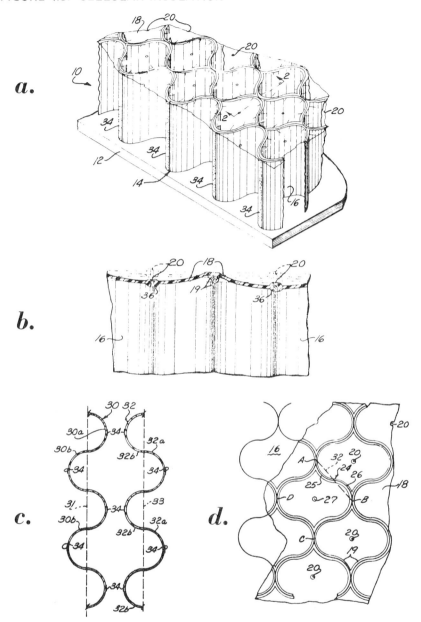

(a)   Perspective view of a portion of the capillary insulation
(b)   Section view along line 2–2 of Figure 4.3a
(c)   Plan view schematically illustrating the strips from which the cellular material
       is constructed
(d)   Top plan view of the assembled capillary insulation with a portion of the
       capillary cover removed to expose the cellular construction

Source: U.S. Patent 3,755,056

the gas column so that the gas columns function as insulators. In addition, the gas columns provide support for the liquid so that the relatively weak cover 18 need not support the liquid.

The cellular material 14 is constructed such that an excess of wall material is available to permit expansion and contraction of individual cells independently of adjacent cells. This is accomplished by forming each quarter or 90° segment of the cell wall in generally S-shaped configuration. As shown in Figure 4.3d, each cell is defined by four quarter segments of the cell wall. These quarter segments extend respectively between points A, B; B, C; A, D; and D, C. It will be noted that the cross-sectional configuration of each of the four quarter segments of the cell wall, extending between any two of the points A, B, C and D, is generally S-shaped in configuration.

For example, the quarter segment of the cell wall extending between points A and B and designated 24, includes a first arcuate portion 25 which bows inward of the cell and a second arcuate portion 26 which bows outward of the cell. Each point along each of the portions 25, 26 has a radius as measured from the longitudinal axis 27 of the cell which differs from the radius of the next adjacent point in that portion of the wall. The distance between the two points, A, B along the dotted straight line 32 is substantially less than the distance between the same two points as measured along the curving surface of that segment 24 of the cell wall.

Accordingly, it should be apparent that more material is provided in the segment 24 due to the curved configuration thereof than if the segment 24 was straight. The excess material thus provided constitutes means which permits the wall 24 to deflect transverse to the axis of the cell without altering the distance between points A and B.

The advantage provided by the excess material can be best appreciated by considering what occurs when a low boiling point liquid is poured into a tank having the insulation. When the insulation is installed within a tank and the low boiling point liquid is poured into the tank or container and contacts the insulation, the temperature of the insulation, particularly that portion which engages or contacts the liquid immediately decreases. As a result, the material of which the cellular structure is made tends to contract. Due to the excess material provided in the cell walls, resulting from the S-shaped configuration of a quarter segment of the cell wall, the contraction does not put any undesirable stress or strain on the material, notwithstanding the fact that the entire bottom edge of each cell is fixedly attached to the container wall; rather, the quarter segment tends to straighten out from the full-line position shown in Figure 4.3d. There is a minimum tendency for the material upon contraction to pull points A and B, for example, together which would result in a large total deflection of the insulation structure due to the fact that this would be magnified from cell to cell.

A cellular structure having S-shaped walls may be fabricated in various ways, but the preferred form is achieved by fabricating the structure from a plurality of individual strips or ribbons of material which are then assembled in the configuration shown in Figure 4.3d. For example, referring to Figure 4.3c, two strips or ribbons 30, 32 as they are shaped in the assembled structure are illustrated. Each of the strips is in the general form of a sine wave defined by loops 30a, 30b and 32a, 32b extending in opposite directions on either side of the neutral axes **31, 33** of the strips. The strips are assembled by offsetting one strip from the other in the manner shown in Figure 4.3b so that every loop **30a** of one strip extending in one direction engages every loop **32b** of the other strip extending in the opposite direction. Adhesive means **34** may be used to securely fasten the two abutting loops together. However, it should be noted that the area over which the adhesive **34** is applied is relatively small so that the cell walls immediately adjacent the adhesive remain free to deflect.

In addition to providing excess cell wall material to accommodate stresses imposed on the cellular structure, similar provision is also made for relieving the stresses imposed on the capillary cover **18**. To this end, an excess of cover material is provided for each of the cells **16**. As shown in Figure 4.3b, the capillary cover **18** is illustrated as being nonplanar, having a concave or dish-shaped configuration over each of the cells **16** so that the distance across the cell as measured along the surface of the cover **18** is greater than the corresponding width of the cell. This excess of material permits the capillary cover for each cell to deflect between the positions shown in solid lines and the position shown in dotted lines in Figure 4.3b. This enables the capillary cover to accommodate strains imposed on each individual cell without transmitting these strains or stresses to adjacent cells.

When the liquid having a low boiling point is poured into the container and contacts the cover **18**, the cover, as discussed above in connection with the material of the cell walls, also tends to contract due to the temperature change within. The cover **18** is made of polyethylene terephthalate film marketed under the trademark Mylar. This contraction results in the Mylar cover deflecting from its full-line position, shown in Figure 4.3b, to the dotted position shown in the same figure. Due to the fact that the cover deflects in this manner, which is possible because of the excess material provided, a minimum of stress is imposed in the cover due to the temperature change therein.

In addition to minimizing the effects of stresses imposed on the capillary cover, there is an additional important benefit achieved by this construction. In particular, the dimpled cover construction insures that the size of the capillary opening **20** remains substantially constant even when the capillary cover is subjected to stresses and strains resulting from contact with the liquid. This benefit is best understood by considering the capillary cover as being stretched taut across each cell **16**. Under those conditions, any temperature change in the cover **18** results in stresses which may alter the size of the opening **20**.

Since the dimensions of the capillary openings **20** are of critical importance to the maintenance of the proper capillary interface between the liquid and gas, any change in the size of the opening will have an adverse effect on that interface. However, by providing the dimpled or concave construction of the capillary cover, sufficient excess cover material is provided to permit deflection of the cover without causing undue stretching of the cover.

Additional provision is made to preclude any enlargement of the capillary openings **20** due to stresses imposed on the capillary cover **18**. This additional provision is illustrated in Figure 4.3b and comprises reinforcement means **36** which extend around each of the openings **20**. As shown, this reinforcement means comprises a thickened section of the cover which, in cross section, is in the form of a torus. The toroidal reinforcement **36** thus provides additional strength at the capillary openings so that any tendency of the openings to enlarge or tear when the cover is under stress is effectively resisted.

WALL CONSTRUCTION

K. Yamamoto, K. Obata and T. Sata; U.S. Patent 3,694,986; October 3, 1972; assigned to Bridgestone Liquefied Gas Company, Ltd., Japan describe a heat insulating wall for a membrane-type low temperature liquefied gas tank in which laminated wood is composed of an inner layer and an outer layer stuck together with adhesives with the joints of the respective layers positioned out of alignment with each other so that the laminated wood which serves as a secondary barrier may have sufficient strength against the internal stress caused by temperature variations.

The heat insulating wall includes a plurality of support members fixed to the outer vessel at one end and extending in perpendicular thereto. The ends of the support members on the inner side are connected with one another by means of a wooden framework. On the wooden framework is provided laminated wood composed of inner and outer layers with the joints of the respective layers positioned out of alignment with each other. Heat insulating material is disposed in the space between the laminated wood and the other rigid vessel.

With the heat insulating wall constructed as described above, the heat insulating material is not required to have compressive strength since it only fills up a space formed with rigid support members. Since the laminated wood serves as a secondary barrier to prevent the leakage of the liquefied gas, there is no need to provide a separate secondary barrier made of metal or the like. Thus, the heat insulating wall of the process is of simple construction, yet can be easily manufactured at low cost. Complete details of preferred structures are presented.

INTERMEDIATE ENVELOPE

A. Marquet; U.S. Patent 3,902,290; September 2, 1975; assigned to B.S.L. (Bignier Schmid-Laurent), France describes an insulating arrangement for a reservoir for liquids or liquefied gases at low temperature. An intermediate envelope is arranged to lie between the external envelope and the reservoir and spaced from both the external envelope and the reservoir. The intermediate envelope comprises a plurality of panels, each panel being secured to the external envelope by securing means connecting the panel to a region of the external envelope which is located substantially opposite to the panel.

Thus the forces exerted on each panel and particularly its self-weight and the action of the insulating material are resisted by the external envelope in the region of the panels, and the displacements of each panel in relation to the corresponding region of the external envelope are reduced to a minimum. This arrangement particularly avoids any concentration of vertical forces of substance such as may be produced on a support structure and enables any accumulation of differences in expansion movements and deformations to be avoided which might otherwise be set up between an external envelope and a support structure.

The process is applicable to any form of reservoir. The reservoir may be spherical, cylindrical with a flat or dished base and with a top which is dished. The external envelope may be of a different shape than that of the reservoir, the intermediate envelope then preferably substantially conforming to the shape of the external envelope.

TUBE HOLDER DESIGN

A process described by *U. Hildebrandt and A. Hofmann; U.S. Patent 3,861,022; January 21, 1975; assigned to Linde AG, Germany* relates to a method for the production of thermally insulated ducts, especially for the flow of low-temperature fluids such as liquefied gases, and more particularly to the formation of ducts having an inner metal tube, an outer metal tube or casing, a layer of insulation between the tubes and holders bridging the tubes and traversing the insulating layer.

The process involves providing a layer of thermal insulation around an inner tube or casing of metal (through which the low-temperature medium is transported) and piercing the layer of thermal insulation from the exterior at spaced-apart locations to form passages in which holders bridging the two tubes or casings are anchored. Preferably the passages are burnt through the layer and the spacers, holders or anchors are affixed at one end to the inner casing or tube or to a sheath surrounding same within the insulating shell provided for the purpose of attaching the holder. At its outer end, each holder may be in direct or indirect engagement with the outer casing or shell.

The duct system produced in this manner, which is effective for extremely cold fluids such as liquefied helium, has a number of advantages. Firstly, the application of the thermal insulation is not limited by the holders to be used or the system for anchoring the holders between the casings. The thermal insulation can, for example, comprise alternating layers of a low-conductivity fabric or fiber materials and high-heat reflectivity metal foils which may be applied in a continuous sheath, one or more helically wound layers in the same or opposite senses, or in any other form totally independently of the holder arrangement. In prior art systems the application of the insulating layer required careful consideration of the holder system and was limited in the nature of the insulation and the manner of its application by the holder arrangement.

Secondly, the process can be effected using apparatus of low capital cost and with the formation of negligible thermally conducting bridges between the inner and outer casings or tubes.

As soon as the thermal insulating layer is completely applied, the passages running from the exterior to the interior of the insulating layer are formed by burning them through the insulating layer. The burning of the passages has the advantage that the sensitive insulating layer receives no significant pressure during the piercing operation and thus does not have diminished insulating effectiveness in the region of the passages.

The passages preferably are of a circular cross section and of a diameter which approaches the diameter of the spacers or holders which may be even perpendicular to the longitudinal axis of the duct system or, when axial stresses are contemplated, may be inclined at angles thereto. The spacers in any event preferably lie in axial planes of the duct. The passages may be formed by rod-like burners generating an open flame or rod-like heated pistons or plungers.

As soon as the passages are burnt through the thermal insulation, the spacers or holders are inserted and anchored at one end with the inner casing. Preferably the spacers are only stressed under tension so that they may be relatively thin wires of low thermal conductivity. The use of spacers only under tension provides the advantage that the diameter of the passage can be held small, thereby reducing cold losses by thermal radiation through the passages. Where conventional systems provided spacers under compressive stress, larger cross-section passages were required and thus greater cold loss was the rule.

The holders or spacers can be affixed to the inner casing by any of a number of techniques. For example, the spacers may engage small fastening elements, such as hooks, eyes or loops formed directly on the exterior of the inner casing. Alternatively the fastening elements in which the holders are hung can be provided on a carrier especially provided along the exterior of the inner shell. In this case, the fastening elements may correspond in number and distribution to the number and distribution of the holders or spacers. This system has the slight disadvantage that the burning-piercing of the passages must be carried out carefully to ensure alignment of the passage with the fastening element.

This disadvantage can be obviated when the burning-piercing of the insulating layer is effected to a template which is applied to the exterior of the insulating layer and has markings, e.g., openings, aligned with the fastening elements of the inner shell.

It has been found to be advantageous, when a large number of holders are to be used, to distribute the fastening elements uniformly over the exterior of the inner casing or tube so that the probability of registry of a fastening element with a passage burnt through the insulation is substantially 100%.

This can be insured when the carrier is a perforated sheath surrounding the inner casing or tube and having a relatively high hole density or distribution and the cross section of the passage burnt in the insulating layer is approximately equal to that which will insure that the holder or spacer member can be hooked around the edge of a hole in the sheath. There should be 100% probability that the edge of a hole engageable by a holder will be exposed through any passage formed in the insulating sheath.

The carrier or sheath may be a cylinder or a helically wound perforated band of metal whose turns are laterally contiguous. Instead of a perforated cylinder or band, however, a wire-mesh structure may be employed. It has also been found to be advantageous to entrain the holder with the piercing member during the piercing operation through the insulating layer and then to affix the holder to the inner casing or tube when the passage is completely burnt through. This results in a saving of one operation.

The holders or spacers may be welded or cemented to the outer wall of the casing or to a carrier sheath and it has been found to be advantageous to provide a heating action by the open flame of the burner, subsequent to the piercing of the insulating layer, thereby melting a portion of the casing wall and forming a pool of metal in which the holder or spacer is seated for bonding

to the casing or a carrier disposed therearound. Welding may also be effected
by passing the holder or spacer through a hollow electrode and spot welding
it in place to the inner casing or a protective sheath therearound. Outer carriers
can then be applied to the spacer members and the outer casing mounted.

SANDWICH STRUCTURE UNITS

*T. Shimomura, H. Saito, M. Sueyoshi and H. Tsuda; U.S. Patent 3,972,166;
August 3, 1976; assigned to Ishikawajima-Harima Jukogyo KK, Japan* describe
a heat insulation structure for a liquefied gas storage tank which comprises a
plurality of heat insulation sandwich structure units attached to the outer tank
shell. Each sandwich structure unit comprises a plurality of core members or
ribs arrayed in spaced apart relation and sandwiched by top and bottom plate
members, and insulating materials filled into the spaced defined by the core
members and the top and bottom plate members. Each core member or rib
comprises a pair of plywood sheet members spaced apart from each other by
a predetermined distance by spacers, and insulating materials filled into the
space or spaces defined by the sheet members and spacers. The above construction not only gives much strength to the heat insulation structure in addition to the improved ability to interrupt heat conduction, but also reduces the
fabrication cost.

Referring to Figure 4.4a, a heat insulation structure 3 is shown as being interposed between an inner tank shell 1 and an inner plating 2 of a ship's hull,
and comprises a plurality of sandwich structure units 11.

Each joist 7 of the sandwich structure units 11, as shown in Figure 4.4b,
comprises a pair of plywood sheets 4 spaced apart by a suitable distance by
spacers 5, and an insulating material 6 filled into the space between the sheets
4. The joists 7 of the above construction are arranged and spaced apart from
each other by a suitable distance upon a bottom plate 8, and a top plate 10 is
attached upon them after an insulating material 9 is filled into the spaces defined by the joists 7 and the bottom plate 8, as shown in Figures 4.4c, 4.4d
and 4.4e. A plurality of sandwich structure units 11 of the above construction
are attached to the inner plating 2 of the ship's hull so that the heat insulation
structure 3 may be assembled.

Following is a description of the method for attaching the sandwich structure
units 11 to the inner plating 2. As shown in Figure 4.4f, the units 11 are
attached to the inner plating 2 with bolts 13 extended through project parts
12 of the units 11. In order to permit the heat insulation structure 3 thus
constructed to carry the transverse loads acting upon it in parallel with the inner plating 2, the side edges of the bottom plates 8 are pressed against or made
into close contact with flat bars or any other suitably formed members 14 attached to the inner plating 2. If the above arrangement is impossible, suitable
filler materials may be filled into the spaces between the bottom plates 8 and
the flat bars 14. Thus the inner plating 2 may carry through the flat bars 14
the transverse loads acting upon the heat insulation structure 3.

If it is desired to install a secondary barrier made of plywood sheets between the
inner tank shell 1 and the heat insulation structure 3, the side edges of the top
plates 10 of the sandwich structure units 11 are notched or stepped as shown

FIGURE 4.4: INSULATION STRUCTURE FOR LIQUEFIED GAS STORAGE

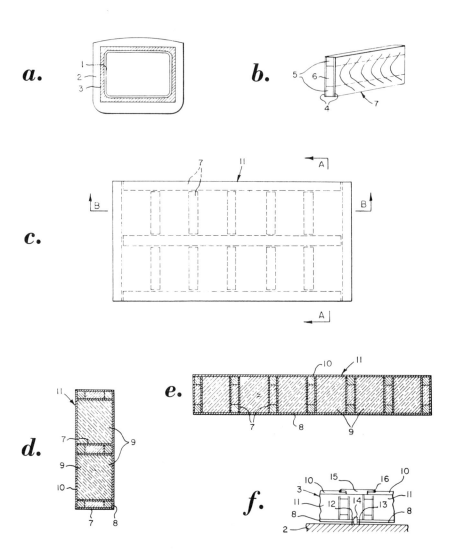

(a)   Diagrammatic sectional view of a liquefied gas carrier in which a heat
      insulation structure is installed between its hull and inner tank shell
(b)   Perspective view of a joist used in the heat insulation structure
(c)   Diagrammatic top view of a heat insulation sandwich structure unit
      making up the heat insulation structure
(d)   Sectional view taken along the line A–A of Figure 4.4c
(e)   Sectional view taken along the line B–B of Figure 4.4c
(f)   Methods for joining the adjacent sandwich structure units and attaching
      them to the inner plating of the ship's hull

Source: U.S. Patent 3,972,166

in Figure 4.4f, and a connecting plate 15 having the mating stepped portions is attached to the adjacent top plates 10 to bridge them. Thereafter sealing materials 16 are applied to the joists between the connecting plates 15 and the top plates 10. Thus, the top plates 10 of the units 11 may make up the secondary barrier.

The loads from the inner tank shell 1 or the secondary barrier are transmitted through the top plates 10, the joists 7, and the bottom plates 8 to the inner plating 2. Therefore the hull of the ship carries all of the loads from the tank shell 1.

When the transverse loads are exerted to the heat insulation structure 3, the joists 7 are subjected to the shearing forces, which are transmitted to the bottom plates 8 so as to tend to cause the lateral displacement thereof, but the bottom plates 8 are securely attached to the inner plating 2 by the flat bars 14 so that the transverse loads are also transmitted to and distributed over the inner plating 2.

In the fabrication of the joist 7 and the sandwich structure units 11, artificial foam material such as polyurethane may be directly filled into the joists 7 and the units 11 through the holes formed in the plywood sheets 4 and the top or bottom plates 10 or 8 so that it may be foamed on the fabrication site.

The spacing or distance between the joists 7 for supporting the bottom plate 8 and the top plate 10 may be suitably selected depending upon the strength of the inner tank shell 1 and/or secondary barrier and upon the degree of cracking of the insulating materials 6 and 9 caused at low temperatures. The spacing or distance may be varied also depending upon the difference in loads exerted from the inner tank shell 1 (the loads exerted on the heat insulating structure 3 being different, for instance, between the top and bottom of the inner tank shell 1) so that the joists 7 having the same cross-sectional area may be used.

CONTINUOUS PROCESS FOR THREE-DIMENSIONAL FIBER INSULATION

Liquid natural gas can be shipped in far less space than when it is in a gaseous state. However, such shipments must be at cryogenic temperatures, requiring insulated containers. To avoid embrittlement of the metal of the containers, the insulation should be inside, thermally shielding the container hull structure from the cold temperature of the liquid natural gas. When liquid gas seeps through the insulation and warms up to a gaseous state it expands and rips insulation from the container walls. For this reason a three-dimensional fabric reinforcement has been developed which contains the insulation foam in place until the pressure has been stabilized.

Current methods of manufacture relate to discontinuous processes that consume large volumes of excess materials, creating a disposal problem of solid waste. Both the use and the disposal add significantly to the cost.

R.L. Long; U.S. Patent 3,881,972; May 6, 1975; assigned to McDonnell Douglas Corp. describes a low cost method of manufacture of three-dimensional fiber reinforced insulation. This is a continuous process requiring a minimum of auxiliary materials. As long as raw materials are supplied, the three-dimensional insulation is of endless length. The channelling of the matrix and polymer during the expanding and setting of the foam minimizes trimming and trimming losses. This process uses a stack consisting of several layers of properly spaced woven, knitted or equivalent fabric that possesses interstices sufficiently large to permit the easy passage of commercial tufting needles and still produce a thread spacing in the fabric sufficient to supply the required reinforcement for the three-dimensional insulation.

Commercial tufting needles can be procured that will pass through at least a five inch stack of spaced and layered fabric. The stack of fabric layers is fed into a moving belt system that grips each side of each layer of fabric simultaneously and at a fixed desired spacing from layer to layer. These layers are held taut as the fabric matrix is carried forward through a trough. Fibers are inserted through the stack of layers normal to the fabric surfaces by bottom tufting and the resultant rigidly held three-dimensional matrix is then foamed into a rigid closed cell polyurethane reinforced insulation structure. The art of foaming, per se, in making such a structure is well known, since mattresses, seat cushions and many other articles are made for cushioning, insulation and other uses with such foaming techniques.

The bottom of the trough may be paper moving independently or riding on a moving belt or the belt alone may be used if made from a nonsticking elastomer. The sides of the trough will consist of the fabric-gripping moving belts. The top of the channel or trough may or may not be a moving belt. This technique controls the size of the block cross-sectional area and will minimize trimming losses. Conventional cutoff and cross-sectional trims can be used as the material leaves the machine.

GLASS FIBER NETS ON TANK SURFACE

K. Katsuta; U.S. Patent 3,970,210; July 20, 1976; assigned to Mitsubishi Jukogyo KK, Japan describes the construction of a heat insulation lined tank for low temperature liquids which is characterized in that the tank wall to be thermally insulated is lined with a plurality of wires in layers assembled crosswise in networks and stretched substantially in parallel with the inner wall surface of the tank, an expanded insulation layer formed on the inner wall surface with the wire layers embedded therein, and a liquid-tight layer or wall that covers the inner surface of the insulation. The spacings of the wires in the netted layers and of the layers themselves are made narrower on the low temperature side or with the approach to the innermost wall or layer of the lining.

In the low temperature liquid tank according to the process, the plurality of wires that consist of glass fiber strings or metal wires are stretched in the form of nets in parallel with the inner wall surface of the tank and are embedded in the insulation layer for its reinforcement. Moreover, the wires in nets and the nets themselves are spaced closer toward the innermost layer of the lining. This construction efficiently precludes the cracking of the insulation layer that might otherwise occur as a result of a change from the normal temperature after manufacture at the factory to the low temperature that is encountered in service.

In Figure 4.5a there is shown the outer wall portion of a tank 1 which may be of any shape suited for the intended use. The plate that forms the tank 1 is,

FIGURE 4.5: HEAT INSULATION LINED TANK

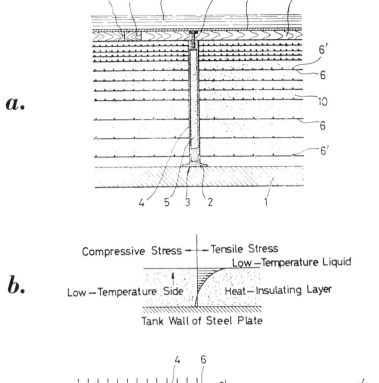

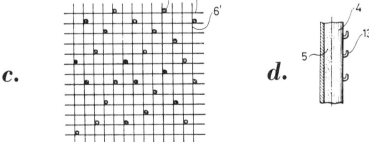

(a)   Fragmentary sectional view of a heat insulation lining for a low temperature liquid tank
(b)   Stress distribution curve of the lining in which rigid polyurethane foam is employed as heat insulating material
(c)   View illustrating the distribution of bolts for supporting the insulation lining on the inner wall surface of the tank
(d)   Enlarged vertical sectional view of one of the supporting bolts

Source: U.S. Patent 3,970,210

for example, the steel plate usually employed as structural material for the construction of ships. A metal anchor or socket 2 serves as seat for supporting bolt 4 and is shown as spot welded at 3 to the inner side of outer wall of the tank 1.

Into this anchor 2 is screwed the lower end of the hollow supporting bolt 4, which in turn is filled with polyurethane foam or other heat insulating material 5. Glass fiber strings or stainless steel wires 6, 6', which are coated with urethane resin, are supported and stretched by the bolts 4 in a plurality of layers in a net-like pattern parallel to the inner wall surface of the tank. The glass fiber strings or metal wires 6 and/or 6' are spaced apart in such a relation that, the farther the strings or wires are away from the inner side of the outer wall of the tank, or the nearer they approach the low temperature side, the greater the density of the arrangement of the string or wire elements.

A sheet of plywood or synthetic resin board 7 is shown as secured to the head of the supporting bolt 4 by the screw 8. The board 7 is formed with an inlet 7', through which an expandable urethane solution is to be injected into the space between the board 7 and the steel 1 so that a heat insulation layer 10 of rigid polyurethane foam can result. Subsequently the inlet 7' is closed fluidtight with a plug 9. The entire surface of the board 7 is covered with a multiply layer 11 of a reinforced synthetic resin built up of several plies of glass fiber and epoxy resin or polyurethane resin alternately laid one on another to keep the seams of the board 7 completely out of contact with the low temperature liquid 12 to be handled.

A tendency generally observed in the use of rigid polyurethane foam as heat insulating material is that, as indicated by the curve in Figure 4.5b, tensile stress is increasingly produced on the low temperature side, whereas the stress diminishes as the temperature rises and slight compressive stress is created on the high temperature side.

When using the rigid polyurethane foam as heat insulating material, particularly as a lining for tanks, it must be noted that without some countermeasure taken the thermal stresses of tension would be intense enough to cause cracks in the insulation layer. With the foregoing in view, the heat insulation lined tank according to this process is designed to prevent the cracking due to thermal stresses of the insulation layer by the networks of glass fiber strings or metal wires 6, 6' having a lower thermal contraction coefficient than that of the insulating material, arranged in parallel with the inner wall surface of the tank.

The vertical spacings among the glass fiber strings or metal wires 6, 6' are made closer on the low temperature side for added strength of the insulation layer 10.

The insulating structure proves particularly advantageous where a large tank, for example having a bottom area of 30 meters by 30 meters and a height of more than 20 meters, is to be lined completely. In the corners of such a huge tank thermal stresses can build up to a serious extent. This problem can be settled by free and ready choice of the structure and construction, for example by the use of glass fiber strings or metal wires 6, 6' stretched crosswise by the supporting bolts 4 and made gradually finer in meshes and shorter in net-to-net spaces inwardly. Since the supporting bolts 4 are themselves good insulators, metal wires stretched thereby would cause transmission of heat merely in the direction parallel to the steel plate of the tank. In the diametral direction the transference of heat takes place only through the insulation layer **10** of rigid polyurethane foam and the supporting bolts **4**. Stretching of the metal wires by and along the supporting bolts **4** is advantageous in that it helps equalize the temperature distribution per unit thickness of the insulation layer **10** and uniformizes the stresses throughout the layer. Desirably the arrangement of the supporting bolts **4** is staggered as shown in Figure 4.5c, so that the spacings among the glass fiber strings or metal wires **6, 6′** can be freely chosen.

As illustrated in Figure 4.5d, hooks **13** may be provided on the supporting bolts **4** to facilitate the stretching of the strings or wires. It is known that polyurethane foam applied by spray foaming, continuous expansion, or other conventional technique can usually serve the purpose of heat insulation for objects at temperatures down to about -100°C. Therefore, a heat insulating effect efficient and economical as a whole will be achieved by providing the insulation for the temperatures down to that level in the conventional way and then providing an insulation for lower temperatures in accordance with the process.

With the features described above, the process permits the use of ordinary steel plates in the construction of the tank outer wall **1**, thus ensuring a substantial saving in cost. Furthermore, the spacings of glass fiber strings or metal wires in the form of nets parallel to the tank wall can be gradually shortened for efficient and sufficient reinforcement of the insulation layer to overcome the problem of thermal stresses to be encountered.

HYDROCARBON-BASED FOAM

L.F. Creaser; U.S. Patent 3,810,839; May 14, 1974; and U.S. Patent 3,682,824; August 8, 1972; both assigned to Shell Oil Company describes an improved thermal insulating material for systems containing liquefied gases at low temperatures. The insulant is a hydrocarbon-based foam having a freezing temperature lower or slightly higher than the liquefied gases.

Gasoline is the preferred hydrocarbon for use as an insulant for LNG, since foams made from it maintain their flexibility to LNG temperatures. Kerosene foams solidify at about 190°K, so they are more suitable for the insulation of materials above that temperature. The foams are of relatively small cell size, less than 2 mm, and preferably less than 1 mm, in average diameter and contain from 30 to 90% of gas by volume.

The foams are gelled to assist in the maintenance of their foamed state. Gelling of the foam may be brought about by any suitable gelling agent, any of which are now known from extensive research in connection with the preparation of safety fuels. A preferred example is aluminum soaps, particularly aluminum octoate. The gelling agent may be employed in any convenient proportion, but it has been found that from 2 to 10%, preferably 3 to 7%, by weight of the hydrocarbon component is generally satisfactory. The following examples illustrate the process.

Example: An insulant suitable for the insulation of LNG storage vessels and conduits (LNG has a temperature of about -160°C) was prepared by mixing together 100 grams gasoline and 50 grams of aluminum octoate in a pressure vessel and injecting nitrogen gas into the mixture at a pressure of about 35 atm, and agitating the mixture vigorously. The pressure vessel was opened, with ejection of a gelled form which was used to insulate a vessel containing LNG. The foam, flexible at 113°K, had a thermal conductivity of 0.53 Btu/ft/°F/hr.

FOAMED SULFUR

According to a process described by *K. Yamamoto; U.S. Patent 3,870,588; March 11, 1975; assigned to Bridgestone Liquefied Gas Company, Ltd., Japan* a heat insulating wall is formed by blowing melted sulfur including a foaming agent at the inner surface of an outer vessel of a rigid structure and having the sulfur foamed and solidified to form a substantially continuous wall. The wall is adapted to include crack-preventing materials.

Sulfur is by itself superior in compression resistance as well as heat insulating characteristics, has a high adhesiveness, and is antiwearing. Furthermore, sulfur is nonhydrogroscopic and is a very stable material. Since sulfur is mixed with a foaming agent and is in a fluidal state blown at the inner surface of a rigid outer vessel thereby to foam and thereafter solidify to form a continuous layer, the heat insulating wall is easily constructed on the site. The foams generated in the layer form a number of independent spaces within the continuous layer, whereby the heat insulating characteristic of the wall is improved, while maintaining the impermeability of the wall to humidity, and at the same time reducing the specific weight of the heat insulating layer.

Since a continuous layer is formed by blowing melted and fluidal sulfur, the wall is provided with a uniform load supporting characteristic. By mixing some fibrous crack-preventing materials such as glass wool in the continuous layer of sulfur, the strength of the wall, especially in its anticracking characteristic, is very much improved.

Sulfur becomes fluidal by being heated up to a relatively low temperature such as 100°C and has a high adhesiveness so that it sticks firmly to the same or foreign materials. Therefore, the surface layer of sulfur can be easily formed by attaching fluid sulfur at the inner surface of the continuous layer in the manner of coating or plastering, whereby the attached layer of sulfur is firmly held there and cannot be removed, even under the application of vibrations or shocks. Thus the safety of the heat insulating layer is further improved by the addition of such an inner surface layer.

The heat insulating layer composed of the foamed sulfur layer and the solid inner surface layer of sulfur provides a sufficiently high impermeability to low temperature liquefied gases so that the low temperature liquefied gases to be stored in the tank can be directly held by the inner surface of the inner surface layer of the heat insulating wall. When the heat insulating wall of the above structure is used as a heat insulating layer for supporting an inner membranous vessel of a low temperature liquefied gas tank of the membrane type, the heat insulating layer provides a smooth supporting surface for the inner membranous vessel and at the same time operates as a secondary barrier wall for provisionally checking leakage of the liquefied gases when a leakage has occurred at the inner membranous vessel.

AEROSPACE APPLICATIONS

REENTRY VEHICLES

Elemental Sulfur and Noble Gas Fluorides

E.K. Weinberg; U.S. Patent 3,725,282; April 3, 1973 describes a composition
which comprises a char-forming polymer, or a porous nonablating material
and a mixture of a solid noble gas fluoride and elemental sulfur, or a solid noble
gas fluoride. The process also involves a method for protecting a spacecraft
from the heat generated upon its reentry into the earth's atmosphere, the spacecraft being provided with a heat shield fabricated from the above-described compositions. In accordance with this method, heat generated by friction between
the earth's atmosphere and the heat shield causes the sulfur and fluoride to react
or the fluoride, when used alone, to decompose, forming gases which are discharged into the aerodynamic flow field of the craft.

The gas or gases so formed and discharged function as a quenchant to protect
the spacecraft from the extreme heat incident to the reentry. The formulation
of gases and their discharge as described can be referred to as "passive wake
quenching." Solid noble gas fluorides that can be used include fluorides of
xenon, krypton and radon. These compounds are excellent fluorinators, forming
sulfur hexafluoride, the gaseous quenchant, when reacted with sulfur. It is usually preferred to utilize one of the xenon fluorides which have relatively high
melting points as shown in the following listing:

Compound	Melting Point, $^\circ$C
XeF_2	129
XeF_4	117
XeF_6	49.5

The following equation shows the reaction that occurs when a mixture of xenon
difluoride and sulfur is employed: $3XeF_2 + S \rightarrow 3Xe + SF_6$. The other solid
noble gas fluorides and elemental sulfur react in a similar manner to form sulfur hexafluoride.

Example: A heat shield is fabricated from a liquid phenol-formaldehyde resin, graphite cuttings and a mixture of xenon difluoride and elemental sulfur. The fluoride and sulfur are present in the mixture in a 3 to 1 mol ratio. The liquid resin is introduced into a vessel provided with a stirrer. The xenon difluoride and sulfur, both in particulate form and thoroughly mixed, are slowly added to the vessel while stirring. Thereafter, the graphite cuttings are slowly introduced into the vessel while continuing to stir. After the added materials are thoroughly dispersed in the liquid resin, the dispersion is poured into a stainless steel matched metal mold.

The mold has an internal cavity corresponding to the shape of the heat shield. The dispersion contains 30 weight percent resin, 35 weight percent of the mixture of xenon difluoride and sulfur, and 35 weight percent graphite cuttings. The mold is heated electrically to 320°F over a period of 6 hours, and then the temperature is reduced to room temperature over a period of 2 hours. After removing from the mold, the heat shield is postcured in an air circulating oven for 15 hours at 300°F and atmospheric pressure. The shield is then machined to provide a smooth outer surface.

The heat shield is subjected to a high temperature, high velocity gas to simulate conditions that might be encountered by a spacecraft upon reentry into the earth's atmosphere. The heat shield ablates at a low rate and quenchant gases are released, thereby reducing the amount of heat carried by conduction through the heat shield.

Reinforced Ablative Foam

R. Gonzalez; U.S. Patent 3,951,718; April 20, 1976; assigned to McDonnell Douglas Corporation describes a woven pile textile construction that has an open weave backing and pile yarns of ⅛ inch centers or more, depending on the strength requirements. The fibers may be synthetic or natural polymers. Polyurethane foam, ablative or nonablative, is then applied to the back side of the pile carpet, which is oriented with the fibers down. After the polyurethane material is applied to the backing during spraying or casting methods, a rigid sheet with a mold release is held against the back to force the polyurethane foaming action through the open weave backing and through the fibers.

Holding the pile fabric with the pile down keeps the fibers in a vertical orientation with relation to the backing. The foam expansion from top down also keeps the fibers straight. Downward foaming, because of gravity, also makes the foam less dense than foaming upwardly. The foam/fiber composite can then be cut to the desired thickness and bonded directly to a fuel tank by applying a cryogenic adhesive to the carpet backing. This insulation material is ideal for a space shuttle drop tank, for example, which requires an efficient cryogenic insulation with high temperature ablative properties to resist ascent thermal environments.

Closed-Cell Ceramic Foam from Cenospheres

A process described by *A.G. Tobin; U.S. Patent 4,016,229; April 5, 1977; assigned to Grumman Aerospace Corporation* relates to a closed-cell ceramic foam material that is prepared by heating cenospheres for a sufficient period of time and at a temperature sufficiently high enough to cause the walls to soften such that the spheres are caused to "foam" by their internal fluid pressure and form a cohesive waterproof material. Cenospheres are cellulated glass microballoons or spherical, hollow glass particles which are a component of the fly ash obtained from the combustion of coal. They are described in detail by Raask, *J. of the Institute of Fuel*, pp. 339-344 (September, 1968).

The closed-pore ceramic foam is prepared by heating cenospheres in the presence of air or any inert or unreactive atmosphere at a sufficiently high temperature for a sufficient length of time to form a coherent material having a bulk density of at least 31 lb/ft^3, and preferably in the range of 40 lb/ft^3 to 60 lb/ft^3. The untreated cenospheres are obtained as a loose-flowing powder that is first placed in a refractory mold that has the configuration of the article which is being prepared. Thereafter, the mold is placed in a furnace or kiln and fired to form the closed-cell ceramic foam.

In the process, the firing temperatures are critical and depend on the chemistry of the system. Firing temperatures in the range of about 2470°F to about 3000°F are preferred. The furnace temperature initially should be raised slowly at a rate of 200°F to 400°F per hour until a temperature has been reached such that any organic binder used is burned off slowly and any carbonates (or other decomposable compounds) are decomposed. This temperature is usually about 1000°F and is maintained for a time period ranging from about 0.5 hour to about 5 hours, preferably for about 1 hour. Then, in what may be termed a second stage of the firing process, the furnace temperature is raised at a rate of 350°F to 550°F per hour and is held at the desired peak temperature until the mass of cenospheres shrink to form a closed pore ceramic foam.

In the process, a firing cycle of about 24 to 30 hours usually is employed and a peak temperature level is maintained for a period of about 0.25 hour to 1.5 hours, preferably 0.5 hour. The cenospheres may be used directly after recovery from fly ash, but it is preferred to first pretreat them by either a decrepitation or separation procedure or both. These steps may be carried out in either sequence.

The decrepitation procedure is essentially a heat treatment to drive off water which is contained in the cenospheres. It is usually carried out by heating the cenospheres in a tray to a temperature at which a crackling noise is heard. This will be accomplished by a heating cycle of 0.5 to 2 hours at a temperature of 600° to 1000°F, preferably at about 900°F. The cenospheres as recovered from fly ash usually have diameters in the range of 20 to 200 micrometers and a shell thickness of 2 to 10 micrometers. The cenospheres obtained from England or domestic sources have the following chemical analysis (wt %).

Constituent	English	Domestic
Al_2O_3	31.97	33.25
SiO_2	60.75	61.60
Fe_2O_3	4.18	3.16
K_2O	1.91	1.44
Na_2O	0.81	0.59

To enhance the physical properties of the closed-cell ceramic foam, the ceno-spheres may be separated to obtain a fraction with diameters in the range of 50 to 100 micrometers, or a bulk density of less than 22 lb/ft^3, preferably about 20 lb/ft^3. The separation may be carried out by placing the cenospheres in a liquid having a density less than water, allowing the heavier fraction to sink and collecting the floating fraction. Suitable liquids include hexane, heptane and other organic solvents with a density of less than about 0.95. This procedure also removes other undesirable impurities or dense particulate matter.

The closed-pore ceramic foam may be used as an insulation panel or structural member for a wide variety of applications. These applications include heat shields for space vehicles, building panels, high-temperature furnace bricks and other ap-plications where a noncombustible insulation is employed. The thermal conduc-tivity of the closed-pore ceramic foam can be varied between 0.9 and 2.7 Btu/in/hr/ft^2/°F depending on composition and density and temperature. The three-point bend strength varies between 400 and 2,500 psi depending on density and chemistry. The compressive strength varies between 2,700 and 9,300 psi de-pending on density and chemistry and the average expansion coefficient is 24 x 10^{-7}/°F from room temperature to 2000°F.

Example: Twenty to thirty pounds of West Virginia cenospheres are placed in ten 8" x 10" steel trays and covered with aluminum foil. The covered trays are placed in an electric kiln furnace which is heated to 900°F over a 2-hour period. The heating is continued at 900°F for 12 hours. The decrepitated cenospheres are removed from the furnace and allowed to cool. The cooled, decrepitated cenospheres are poured into a glass beaker that is filled with n-heptane. After 2 to 3 minutes, the spheres separate into a floating layer and a sinking layer. The floating spheres are skimmed from the surface with a ladle and placed in a Buchner funnel lined with filter paper. The drained cenospheres are removed and allowed to air dry in a vented hood.

1.3 lb of separated and decrepitated cenospheres are mixed with 0.17 lb of co-balt carbonate and sieved several times to separate the –20+40 fraction. To the sieved mixture is then added 0.25 lb of a binder which consists of 20% polyvinyl alcohol and 4% glycerin in water (w/w). The binder is blended to the ceno-spheres and cobalt carbonate for 5 minutes in a Hobart blender at maximum speed. The blended mixture is sieved again to separate the –20+40 fraction and, thereafter, placed in a 9¼" x 9¼" x 1¼" mold with a Formica liner. The mold is placed in a press and a block is formed by applying a pressure of 125 psi. The block is removed from the mold and dried in an oven at 200°F for 8 hours.

After drying the block is placed in a fused silica muffle that is fitted with a thermocouple insert. The base of the muffle is covered with an alumina grog to prevent sticking and the block is fired in an electric kiln or gas fired kiln accord-ing to the following cycle: Room temperature to 1000°F at 333°F/hr; hold at 1000°F for 1 hour; 1000° to 2500°F at 550°F/hr and hold at 2500°F for 0.5 hr. Thereafter, the furnace is shut down and allowed to cool prior to removing the closed-pore ceramic foam.

Carbonized Granulated Cork

R.D. Klett; U.S. Patent 3,914,392; October 21, 1975; assigned to the U.S. Energy Research and Development Administration describes a process which uses natural cork which is the bark of the cork oak, *Quercus suber.* In the process, cork which has had its impurities removed through processes old in the art, is comminuted to a granule size of from between about 150 and 2,000 micron diameters and preferably between about 300 and about 1,200 micron diameters. The comminuting process and granule size achieved serve the purpose of substantially eliminating voids or structural deficiencies which may have been inherent in the original or natural cork slab.

The comminuted particles or cork granules may then be molded or shaped by placing in a mold made of a suitable material such as stainless steel, aluminum, or other metal. This mold or confining means may generally be in the shape of the final product, or may approximate the same so that minimal amount of machining of the finished product may be required. The comminuted granules or particles so confined or molded may be compressed if required to achieve a given density. The pressure used will be dependent upon the density desired. The material will expand during heating due to thermal expansion creating thereby an internal pressure so that external compaction may be necessary.

The mold containing the cork particles, which may be compressed or slightly packed, is then heated to a temperature of between about 260°C and about 310°C for a length of time suitable to satisfactorily cure and bind the cork granules. This temperature is used because it effects flow of the lignin naturally found in cork which acts as a binder. Times which may be necessary are dependent upon the quantity of granules involved, mold size, material thickness, etc. Times which have been generally found to be satisfactory at the recited temperature range are from about 4 to about 6 hours. Heat and compression effectively bind and cure the granules.

Use is made of the natural binder material (lignin) of cork as recited above in the curing step so that an additional additive or adhesive or binder is not required. If it is desired to increase the strength of the final product, a suitable carbonizable adhesive may be added to the granules prior to application of heat and pressure. This effectively increases the strength of materials and reduces grain boundary failure at given loads. Suitable adhesives or binders which have been used are such as refined coal tar pitch, furan resins, lignin, phenolic resins, and epoxides. Generally the amount of binder that may be added is from about 10% to about 25% by weight depending upon the binder composition and properties.

Figure 5.1a is a representation of an electron photomicrograph at 600X magnification of a cork material which illustrates the cell structure of cork. The photomicrograph was taken of a sample that did not have a binder additive other than the natural lignin found in the cork, and that had been heated and slightly compressed to a density of about 9.6 pounds per cubic foot (lb/ft^3). As noted, the application of pressure should be minimal so as to minimize distortion of the cells. Cork is made up entirely of 14 sided, i.e., tetrakaidecahedronal, closed cells (hereinafter referred to as TKD) averaging approximately 0.001 inch in diameter. Figure 5.1a also shows that the compression force exerted was in a direction from top left to bottom right.

After curing and binding of the cork granules, these are carbonized by a suitable carbonizing cycle, such as illustrated in Figure 5.1b, in an inert atmosphere such as argon or helium.

FIGURE 5.1: HIGH TEMPERATURE INSULATING CARBONACEOUS MATERIAL

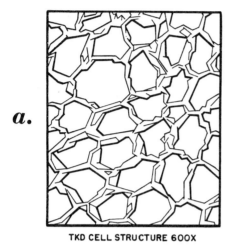

TKD CELL STRUCTURE 600X

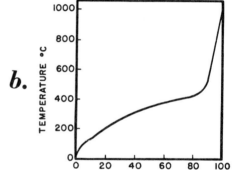

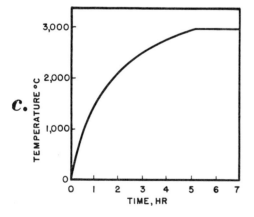

(Continued)

FIGURE 5.1: (continued)

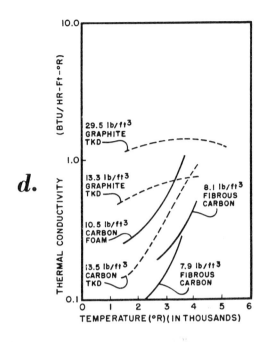

d.

(a) Representation of an electron photomicrograph magnified 600 times illustrating the cell structure of cork
(b) Typical carbonization cycle for carbonizing cork
(c) Typical graphitization cycle for graphitizing carbonized cork
(d) Comparative thermal conductivity curves

Source: U.S. Patent 3,914,392

As used herein, carbonizing cycle refers to the period required to bring the granules to a carbonizing temperature of from about 900°C to about 1200°C. Typical carbonizing cycles will be dependent primarily upon the thickness of the cork. Cork has been successfully carbonized by heating it to a temperature of between about 900° and about 1200°C for a cycle period of from about 30 to about 100 hours. A thin cork material, such as of one inch thickness, will require minimum cycle time, such as about 30 hours, whereas a thick material, such as of 5 inch thickness, will require long cycle times such as about 100 hours in order to achieve uniform temperature within the material, removal of entrapped gases, uniform expansion of cork structure, etc. and thus avoid cracking of the cork.

Heat input should be gradually increased in a manner as shown in Figure 5.1b which illustrates temperature versus percent of cycle rise time in an ideal carbonizing cycle. Thus a carbonizing cycle that attains or incorporates a temperature sequence in relation to a cycle rise time, such as: heating to not more than about 240°C, and preferably 200° to 240°C, at about 20% of cycle rise time, then to not more than 340°C, and preferably between 300°C and 340°C, at about 40% cycle rise time, then to not more than about 400°C, and preferably to between about 360°C and 400°C, at about 60% of cycle rise time, then to not more than about 430°C, and preferably to between about 390°C and about 430°C, at about 80% of cycle rise time, then to not more than about 1200°C, and preferably to between about 900°C and about 1200°C at about 100% cycle rise time, and then holding at the temperature reached of between 900°C and 1200°C for from about 1 hour to about 5 hours at the completion of the cycle rise time, may be preferred.

At any event, 1 to 5 hour hold at about 1000°C at the end of the cycle is preferred to obtain uniform properties. The density and other properties of the material are determined by many factors such as the degree of compression, particle size of the granules, the amount of adhesive or binder used, etc. Materials having densities of from about 5 to about 30 lb/ft^3 have been carbonized using this process. The carbonized cork material may be graphitized, after cooling to ambient temperature or directly from the maximum temperature of the carbonization cycle, at a suitable temperature of from about 2800°C to about 3000°C in a vacuum or an inert atmosphere such as argon or helium. An ideal curve for graphitizing the carbonized material of this process is shown in Figure 5.1c.

Thus a graphitizing cycle that attains or incorporates an increasing temperature sequence in relation to time, such as: heating to between about 1300°C and about 1700°C after 1 hour heating period, and then to between about 2000°C and about 2400°C after 2 hours of heating, and then to between about 2400°C and about 2800°C after 3 hours of heating, and subsequently heating to the graphitizing temperature of between about 2800°C and about 3000°C and holding this temperature for from about 1 to about 4 hours may be preferred. The temperature of about 3000°C is preferably attained by gradual heating as described above, that is, increasing the temperature incrementally to approximate the curve of Figure 5.1c, and is then maintained at about 3000°C for about 1 to 4 hours to achieve best results.

The material of this process, TKD, may be used to insulate heat sources in radio-isotopic thermoelectric generators used in the space programs. The fact that the conductivity of graphite TKD does not increase with temperature may be more important for this application than the low value of conductivity. Graphite TKD is strong enough to support the heat source in the generator and resilient enough to compensate for thermal expansion.

Thermal conductivity data for several densities of TKD graphite and carbon are shown in Figure 5.1d. The carbon foam and fibrous carbon used in the graph and in the following table are the materials which would most closely approximate the TKD structure of this process. Thermal conductivity is proportional to the conductivity of the solid and the density of insulation according to theory. Figure 5.1d shows the carbon TKD has a lower conductivity than carbon foam for all temperatures. The conductivity at low temperatures would be about the same as the fibrous carbon of the same density.

At temperatures above 4000°R the conductivity of the other insulators is increasing proportional to the temperature cubed, but the rate of increase for carbon

TKD is decreasing. The conductivity of graphite TKD is nearly constant and it actually drops at high temperatures as shown in Figure 5.1d. These results enable the use of this product as high temperature insulating material. The table compares the strength of TKD with other high temperature insulators. The strength of carbon TKD is about the same as the higher conductivity foams and is much stronger and more rigid than the fibrous carbons.

Material	Density (lb/ft³)	Crush Stress (psi)	Stress at 10% Deflection (psi)	Stress at 20% Deflection (psi)	Modules of Elasticity	% Compression to crush
Graphite TKD	17.5	596	—	—	45,000	1.32
Carbon TKD	13.5	960	—	—	107,600	.89
Carbon Foam	12.5	876	—	—	71,300	.81
Fibrous Foam	8.2	—	24	35	384	—
Fibrous Carbon	14.9	—	171	245	3,610	—

With regard to tensile strengths, carbon foam at a density of 12.5 lb/ft³ has a tensile strength of 230 psi. Carbon TKD at a density of 13.5 lb/ft³ has a tensile strength of 238 psi which is comparable to carbon foam. Carbon TKD with 1.2% fibers (cellulose acetate) at a density of 13.3 lb/ft³ has a tensile strength of 281 psi. Thus the addition of this relatively small amount of fibers has increased the tensile strength by about 18%.

Ceramic Glaze Coatings

A. Pechman and R.M. Beasley; U.S. Patent 3,955,034; May 4, 1976; assigned to NASA describe ceramic glaze coatings and a method for treating low density rigid, fibrous silica insulating materials, generally adhered to a metallic substrate to thereby render such insulations nonporous and impervious to moisture. Silica insulations coated in accordance with the process are suitable for use in environments in which the insulation must withstand repeated exposures to alternating high and low temperatures.

The fibrous, silica insulation to which the coatings are applied generally exhibit a density in the range of from about 9 to about 20 lb/ft³ and are characterized as fibrous. Such insulations exhibit a low coefficient of thermal expansion, generally in the range of 2.0-5.0 x 10⁻⁷ in/in/°F and a low emissivity at temperatures in the range of 2000° to 2500°F. Being fibrous, such insulations readily absorb moisture and, for this reason, must be coated with a suitable ceramic glaze coating.

Example: In the following example, there are described the preparations of each of the three components of the ceramic coating and a description of the application technique by which they are employed.

The Barrier Coat — A fused silica aqueous slip containing approximately 82% solids was prepared using 99.6% silica (SiO_2) and water. The grain size of the fused silica was such that there was less than 3% retention on a Standard No. 325 sieve. The pH of the aqueous slip was adjusted to a pH of from 5 to 7. The silica slip was applied to a 3 inch x 3 inch x 1 inch tile of fibrous silica having a density of approximately 13 lb/ft³. The dense silica slip was applied by brushing with a soft, lint-free brush to a thickness of from about 4 to about 8 mils. The tile was then dried in an oven for 30 minutes at 250°F after which it was placed in a furnace and fired at 2200°F for 15 minutes. It was then allowed to cool to ambient temperature.

The Emissivity Coat — To a plastic bottle were added 380 parts by weight of a high silica glass (Corning Glass Works No. 7913; 96.5% SiO_2, 3.5% B_2O_3), 20 parts by weight silicon carbide (1,200 grit) and 350 parts by weight of a 0.5% solution of methylcellulose in water. These materials were dispersed with an air-driven mechanical mixer to form an aqueous slurry.

The emissivity coat in the form of the aqueous slurry was applied to the cooled specimen by spraying using a Binks spray gun, Model 18-V fitted with a pressure nozzle. The emissivity coat was sprayed to a thickness of 9 to 12 mils. The coating was allowed to air dry for approximately 10 minutes and was then dried in an oven for 30 minutes at 250°F. Thereafter, the specimen was placed in a furnace and heated to 1000° to 1400°F for approximately 20 minutes after which it was fired at a temperature of 2500°F for 15 minutes. The specimen was then allowed to cool to ambient temperature.

The Overglaze Coat — To a plastic bottle were added 380 parts by weight of the high silica glass of the composition employed in the intermediate coat, 20 parts by weight of a borosilicate glass (Corning Glass Works No. 7740; SiO_2, 80.4%; B_2O_3, 13.3%; Na_2O, 4.3%; Al_2O_3, 2.0%), and 350 parts by weight of a 0.5 wt % solution of methylcellulose in water. These constituent materials were then dispersed with an air-driven mechanical mixer.

The overglaze coat was applied to the cooled specimen by air spraying with the equipment used for the emissivity coat to a thickness of from about 2 to 4 mils. The specimen was allowed to air dry for 10 minutes, after which it was placed in an oven and dried for 30 minutes at 250°F. The specimen was then transferred to a furnace and heated to 1000° to 1400°F for approximately 20 minutes and was then fired at 2500°F for 15 minutes. The specimen was then allowed to cool to ambient temperature.

The test specimen was then subjected to cyclic thermal tests wherein the specimen was subjected to alternating high and low temperatures. In a test, the specimen was exposed to a pulse consisting of radiant heating to simulate the temperatures encountered in a reentry into the atmosphere from a space orbit of approximately 100 nautical miles. The sample was exposed to a peak surface temperature of about 2500°F over a 50 minute interval. The sample was exposed to 3.5 hours of such cycling wherein the time at 2500°F peak was held approximately 3 minutes and the heat-up time was approximately 11 minutes and the cool-down time was 16 minutes.

At the completion of the thermal cycling, the test specimen was then tested for moisture-tightness by placing droplets of water on the coated surface and allowing the specimen to stand for 5 minutes. The surface was then examined by microscope. The test specimen showed no absorption of water and there were no apparent cracks in the coating.

In related work, *A. Pechman and R.M. Beasley; U.S. Patent 3,953,646; Apr. 27, 1976; assigned to NASA* describe low density, fibrous, rigid, silica insulations which are rendered impervious to moisture by application of a ceramic glaze coating comprising a silica barrier layer and an emissivity glaze layer comprising a high silica glass component, an emissivity agent and a borosilicate glass component. The resulting ceramic glaze laminate adhered to the fibrous silica insulation provides a moisture-impervious insulating material which exhibits a high emissivity and is resistant to delamination and spalling at repeated cycles of thermal shock.

Fibrous Zirconia Cement

According to a process described by *B.H. Hamling; U.S. Patent 3,736,160; May 29, 1973; assigned to Union Carbide Corporation* composites comprising fibrous zirconia in a matrix of zirconia are prepared by subjecting fibrous zirconia impregnated with a mixture of a liquid containing a zirconium compound and a refractory powder to a temperature sufficient to insure conversion of substantially all of the zirconium compound to zirconia.

The fibrous zirconia that is employed can be in the form of tow, yarn, woven fabrics, felts, roving, knits, braids and paper. The fibrous zirconia can be prepared by the method described in U.S. Patent 3,385,915. Preferably, the fibrous zirconia is in the form of yarn, felt, especially felt wherein the fibers have been interlocked by needle punching or other method, or woven fabric.

It is preferred that the zirconia fibers be stabilized. The stabilization of zirconia in either the tetragonal or cubic form is known. The preferred stabilizer oxide for use in stabilizing the fibrous zirconia in the tetragonal form is yttria.

The stabilization of zirconia fibers in the cubic form is accomplished by incorporating from about 5.5 mol percent to about 9.6 mol percent of yttria. The refractory powder which is employed can be a refractory metal oxide, zircon, barium titanate, strontium titanate or other refractory powder.

The fibrous zirconia-cement composites are useful as heat shields, flame barriers, corrosion protection barriers, and in other applications requiring resistance to high temperature and/or corrosion resistance. The following examples illustrate the process.

Example 1: Representative Method of Preparing Liquid Component-Chloride Base — Basic zirconium chloride (ZrOOHCl), also called zirconium hydroxychloride, such as that sold by National Lead Company, as an aqueous solution is used. The solution has the following properties:

ZrO_2 content	234 g/l
Chloride content	69.0 g/l
Fe_2O_3 content	0.03 g/l
pH	0.4 at 25°C
Color	slightly amber
SG	1.26 at 25°C
Viscosity	17.5 cp at 25°C

It is concentrated further by evaporation (or boiling) to the following composition:

SG	1.65
ZrO_2	586 g/l
Viscosity	70 cp at 25.5°C

To the concentrated solution is added yttria stabilizer in the form of yttrium chloride, YCl_3. This salt may be purchased commercially, or it can be prepared by reacting the oxide Y_2O_3 with a stoichiometric quantity of HCl. The YCl_3 solution is made up to a SG of 1.43 and has a Y_2O_3 content of 264 g/l. 150 ml of YCl_3 solution is mixed with 1.0 liter of ZrOOHCl solution. A typical solution has the following properties:

SG	1.61
Viscosity	40-50 cp at 25.5°C
pH	Less than zero
Chemical analysis:	
ZrO_2	462 g/l
Y_2O_3	38.4 g/l
Cl	189 g/l
Rare earth oxides	1.8 g/l

Representative Preferred Method for Preparing Liquid Component-Acetate Base — Zirconium acetate solution, such as that sold by National Lead Company as an aqueous solution, is further concentrated. The as-purchased properties are:

ZrO_2 content	22%
Viscosity	31 cp at 25.5°C
pH	3.8-4.2 at 25°C
Color	Clear white-pale amber
SG	1.30

The above solution is concentrated by evaporation or boiling for cement use to the following composition and properties:

ZrO_2 content	379 g/l
Viscosity	120.3 cp at 25.5°C
pH	3.4 at 25°C
Color	Clear white-pale amber
SG	1.44

During concentration by steaming or boiling, the acetate solutions tend to set up as a very thick gel on the bottom of the vessel where heated, requiring frequent mixing. Upon cooling, the thick gel decomposes back to a pourable (lower viscosity) solution. Yttrium acetate crystals are dissolved in the above concentrated solution in sufficient amount to fully stabilize the ZrO_2. The acetate salt is prepared by reacting yttrium oxide powder (99% purity) with hot acetic acid (50% concentration in H_2O).

Yttrium acetate has rather low solubility in water and is collected as crystals on cooling down the reaction solution. The crystals are dried and have a yttria content of 44.1%. The crystals are added to hot Zr acetate solution in which they dissolve readily. Alternatively, the salt can be recovered from the reaction solution simply by drying the solution or the solution can be added directly to the zirconium acetate solution, followed by concentration.

Example 2: A representative method of producing yttria-stabilized zicronia powder suitable as a filler material is the following: (1) Contact sheets of wood pulp, by immersion, in an aqueous solution of zirconium oxychloride and yttrium chloride and having a specific gravity of 1.35 and containing 250 g/l ZrO_2, 20 g/l Y_2O_3 and rare earth metal oxides, and 160 g/l chloride ion.

(2) After thorough saturation of the solution into the wood pulp (time may vary from several minutes to a day or more), the pulp is centrifuged of excess solution, i.e., solution not absorbed into the pulp.

(3) The wet, salt-loaded pulp is next burned in any convenient manner such as an incinerator, kiln, or the like. During burning the material reaches a maximum temperature of around 1500° to 1800°F for several minutes.

(4) After the charge has completely burned, the white ash is collected. The ash at this point is a soft, fluffy material composed of loosely agglomerated crystallites of fully stabilized zirconia. Particle sizes of the crystallites, as determined by x-ray diffraction line broadening analysis and electron microscopy, are in the 200 to 500 angstrom range. The ash is next broken down to about 100 mesh size in a blender or pulverizer and wet milled for 4 to 8 hours. Zirconia beads have been used as the grinding media in small preparations, but other hard grinding media are acceptable. The wet milled powder passes (more than 98%) through a 10 micron sieve and has a mean particle size under one micron. Typical analysis of fully stabilized wet-milled powder is:

	Weight Percent
ZrO_2	92.70
Y_2O_3	3.53
Rare earth metal oxides	2.42
Fe_2O_3	0.16
Cl	0.09
Moisture	0.11
Loss on Ignition	0.24

For the preparation of submicron sized metal oxide powders other than zirconia for use in the zirconia cement, compounds of other metals can be used to impregnate the wood pulp or other preformed polymeric material such as cotton linters. For instance, aqueous solutions of one or more compounds of yttrium, thorium, beryllium, cerium or other rare earth metal, hafnium, or the like, can be used. Thereafter, the process employed is analogous to that illustrated above wherein the loaded polymer is ignited to burn off the polymer and produce fragile agglomerates of submicron sized metal oxide particles. The following example describes the fabrication of yttria-stabilized fibrous zirconia composites utilizing zirconia cement, as described above, to bond zirconia fibers.

Example 3: Fabrication and Properties of a Rocket Nozzle Throat Insert — A composite structure was fabricated suitable for use as a throat insert in a rocket nozzle where a constant throat area (no erosion) is required under an oxidizing environment. The composite was fabricated with zirconia cloth which contained 8 weight percent yttria and which had the following properties:

Cloth construction	Five-harness satin weave
Thickness	28–31 mils
Weight	22 oz/yd^2
Bulk density	63 pcf
Porosity (void content)	83%
Breaking strength	6 lb/in width
Elongation at break	8%

The cement used was prepared by mixing together a ratio of 100 ml of liquid binder to 150 grams of zirconia powder. The liquid binder contained 35.6 weight percent ZrO_2 in the form of dissolved ZrOOHCl and 2.75 weight percent Y_2O_3 in the form of dissolved YCl_3 (8 weight percent Y_2O_3 relative to ZrO_2). The ZrO_2 powder also contained 8 weight percent Y_2O_3 stabilizer and was prepared as described above by ashing and wet balling $ZrOCl_2$ + YCl_3 loaded wood pulp. The ZrO_2 cloth was saturated with cement, thoroughly padding cement into both sides of the cloth and removing excess cement from the cloth prior to lamination.

The lamination was prepared by cutting round discs 2 inches in diameter from the the cement-filled cloth and stacking 100 layers high. The stack was placed in an aluminum die which had been lined with Mylar film. The laminate was pressed at 150 psi, the temperature being raised over a period of three hours to 250°F. After holding for 3 hours at 250°F, pressure was released and the laminate removed from the die. (At this point the dried laminate was hard and strong. The cement had been dried and formed a strong, green body.)

The laminate was next cured in a gas fired kiln, which was raised to 3000°F over a period of 6 hours and held at 3000°F for ½ hour. The cured laminate measured approximately 1¾" high x 1¾" diameter. A ⅜" diameter hole was machined in the center of the laminate using diamond grinding media, such that the cloth layers were normal to the axis of the hole. The finished specimen had a bulk density of 280 pcf (4.5 g/cc). In separate tests on zirconia-bonded satin weave cloth laminates prepared and fired in the manner described above (except that maximum firing temperature was 2800°F), the properties of such laminates are as follows:

Porosity	32%
Flexural strength	6,000–8,000 psi
Modulus of elasticity	$3-5 \times 10^6$ psi
Coefficient of thermal expansion (25°-1000°C)	105×10^{-7} in/in/°C
Melting point	4800°F

Intra-Cell Thermal Resistance in Honeycomb Cores

Since the core of a honeycomb panel is ordinarily of extremely thin sheet material, the cell walls provide very low heat conductance from one facing sheet of the panel to the other. While there is little radiation through the core cells from one facing sheet to the other at low temperatures, when the temperature of the hotter side of the panel is of the order of 1000°F or more, such radiation becomes the principal source of such heat transfer.

R.J. Scanlon; U.S. Patent 3,802,145; April 9, 1974; assigned to Rohr Corporation describes a technique to increase the thermal resistance of a honeycomb panel by providing a heat barrier in the form of one or more transversely extending layers of light, thin, heat shielding material within each of the cells of the honeycomb core of a honeycomb panel. Figure 5.2a illustrates a fragment of a well-known type of honeycomb panel **A** comprising parallel, spaced apart facing sheets **10** and **11**, attached by suitable means such as brazing, welding, diffusion bonding or adhesion, to opposite ends of a square cell type honeycomb core **12**.

FIGURE 5.2: HONEYCOMB PANEL CORES

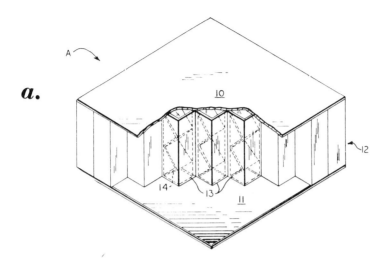

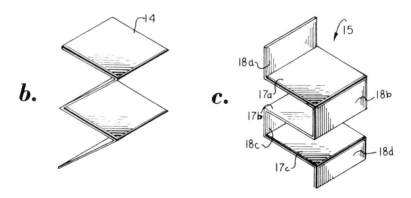

(a)   Perspective view of a fragment of square cell honeycomb panel.
(b)   Enlarged perspective view showing a zig-zag heat shield insert
       of the type used in the panel of Figure 5.2a
(c)   Similarly enlarged perspective view of a modified form of heat
       shield.

Source: U.S. Patent 3,802,145

Fitted snugly into each cell **13** of the core **12** is a zig-zag bent insert **14** formed by the zig-zag bending of a strip of thin, suitable sheet material, preferably metal foil, of a metal capable of withstanding without melting, substantial softening, or contamination of the panel A, the maximum temperature to which the panel is to be exposed, either during attachment of the facing sheets **10** and **11** to the core **12**, or in subsequent use.

Honeycomb cores vary widely in the cross sectional configuration of their cells, for example, square, triangular, hexagonal, etc. Obviously, therefore, each cell insert will be so shaped that the transversely extending portions thereof fit as snugly as practicable within its respective cell so as to intercept a maximum amount of radiant heat rays emanating from the hotter of the facing sheets and toward the other. Since such shaping of the inserts is well within the capabilities of any routine worker familiar with honeycomb panel manufacture, such alternate shapes for different types of core cells are omitted.

In the modified form of the process shown in Figure 5.2c, an insert **15** is formed by bending a strip of suitable material in square, zig-zag manner so that portions **17a**, **17b** and **17c** thereof extend transversely across a cell of a honeycomb core, such as the cell **13** of Figure 5.2a into which the insert is fitted, while the other positions **18b**, **18c** and **18d** of the insert extend along the walls of such cell to position and support the insert therein.

Honeycomb panels of the process may be used in numerous applications, particularly in the fields of aviation and aerospace. For example, in a jet engine tailpipe the use of the honeycomb panels will reduce external temperature of the tailpipe and thereby reduce radiation to external heat sensitive structures or components. Also, in supersonic aircraft, space shuttles, space modules and reentry capsules the material can be used to reduce heat transference between the outer and inner skins of the vehicle.

GENERAL PROCESSES

Polysulfide-Epoxy Compositions

A process described by *F.A. Marion and H.J. McSpadden; U.S. Patents 4,001,126; January 4, 1977; and 3,714,047; January 30, 1973; both assigned to Universal Propulsion Co.,* relates to materials having properties of withstanding temperatures in the thousands of degrees for extended periods of time without gross decomposition.

The materials include binders and fillers mixed with the binders. The materials form a char when subjected to the elevated temperatures. The surface of the char glows in the luminous to incandescent range and rejects incoming heat by thermal radiation. The fillers have the properties of decomposing through a number of successive steps, each step advanced in temperature from the previous step by a relatively small amount. The gaseous products of such successive decompositions flow through the char, which is porous so as to provide transpirational heating. The resultant gases produced at the surface of the material provide a gaseous barrier to absorb heat and to transport heat away from the surface of the material.

A coating of the material of the process was applied in a thickness of approximately 0.6 inch to a steel plate having a thickness in the order of 0.050 inch and dimensions in the order of 6 x 6 inches. The flame from an air-acetylene torch was applied to the surface of the coating to provide a temperature of approximately 3000°F and a heat transfer of approximately 60 Btu/ft²/sec. At the end of approximately one hour, the temperature of the steel plate was approximately only 200°F and at the end of approximately 30 hours the temperature of the steel plate was approximately only 350°F.

The material can be bonded to any desired backing member by application in liquid, plastic or solid form of any desired thickness. The material can be applied and cured at ambient temperatures between approximately +50°F and +125°F. The material has the properties of retaining its properties even after prolonged exposure to ambient temperatures and even when exposed to various solvents such as water, oil, gasoline and toluene.

The material constituting this process consists of a binder and filler material in the binder. The binder may be formed from a mixture of a polysulfide and an epoxide in a ratio in the order of 9 parts to 1 part by weight of polysulfide to from 1 part to 9 parts by weight of epoxide. A particularly preferred range is 70 parts to 30 parts by weight of polysulfide to 30 parts to 70 parts by weight of epoxide. Typically the two may be approximately equal in weight or the epoxide may have a percentage by weight in the order of 40 parts to 45 parts per 100 parts of binder and the polysulfide may comprise the balance of the binder.

An especially preferred amount is 57.5% by weight of polysulfide and 42.5% by weight of epoxide. The ratio is dependent upon the properties desired for the binder. For example, the epoxide tends to be brittle and glass-like with high tensile strength and low elongation and is subject to shock.

The polysulfide tends to provide a high elongation and low tensile strength. It is flexible and rubbery and neither glass-like nor plastic. Variations in the ratios of the polysulfide and the epoxide will affect the physical and structural properties of the cured binder matrix and the rate of cure. Typical compositions of the process are shown below.

Example 1:

Material	Parts by Weight
Polysulfide*	55.0
Curative**	4.5
Ammonium biborate	47.0
Epoxide***	45.0
Sodium phosphate monobasic	34.4

    *Thiokol LP-33, liquid polysulfide resin having terminal mercaptan groups and a molecular weight of about 1,000.
   **Rohm & Haas DMP-30, 2,4,6-tri(dimethylaminomethyl) phenol
 ***Araldite 6020, epoxy equivalent 208 grams per gram mol of epoxide

Example 2:

Material	Parts by Weight
Polysulfide (Thiokol LP-33)	55.0
Anhydrous sodium borate	36.0
Monobasic ammonium phosphate	32.0
Curative (Rohm & Haas DMP-30)	4.5
Epoxide (Ciba Araldite 6020)	45.0

The coating formed from the above material had a thickness of approximately 0.9 inch. It was bonded to a steel plate having a thickness of approximately 0.05 inch. The temperature at the back side of the steel plate was only approximately 300°F after the application of the temperature of approximately 3000°F continuously to the front of the coating for about 50 hours.

Graphite Flakes and Fibers

Z.L. Ardary, D.H. Sturgis, and C.D. Reynolds; U.S. Patent 3,793,204; Feb. 19, 1974; assigned to the U.S. Atomic Energy Commission describe a process which relates generally to fibrous thermal insulation and more particularly to such insulation where graphite flakes are incorporated for increasing the opacity of the insulation to radiant heat transfer.

In the process, the fibrous insulating composite comprises a plurality of discrete fibers of a carbonaceous or refractory material joined together in random orientation by a binder consisting essentially of carbonized starch which has been previously gelatinized, the improvement being in the incorporation of thin graphite flakes in the composite with the maximum dimension of each of the flakes being disposed essentially orthogonal to the expected direction of heat flow through the composite.

The fibrous thermal insulation is prepared by employing the steps of mixing together fibers exhibiting low thermal conductivity and high heat resistance such as provided by refractory or carbonaceous materials with thin graphite flakes of a discoidal configuration and a mixture of water and starch particulates which provide the binder for joining together the fibers in a random orientation and securing the flakes in the resulting composite. The graphite flake-fiber-water-starch mixture or slurry is vacuum formed into the desired configuration, with the starch particulates and flakes uniformly dispersed throughout the fibers. The graphite flake-starch-water-fiber mixture is then subjected to a sufficient quantity of heat to effect a reaction between the starch and the water to gelatinize the starch.

After gelatinization and prior to hydrolysis the formed mixture is dried and then heated to a temperature sufficient to convert the starch to carbon for joining together the fibers. The formation of the fibrous insulation into composites or structures of the desired configuration while providing relatively smooth surface finishes and uniform wall thicknesses of 1.5 inches or more has been satisfactorily achieved by employing a vacuum molding process. Basically, a vacuum or pulp molding process found suitable comprises the steps of separating the fibers from intertwined clumps or agglomerates of the fibers by a conventional screening technique, mixing the fibers and graphite flakes in water containing the desired quantity of starch for a period of about 30 minutes to assure a homogeneous distribution of the solids, feeding the resulting slurry from a mixing tank into a molding assembly containing a perforated mold or mandrel which is under the influence of vacuum at a pressure in the range of about 500 to 760 mm, and drawing off excess water after all of the solids have been deposited on the mandrel. During the vacuum drawdown the graphite flakes tend to orient themselves flat against the composite mat being formed by the molding or filtering process resulting in the flat surfaces of the flakes, i.e., either flat side of the flake as opposed to the edge of the flake, being disposed in substantially parallel planes which are at least substantially normal to the source of vacuum defined by the perforated mold.

With the graphite flakes so oriented in the composite the flat surfaces are disposed in parallel planes that are at least substantially orthogonal to a plane projecting through the thickness of the composite which is parallel to the path of radiant heat transfer in the composite when the latter is employed as the thermal insulation.

The gelatinizing of starch and pyrolysis of the latter in the composite may be achieved by placing the composite containing mold assembly in a heated, water-saturated atmosphere at a temperature of about 95°C for a duration in the range of about 8 to 20 hours to gelatinize the starch, setting the starch by continuing the drying operation at low humidity and at about the same temperature for approximately an additional 20 to 32 hours (if desired, higher temperatures, e.g., up to about 150°C may be used for the drying step to complete drying more rapidly), and thereafter removing the rigid composite from the mandrel and pyrolyzing or carbonizing the starch at a temperature in the range of about 900 to 1200°C in an inert atmosphere for about 16 to 36 hours to effect carbon-bonding of the fibers. The following example illustrates the process.

Example: A 14-inch diameter by 1-inch thick disc of insulation was vacuum molded at a pressure of 700 mm from a dilute water slurry on a perforated plate. The slurry contained 250 grams of carbon fibers prepared by pyrolyzing rayon fibers of 0.010 inch length and 3.7 micron diameter, 375 grams of tapioca starch made from cassava plants, 250 grams of graphite flakes having a surface area of about 19.2 m²/g, and 150 liters of water. The prepared fibrous mat was processed through curing, drying, and heat treatment operations to effect a semirigid, carbon-bonded structure. The curing and drying operations were performed at 90°C in a high-humidity environment.

The structure was heat treated at 1000°C in an inert atmosphere to pyrolyze the starch binder. Analysis of the disc indicated a density of 17.0 pcf, a compressive strength at room temperature of 132 psi at 10% deflection and 155 psi at 20% deflection, a compressive strength at 2340°C of 45 psi at 10% deflection and 60 psi at 20% deflection, a modulus of elasticity at room temperature of 4920 pounds per square inch, and a coefficient of thermal conductivity in a vacuum of 3.0 Btu/in/hr/°F/ft² at 3860°F mean temperature.

Heat Reflective Metal Coated Spheres

N.H. Deschamps and G.L. Bernier; U.S. Patent 3,769,770; November 6, 1973; assigned to Sanders Nuclear Corp. describe a thermal superinsulation which utilizes highly reflective, thin layers on high-temperature-resistant bases to increase reflection and decrease heat transfer by radiation. Preferably the thermal

superinsulation comprises a plurality of spheres of a hard low thermal conductivity material grouped together with point contact between adjacent spheres and
a vacuum is created in the interstices between the spheres thus forming a heat
barrier. The spheres are individually coated with a thin layer of highly heat reflective, low emissivity material which is preferably a reflective metal. Preferably
the spheres are uniform in size and are formed of a peripheral wall of an inorganic insulating material defining a space therein which may be filled with air,
other gas or evacuated. In the preferred case, the spheres each have a diameter
no greater than 250 microns, the sphere walls have a thickness no greater than
10 microns and the heat reflective layers each have a thickness no greater than
5 microns.

In insulating an object in accordance with the process, the spheres described
above are formed into a superinsulation heat barrier by positioning a plurality
of the spheres in point contact with each other in a compact mass defining interstitial spaces therebetween about the object to be insulated. The spaces formed
between the spheres are maintained at a vacuum preferably in the order of 10^{-4}
torr. Because of the point contact between spheres, heat transfer by conduction
is substantially eliminated. Since high vacuum conditions are used, convective
heat transfer is substantially eliminated. The large surface area of the heat reflective layers acts to retard heat transfer by radiation.

The process also encompasses the use of a fiber base rather than individual
spheres. Thus, fibrous material having good insulating values, for example,
MIN-K is used with the fibers being coated with a thin layer of highly heat reflective, low emissivity material. The layer of reflective material is preferably a
radiation reflective metal. The metal is preferably applied by a vacuum sputter-
ing technique to deposit individual films around each fiber in a mass.

Amorphous Carbon Layered Composite

*R.D. Allen and W.M. Lysher; U.S. Patent 3,715,265; February 6, 1973; assigned
to McDonnell Douglas Corporation* describe a composite thermal insulation comprised of separate layers of particulate material selected from the group consisting
of (a) materials characterized by low thermal conductivity above 1000°C and (b)
material characterized by low thermal conductivity below 1000°C arranged alternately with and separated by one or more layers of metal sheets, and wherein
the first and last layers are comprised of particulate material.

The composite thermal insulation provides very low thermal conductivity up to
2000°C. This is partially due to the fact that while most prior art materials employ particles which are physically or chemically bonded together and thus conduct heat readily, the composites of the process employ particulate unbonded
particles. Further, the particulate layers of the composite are separated by metal
sheets which prevent commingling or reaction between adjacent layers. The
metal sheets provide additional advantages by acting as radiation shields, as barriers to thermal conduction, as barriers to transverse fracturing, and as internal
structural support for the separate particulate layers.

Referring to Figure 5.3a there is shown a three layer composite having a first
layer **10** of particulate material characterized by having low thermal conductivity
above 1000°C, e.g., amorphous carbon black, the particles of such material being
indicated at **11**, a second layer adjacent the first layer and substantially coexten-

sive therewith, comprised of a metal sheet **12**, e.g., tantalum foil, and a third layer **14** of particulate material characterized by having low thermal conductivity below 1000°C, e.g., zirconia, the particles of which are indicated at **15**, disposed adjacent to the metal sheet so that the metal sheet separates the layers **10** and **14** of particulate material. Figure 5.3b shows a three layer composite thermal insulation similar to that of Figure 5.3a, except that here the composite has a hemispherical cross-sectional configuration.

FIGURE 5.3: COMPOSITE THERMAL INSULATION

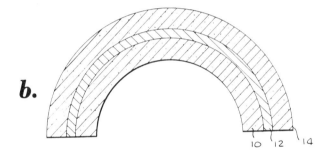

(a)   Schematic, partially fragmented, cross-sectional representation of a three layer composite thermal insulation
(b)   Three layer composite having a hemispherical configuration

Source: U.S. Patent 3,715,265

Example 1: A thermal insulation composite substantially as shown and described in Figure 5.3b is fabricated using amorphous carbon black as material (a) in layer **10**, zirconia as material (b) in layer **14**, and tantalum foil as the metal sheet **12**. The particulate amorphous carbon black having a particle size range of from about 0.1 to about 0.2 micron, and substantially all of which passes through a 325 mesh screen, is placed in a hemispherical mold to a thickness of 0.35 inch, a 1 mil thick sheet of tantalum foil is placed on top of the first layer, followed by a 0.4 inch layer of zirconia having a particle size range of from about 1 to about 15 microns, and substantially all of which passes through a 325 mesh screen. The resulting assembly is pressed at 5 psi to form the composite.

The thermal conductivity of the above composite is tested by means of a technique utilizing electron beam heating in conjunction with a hollow hemispherical Armco iron calorimeter containing the specimen, a hemispherical tantalum target at the specimen's center and a hemispherical insulated container for the calorimeter. The target, calorimeter and insulated container are arranged concentrically. A vacuum environment and thermoelectric temperature measuring devices are provided.

A given electron beam power setting produces a constant temperature in the hemispherical surface of the tantalum button. Heat generated in the tantalum button by the electron beam passes radially through the hollow hemispherical test material to the Armco iron heat sink. Calorimetric measurement of total heat added to the heat sink provides values which are used to calculate the thermal conductivity of the test specimen. The composite of Example 1 is placed in the hollow hemispherical iron calorimeter with the layer of amorphous carbon black facing the tantalum button. The thermal conductivity obtained for the composite at varying temperatures is given below in Table 1.

TABLE 1

Thermal Conductivity (cal/cm/sec/°C x 10^4)	Temperature (°C)
0.405	799
0.676	1180
1.27	1552
1.67	1905

For purposes of comparison, the thermal conductivities of various single phase granular powders as well as that of a composite of layers of amorphous carbon black separated by layers of pyrolytic graphite, and having comparable thickness to the above described composite, are determined by means of the electron-beam method described above. The results are presented in Table 2. In the table "first sequence" refers to values obtained initially and "second sequence" refers to values obtained after a period of heating and shows changes due to sintering.

TABLE 2

Insulation	Thermal Conductivity (cal/cm/sec/°C x 10^4)	Temperature (°C)
Amorphous carbon black,	0.793	810
particle size 0.1 to 0.2	1.01	920
microns	1.22	1010
	0.998	1025
	1.38	1120
	1.69	1181
	1.55	1220
	2.13	1320
	2.15	1420
	2.66	1515
	3.28	1615
	2.54	1622
	4.20	1720
	4.46	1830
	4.79	2000

(continued)

TABLE 2: (continued)

Insulation	Thermal Conductivity (cal/cm/sec/°C x 10⁴)	Temperature (°C)
. . . . First Sequence		
Zirconia, particle size	0.724	1120
range 1 to 15 microns	0.775	1263
	1.40	1582
	1.93	1762
	2.72	1942
	3.36	2052
. . . . Second Sequence		
Zirconia, particle size	1.21	792
range 1 to 15 microns	1.56	988
	1.93	1204
	2.26	1417
	2.84	1615
	3.27	1846
	3.98	1956
. . . . First Sequence		
Zirconium carbide,	0.813	796
particle size range	0.972	910
5 to 40 microns	0.882	1012
	1.32	1120
	1.94	1236
. . . . Second Sequence		
	1.12	904
	1.75	1122
	1.85	1237
	2.35	1341
	3.35	1495
	4.34	1621
Composite of amorphous	0.496	800
carbon black/pyrolytic	0.681	1180
graphite/amorphous car-	1.364	1552
bon black/pyrolytic	1.933	1905
graphite/amorphous car-		
bon black/pyrolitic		
graphite/amorphous car-		
bon black		

A comparison of the above thermal conductivity values for the composite thermal insulation of the process in Table 1 with the various insulating powders and with the carbon-pyrolytic graphite composites of Table 2 shows clearly the superiority and effectiveness of the process composite. For example, at about 1900°C the process composite having only 3 layers has a thermal conductivity of 1.67 (cal/cm/sec/°C) x 10⁻⁴ while the seven-layer carbon-pyrolytic graphite composite has a thermal conductivity of 1.933 (cal/cm/sec/°C) x 10⁻⁴. None of the powdered insulation materials used alone has values anywhere near that of the process composite. Furthermore, most of these powders are subject to at least moderate sintering while the process composite shows no sintering or radial fracturing.

Example 2: In the manner described in Example 1, a three layer composite is formed having a first layer of boron nitride powder of particle size range of 0.5 to 6 microns, a second layer of tungsten foil of 1 mil thickness, and a third layer of magnesium oxide powder having a particle size range of 5 to 40 microns. The powder layers each have a thickness of 0.3 inch. The resulting composite exhibits low thermal conductivity and other properties comparable with the composite of Example 1.

Substantially the procedure of Example 1 is repeated to produce a five-layer composite having first and third layers of amorphous carbon black of particle size range 0.1 to 0.2 micron, second and fourth layers of 1 mil thick tantalum foil, and a fifth layer of zirconia having a particle size range of 1 to about 15 microns. The powder layers each have a thickness of 0.2 inch. The resulting composite exhibits thermal conductivity which is somewhat lower than the composite of Example 1. Other properties are comparable.

Alkaline Earth Sulfate and Carbonaceous Material

R.M. Gunnerman; U.S. Patent 3,817,925; June 18, 1974 has found that an effective heat barrier can be produced in situ upon the surface of a composition consisting essentially of an intimate mixture of alkaline earth sulfates and finely divided carbonaceous material. Upon application of heat to a surface of such a composition, the alkaline earth metal sulfate and carbon react, primarily at the surface of the composition, whereby the sulfates are reduced to sulfides to provide an effective thermal barrier on such surface.

It has been found that the alkaline earth sulfide coating formed upon the surface of the products is so effective as a heat barrier that even under application of extreme heat thereto by arc plasma jet flames, heat penetration into the body of the product is inhibited to such a degree that the back surface layer of the coated product does not rise materially in temperature. The compositions may be formed by using calcined alkaline earth sulfates, i.e., those sulfates which are not fully hydrated. The sulfates may be mixed with water to form a slurry to which carbonaceous material in the condition of fine subdivision is added and thoroughly admixed.

The mixture is allowed to set due to hydration of the alkaline earth sulfates so that there is formed a composition in which the alkaline earth sulfates' crystals form a matrix in which the particles of finely divided carbonaceous material are held in intimate mixture. The composition, before setting, can be formed into sheets or blocks or can be applied directly to the surface to be protected from heat. When high temperature is applied thereto, there results the formation upon the surface of an alkaline earth sulfide due to the reduction of the sulfate to the sulfide by the carbon which is available at such surface.

The term "high temperature" as used herein means a temperature in the order of 900°C or greater, i.e., at least that temperature at which the sulfates react with carbon to form sulfides. The carbonaceous materials which are useful include all materials which will provide available carbon under the influence of sufficient heat to effect the reaction of carbon with sulfates according to the representative equation: $MSO_4 + 2C \rightarrow MS + 2CO_2$, where M is an alkaline earth metal. One class of suitable materials is that known under the trademark Microballoons as described in U.S. Patent 2,797,201. The following examples illustrate the process.

Example 1: Calcium sulfate, i.e., calcium sulfate hemihydrate $(CaSO_4 \cdot \frac{1}{2}H_2O)$ is mixed with water to form a slurry using the same proportions of calcium sulfate hemihydrate and water as is ordinarily used in forming a slurry for casting purposes. The slurry thus formed is allowed to stand for such period of time as will permit the addition of phenolic resin either in finely ground condition or as Microballoons without the tendency of the phenolic resin to separate during the subsequent setting, which thereby forms a matrix holding the phenolic resin intimately dispersed with the calcium sulfate throughout the composition.

The composition which was used in this example was 30% by volume of $CaSO_4 \cdot \frac{1}{2} H_2 O$ and 70% by volume of finely divided phenolic resin of 500 micron maximum average particle size and the amount of water added to form the slurry is in accordance with good practice that amount which will permit the formation of a thick cream slurry, i.e., in most cases employing the minimum of water which will permit the formation of a uniform intimate mixture of the $CaSO_4 \cdot \frac{1}{2} H_2 O$ and phenolic resin.

Example 2: In a similar manner one produces the slurry of the alkaline earth sulfate and water and thermosetting plastic and adds a suspension of thermoplastic material. It has been found that the composition thereby produced is flexible as distinguished from a rigid composition. Specifically, one mixes 100 parts by volume of $CaSO_4 \cdot \frac{1}{2} H_2 O$, sufficient water to form the slurry as above described, to which is added 200 parts by volume of phenolic resin, and 100 parts by volume of the approximately 50% suspension of polyvinyl acetate in water.

Upon setting, the resultant composition was flexible as distinguished from the rigid composition produced by the first stated example. The operation apparently performed in this example is that the addition of the vinyl plastics to the composition in excess of that set forth in the first example given has prevented the complete interlock of the crystals of the alkaline earth sulfate, leaving the composition flexible.

Heat-Barrier Tests — A test block of the composition was prepared in the manner set forth in Example 1 in which proportions were 70% by volume of phenolic resin Microballoons (maximum average particle size 500) to 30% by volume of $CaSO_4 \cdot \frac{1}{2} H_2 O$ with water which was allowed to set to form test blocks which were 2" x 2" square and of ½" thickness. An arc plasma jet flame was applied to the surface of such block under the following conditions:

Heat flux	1,000 Btu/ft²/sec
Gas enthalpy	6,050 Btu/lb
Test duration	24 sec
Stagnation pressure	1.242 psig
Gas temperature	10,025°F
Gas velocity	1,913 ft/sec

The results were as follows:

Decrease in weight	4.7874 g
Decrease in weight	3.4%
Depth of erosion in the surface so exposed	0.199 inch
Final temperature	
(a) Back face	140°F
(b) Front face (optional)	4820°F
(c) Front face (total radiation)	4340°F

Further tests were conducted in blocks of the same composition to determine the rate of erosion of the surface with the following results:

Specimen	Heat Flux (Btu/ft^2/sec)	Test Time (sec)	Erosion Rate (mils/sec)
A	100	137	3.27
B	500	24	3.96
C	500	44	7.70
D	1,000	24	8.30

Further tests were conducted employing larger sheets of the same composition of like thickness which were tested in an oven over propane gas burners which were spaced 18" on center and where the burners were placed 30" from the face of the test panel. Propane gas was burned in the burners. The temperatures at the face of the panel, toward which the flames were directed were measured by means of eight Chromel-Alumel thermocouples encased in ½" iron pipe and placed approximately 6" from the panel face. The vertical and lateral placement of the thermocouples was determined by experimentation and visual observation of the flow pattern of the burning gases.

Thermocouple leads were connected through a rotary switch to a Techniques Associates Pyrotemp Model 9-B pyrometer from which temperature readings were obtained manually. The backside temperatures were taken by placing seven iron-constantan thermocouples on the cross framing of the panel, with one thermocouple placed directly on the back surface of the composition. The following temperature readings in degrees Fahrenheit were taken from the panel face:

	.... Temperature Readings, °F, Panel Face								Required Temperature (°F)
Minutes	.... Station Number.								
	1	2	3	4	5	6	7	8	
2	250	200	164	164	204	174	115	158	—
5	490	—	—	—	—	—	—	—	1000
7	760	605	334	273	630	555	300	280	1150
12	1020	870	590	—	—	—	—	—	1340

The pressure of the gas flowing to the burners at the start of the test was five pounds per square inch which was increased to ten pounds per square inch after five minutes at which gas pressure the test was continued for the full duration of twelve minutes. The temperatures at the back of the panel measured by the thermocouples as above described were as shown by the following table:

	Temperature Readings, °F, Panel Back	
Minutes	Minimum	Maximum
0	87	102
2	87	102

(continued)

Minutes	Temperature Readings, °F, Panel Back	
	Minimum	Maximum
4	90	103
6	90	110
8	87	109
10	92	112
12	92	113
14	93	124
16	94	126
18	76	108

Further tests utilizing substantially the same equipment and test samples of the same character have been conducted wherein for a total elapsed time of two hours and fifteen minutes the average surface temperatures taken over the hot side surface have been in excess of 1900°F and the temperature of the back face taken likewise over the distributed points as above indicated has been less than 180°F.

Molding Technique

A process described by *J.J. Asbury and J.M. Googin; U.S. Patent 3,634,563; January 11, 1972; assigned to the U.S. Atomic Energy Commission* relates to an inorganic thermal insulating product of a density less than 25 pounds per cubic foot and a thermal conductivity factor of less than 0.5 Btu/in/ft^2/hr/°F at a temperature of 1500°C. The product is prepared by mixing minute silicon oxide particles with inorganic fibers, an organic liquid which wets the surface of the particles and inhibits crystallization and hydration of the particles, and a quantity of octanoic acid.

The organic liquid and a portion of the octanoic acid are evaporated, with the remaining octanoic acid being retained. The mixture is ground and then pressed into a desired configuration of essentially final dimensions, with the octanoic acid flowing during the pressing step to act as a binder for the silicon oxide particles. After pressing the compact is lightly sintered in an inert atmosphere during which virtually all, if not all, of the remaining octanoic acid volatilizes. Octanoic acid forms a thixotropic-like system with the oxide particulates in that the acid flows under the influence of pressure and has been found to provide a unique function in the formation of the thermal insulator in that, while providing a binder for holding the powders together during the pressing operation, it also functions as a lubricant for enhancing the pressing of the powders.

Upon "drying" the slurry with the wetting agents, the portion of the octanoic acid remaining is uniformly dispersed throughout the mass of silicon oxide and upon application of the stress during the pressing operation, the octanoic acid flows to function as the binder and lubricant or pressing aid for the pressing operation. Once the pressing is completed and the stress removed, the octanoic acid-silicon system again returns to a gelatinous state. The octanoic acid remaining in the pressed compact is virtually, if not entirely, volatilized during the sintering of the compact.

Example: A mixture of 32.2 grams of silicon oxide of a size of about 100 to 150 angstroms, 7.9 grams of titanium oxide of –325 mesh, and 4.5 grams of

½" long, 10 mil diameter silicon oxide fibers is thoroughly blended. The blended particulate mixture is added to 225 cc of ethyl acetate and 75 cc of octanoic acid. The mixture of the particulates and the wetting agents is slurried for about 5 minutes in a blender and then spread as a thin layer on an evaporating dish and dried in air at room temperature for 36 hours to remove most of the ethyl acetate and a portion of the octanoic acid. The dried mixture is again blended for 15 seconds to break up the larger flakes resulting from the drying.

The ground or blended mixture is placed in a die-forming assembly and pressed at 3,000 psi into the desired shape. The formed part is then slightly sintered at 900°C for 6 hours. The sintered density of the product is 22.78 lb/ft³. The thermal conductivity or K factor of the product is approximately 0.28, 0.32, and 0.40 Btu/in/ft²/hr/°F at 700°, 1000°, and 1500°C, respectively.

PIPE COATINGS, AUTOMOTIVE
AND OTHER APPLICATIONS

PIPE COATINGS

Tripoli Rock and Glass Fiber

J.L. Helser and R.F. Shannon; U.S. Patent 3,895,096, July 15, 1975; assigned to Owens-Corning Fiberglas Corporation describe a process of manufacturing hydrous calcium silicate products. The products are prepared from a molded aqueous slurry of reactive cementitious materials, cellulosic reinforcing fibers and glass reinforcing fibers. By employing tripoli as one of the reactive siliceous materials, a marked improvement in the product is achieved.

Tripoli is a porous, siliceous rock resulting from the natural decomposition of siliceous sandstone. The various grades of tripoli, according to fineness, are rose, cream and white. Tripoli often is referred to as rotten stone. Tripoli is a sort of diatomaceous earth, but is not to be confused with diatomaceous earths which consist chiefly of amorphous silica derived from fossil diatoms. Deposits of tripoli are found in the Missouri-Oklahoma district and in southern Illinois. The terms microcrystalline silica and cryptocrystalline often are used in reference to tripoli.

Alkali resistant glass fibers that can be employed with the cellulose material include those described in British Patent 1,243,972 and 1,290,528 and in U.S. Patent 3,840,379. Certain ZrO_2- and TiO_2-containing compositions provide a unique combination of alkali resistance, low liquidus temperature and desirable viscosity for the fiberization of glass compositions and for the reinforcement of cementitious materials. The glass compositions of U.S. Patent 3,840,379 have the following range of proportions by weight: 60 to 62% SiO_2, 4 to 6% CaO, 14 to 15% Na_2O, 2 to 3% K_2O, 10 to 11% ZrO_2 and 5.5 to 8% TiO_2.

Example 1: A dispersion of the following materials was made in water with the ratio of water to solids as noted. The dispersion was produced in a hydropulper, placed in U-shaped mold forms and prehardened in a steam atmosphere at a temperature of 190°F. These half-section insulation pieces are pipe covering with

185

a thickness of 2 inches. The U-shaped molds filled with the dispersion were then placed in an autoclave. After the autoclave was sealed, the pressure in the autoclave was raised to 250 psi over a 15 minute cycle, following which the molds were subjected to saturated steam at this pressure for 60 minutes to indurate the dispersion. The temperature in the autoclave was then raised by heating coils to 575°F to produce superheated steam which slowly indurated and dried the insulation over a further 175 minute period. The autoclave was then depressurized for a half-hour period, and the molds were removed from the autoclave.

The pipe insulation so produced had a free moisture content of at least 10% by weight of solids, a modulus of rupture of 75 psi, and a hardness of 44 mm penetration. In the examples, the modulus of rupture is determined according to ASTM Specification C446-64 and hardness is determined according to ASTM Specification C569-68.

Materials (Pan Batch)	Dry Wt %
Glass fiber	1.8
Wood pulp	8.4
Quicklime	31.9
Tripoli	21.9
Diatomaceous earth	16.0
Filler (calcium silicate dust)	9.7
Bentonite clay	3.9
Limestone	3.9
Liquid sodium silicate	2.6
Calcium/silica ratio	0.77/1
Water/solids ratio	4.8/1

Example 2: The process of Example 1 was repeated except that the following batch composition and water-to-solids ratio were employed:

Materials (Pan Batch)	Dry Wt %
Glass fiber	2.0
Wood pulp	9.2
Quicklime	19.6
Diatomaceous earth	32.9
Filler (calcium silicate dust)	10.5
Bentonite clay	4.2
Portland cement	18.4
Liquid sodium silicate	3.2
Water/solids ratio	5.0/1

The pipe insulation so produced had a free moisture content of at least 10% by weight of solids, a modulus of rupture of 42 psi and a hardness of 85 mm penetration. A comparison of the modulus of rupture and the hardness of Examples 1 and 2 reveals the marked improvement of employing tripoli in the production of crystalline hydrous calcium silicate insulation products reinforced with a combination of glass fibers and organic fibers.

Alkali Resistant Glass Fibers

J.L. Helser and R.F. Shannon, U.S. Patent 3,902,913, September 2, 1975; assigned to Owens-Corning Fiberglas Corporation describe hydrous calcium silicate insulation products which are reinforced with a particular combination of organic fibers and alkali resistant glass fibers. Thus, it has been discovered that a much stronger product is achieved when a combination of organic fibers and alkali resistant glass reinforcing fibers are incorporated into the cementitious slurry.

Alkali resistant glass fibers that can be employed with the cellulose material include those described in British Patents 1,243,972 and 1,290,528. The ZrO_2- and TiO_2-containing compositions of U.S. Patent 3,840,379 provide a unique combination of alkali resistance, low liquidus temperature and desirable viscosity for the fiberization of glass compositions and reinforcement of cementitious materials.

The organic materials generally have a fiber diameter of 30 microns or less, as in the case of cotton fibers, and may average less than 1 micron in fiber diameter as in the case of wood pulp. The glass fibers will generally have a diameter of less than 0.001 inch. The glass fibers have a length from 0.25 to 2.0 inches, desirably from 0.5 to 1.25 inches, and preferably from 0.625 to 1.00 inch. The hydrous calcium silicate insulation materials generally will have a density between 10 and 20 pounds per cubic foot and comprise the following materials in percent by weight of solids:

Materials	Percent
Organic fibers	1.0 to 20.0
Glass fibers	0.1 to 10.0
Hydrous calcium silicate	60.0 to 95.0
Fillers	0 to 20.0

Preferably, the amount of organic fibers ranges from 5 to 10% by weight of solids and the amount of glass fibers ranges from 0.5 to 2.5% by weight of solids. The following example illustrates the process.

Example: A low density hydrated calcium silicate heat insulation material is made from the following materials in part by weight of solids:

Materials	Parts by Weight
Hydrated lime	45.2
Diatomaceous earth (86 ft^2/g)	22.6
Diatomaceous earth (54 ft^2/g)	22.6
Wood pulp (sulfate type pulp)	8.5
Chopped glass fibers	1.0
Aluminum sulfate	*

*add as required to control alkalinity

A dispersion of the various materials is made by dispersing the wood pulp and the aluminum sulfate in 350 parts by weight of water heated to 200°F in a hydropulper to produce a dispersion. This dispersion of the wood pulp is then added to a premixer wherein the hydrated lime is added and mixed for 1 minute. Another 900 parts by weight of water at 200°F is added to another premixer and the diatomaceous earth is added and mixed for 1 minute. Thereafter the contents of the two premixers are added to a gel tank, where the materials are thoroughly mixed for 10 minutes.

The resulting slurry or suspension of ingredients is then permitted to gel quiescently for 10 minutes and is then slowly stirred in the gel tank for 2 minutes.

The partially formed gel is again allowed to remain quiescent for a period of 10 minutes, followed by another period of slow stirring of approximately 2 minutes. The gel so produced is then allowed to sit for 80 minutes before being drawn off in small quantities to a volumetric tank in precise quantities for charging a precision type filter mold shaped to make 3 inch annular pipe insulation of 1½ inch wall thickness and a length of 36 inches. The ram of the mold compresses the gel to force the water out through the cylindrical filter forming the inside surface of the pipe insulation to leave a pipe insulation which is self-sustaining and handleable.

The pipe is removed from the filter mold and is stacked in a rack which, when filled, is rolled into an autoclave for induration. After the autoclave is sealed, the pressure in the autoclave is raised to 175 psi over a 30-minute cycle and the pipes are subjected to saturated steam at this pressure for 1½ hours. Thereafter, the temperature in the autoclave is raised by heating coils to 600°F to produce superheated steam which slowly dries the pipes over another 2-hour period. The autoclave is then depressurized, over a ½-hour period, and the racks which hold the insulation block are removed. The material so produced has a modulus of rupture of 115 psi and a density of 12.5 pounds per cubic foot.

By way of comparison, a prior art material made from asbestos using the same parts by weight of diatomaceous earth and hydrated lime, and devoid of the cellulose fibers and the soluble aluminum material, has a modulus of rupture of only approximately 95 psi. The modulus of rupture is determined according to the ASTM Specification C 446-64.

The glass fibers used in the above product are resistant to calcium hydroxide attack. As far as is known, only zirconia fibers (i.e., fibers having 2½% or more of ZrO_2) will withstand autoclaving with lime. A ZrO_2 content of 2½% may be acceptable in low pressure autoclaving (i.e., 175 psi or less), in high temperature autoclaving (i.e., above 450°F) 5% or more may be necessary. Examples of zirconia glasses are described in British Patent 1,243,972.

The fibers were made by drawing molten glass from a bushing having 204 orifices into individual fibers having a diameter of approximately 0.00055 inch. These individual fibers were coated with an aqueous solution of a water-soluble polyvinyl acetate, and were gathered into a strand that was coiled into a package and dried. The strand had approximately ½% of polyvinyl acetate thereon, based on the weight of the dried coated fibers. These fibers were chopped into 1-inch lengths before dispersing in the water.

The wood pulp used is a chlorine bleached sulfate pulp. Sulfate pulp generally includes approximately 0.1% of aluminum sulfate. The soluble aluminum sulfate is used in a quantity sufficient to change the basic negative charge on the fibers to a positive charge. This is highly beneficial in that the positively charged fibers disperse readily in water, and are drawn to the negatively charged particles of diatomaceous earth to surround the same. In addition, the positively charged fibers repel the positively charged calcium ions and leave the hydroxyl ions free to migrate to the particles of diatomaceous earth.

Cellulose fibers treated with soluble aluminum compounds are, therefore, a highly desirable form of cellulose fiber for use in the process. Other types of cellulose fibers which are treated to become positively charged are also a preferred fiber material. In those instances where the cellulose fibers are not pretreated before dispersing in the water, cations may be used in the batch formulation to accomplish generally the same result.

Composite Laminate with Woven Glass Fiber Interlayer

W.C. King and R.J. Medvid; U.S. Patent 3,959,541; May 25, 1976; assigned to Pittsburgh Corning Corporation describe a composite laminate insulating material which retains its insulating ability even when the component layers have fractured under excessive thermal stress. The insulating material is comprised of layers of a substantially rigid cellular ceramic of siliceous material with a layer of insulating material, preferably a woven glass fiber reinforcing material, sandwiched between. The reinforcing layer of insulating material sandwiched between the layers of rigid cellular materials prevents the cracks or fissures formed within the ceramic layer adjacent the heat source from propagating into the outer layer, and limits their formation in the outer layer of material. The glass fiber insulating and reinforcing layer also maintains the integrity of the fractured pieces in contact with the body to be insulated, sustaining the insulation ability of the fractured layer and hence of the composite body.

In a preferred instance, a composite laminate insulating material is provided, comprising inner and outer layers of a substantially rigid cellular glass with the outer layer having a thickness of less than 2 inches. The inner and outer layers are adhesively secured to a layer of woven glass fiber insulating and reinforcing material sandwiched between. Preferably, the adhesive bond is provided by a quick setting inorganic adhesive, most preferably a calcium sulfate adhesive.

The process is described for forming a composite laminate insulating material including the steps of preshaping a layer of cellular ceramic material to fit the surface to be insulated; preshaping a layer of glass fiber reinforcing material to fit such surface; adhesively securing the layer of cellular ceramic material to the inner surface of the reinforcing layer by use of a bonding agent; and adhesively securing a second cellular ceramic layer to the outer surface of the reinforcing layer.

Figures 6.1a and 6.1b illustrate the basic construction of the laminate insulating material of this process as a pipe covering. Figure 6.1a shows cylindrical pipe insulation formed from composite laminate insulating material; while Figure 6.1b is a view in section taken along the line II–II of Figure 6.1a, illustrating the fissures formed in the inner layer of cellular ceramic insulation and the manner in which the reinforcing layer limits the formation of fissures in the outer layer of cellular ceramic insulation.

The numeral **10** generally refers to the cylindrical composite insulating body that encapsulates a pipe **12**. The composite laminate insulating material has a cylindrical configuration formed in semicylindrical parts **14** and **16** that are suitably secured to each other around the pipe. The semicylindrical halves are of similar construction and each includes an inner layer **18** of rigid cellular ceramic material such as cellular glass having a low coefficient of thermal conductivity and a low density.

The inner cellular layer has an inner semicylindrical surface **20** and an outer semicylindrical surface **22**. The inner semicylindrical surface has substantially the same configuration as the outer surface of the pipe and is arranged to be placed in abutting relation with the outer surface of the pipe. The semicylindrical inner cellular layer may be formed in a conventional manner by cutting and shaping the semicylindrical layer from a block of cellular glass.

FIGURE 6.1: PREFORMED LAMINATE

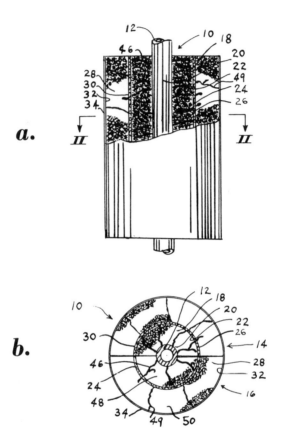

Source: U.S. Patent 3,959,541

A glass fiber insulating and reinforcing layer 24 is precut to conform to the outer surface 22 of the inner cellular layer 18 and is positioned in overlying relation with the inner cellular layer outer surface. A bonding agent is applied to the outer surface 26 of the glass fiber reinforcing layer by any conventional method, as by spraying or brushing a slurry of the bonding agent onto the surface.

An outer layer of cellular ceramic material 28 is shaped in a similar manner and has an inner semicylindrical surface 30 and an outer semicylindrical surface 32. The inner semicylindrical surface has substantially the same configuration as the outer surface 22 of inner cellular layer 18. The outer semicylindrical layer 28 is positioned with the inner surface 30 in overlying relation with the glass fiber reinforcing layer outer surface 26. A pressure is applied to the outer layer 28 to displace the air between the glass fiber reinforcing layer 24 and the adjacent surfaces 22 and 30 of the inner and outer cellular layers 18 and 28. The bonding agent thus secures the inner cellular ceramic layer 18 to the glass fiber reinforcing layer 24 and the outer cellular ceramic layer 28 to the glass fiber reinforcing layer 24.

The semicylindrical halves 14 and 16, after assembly as described, may be positioned around the pipe and secured by a suitable cylindrical outer covering 34. The laminate insulating material 10 thus surrounds and insulates the pipe.

The preferred bonding agent comprises a totally inorganic quick-setting adhesive, such as calcium sulfate (gypsum) cement. A suitable gypsum cement is Hydrocal A-11, which is obtained as a powder having a size less than 200 mesh Tyler Standard screen. The bonding agent comprises $CaSO_4 \cdot \frac{1}{2}H_2O$. The pulverulent bonding agent is preferably mixed with water on a weight basis of 2.2 parts dry powder to 1 part of water to form a slurry. The slurry is applied to a thickness of 33 pounds of slurry per 100 square feet of cellular ceramic insulation. Other cements, such as Hydrocal B-11, gypsum plaster and plaster of Paris, have also been found suitable.

The glass fiber reinforcing layer 26 preferably comprises a relatively thin mat having a nominal thickness of less than ¼ inch. The glass fiber reinforcing layer preferably comprises a scrim fabric with a Leno-Type weave. The glass fiber has a thread count of 20 warp and 10 fill, a weight of 1.65 ounces per square yard, and a thickness of 0.004 inch. The fabric has no organic binders and coatings.

It has been found that randomly woven glass fibers mixed with a bonding agent and applied as a reinforcing layer have proven unsuitable for resisting thermal stresses exerted by the cellular layers 18 and 28. Other suitable woven glass fiber fabrics include fabrics having a plain weave with a thread count of 20 x 20 and a Leno-Type weave with a thread count of 10 warp x 10 fill. The glass fiber fabrics having both a plain and Leno-Type weave are relatively uniformly woven mats and are relatively thin.

Example: A foamed cellular glass material (Foamglas) was chosen as the cellular insulating material. The cellular glass has a density of approximately 9 lb/ft^3 and a thermal conductivity of about 0.40 Btu/hr/ft^2/°F/inch at a temperature of 72°F. This material is corrosion resistant, water impervious and capable of resisting thermal stress exerted by temperatures as high as 800°F.

Two semicylindrical segments of cellular glass were preshaped with an arcuate inner dimension substantially equal to the outer dimension of the body desired to be insulated. The segments were joined lengthwise using Hydrocal A-11 gypsum cement as a slurry formed of 2.2 parts Hydrocal A-11 and 1 part water by weight to form a cylinder of cellular ceramic material. The cement slurry was next evenly applied to the outer surface of the cylinder of cellular ceramic material. Woven glass fiber [Scrim Fabric No. 1659 (J.P. Stevens & Co.)], 0.004 inch thick, was cut to substantially the length and surface dimensions of the ceramic cylinder and applied in place thereon. Pressure applied to the glass fiber secured it to the ceramic cylinder. The gypsum cement was applied evenly to the surface of the glass fiber thus secured and to a second pair of cellular ceramic segments appropriately preshaped with inner semicylindrical dimensions substantially equal to the outer semicylindrical dimensions of the insulated body formed to this point. The semicylindrical segments were then brought into contact with the cement coated surface of the glass fiber reinforcing layer. Sufficient pressure was applied to securely bond the outer cellular cylinder to the inner cellular cylinder and the glass fiber reinforcing material therebetween.

After setting, the composite laminate insulating material thus formed was subjected to thermal stress by contacting the inner surface of the inner cellular cylinder with a heated rod of substantially the same outer diameter and maintained at 1000°F. The heated rod was allowed to remain in contact with the composite insulating body for 24 hours.

At the close of the heating period, it was noted that small cracks or fissures had formed throughout the thickness of the first cellular ceramic layer and to a lesser amount in the outer cellular ceramic layer. The cracks in the outer layer were not direct propagations of the cracks in the inner layer. The composite insulating body of this process thus retained its insulating ability even when cracks occurred under thermal stress.

Perlite and Metallic Phosphate Binder

J.M. Venable; U.S. Patent 3,886,076; May 27, 1975; assigned to Pamrod, Inc. describes a compression-molded, corrosion-inhibiting thermal insulating product for use at temperatures up to about 1800°F. The product is formed from a mixture of 35 to 52% water, 20 to 32% expanded perlite, 1.75 to 3.0% sodium tetraborate, 2.5 to 3.0% sodium silicate, 1.0 to 3.5% chopped inorganic fiber, 0.33 to 2.0% silicone, and 13.5 to 21% of a binder consisting of a metallic phosphate, for example, an aluminum phosphate such as Alkophos.

Example: In preparing the material of this example, which illustrates the preferred composition of the process, two mixing chambers were provided, one for a wet phase and one for a dry phase. In the wet phase, 138.75 pounds of water containing a maximum grain hardness of 5 grains was introduced into the wet phase mixing chamber. The grain content of the water has been found to be critical, as metal ions which are dissolved in water can cause the Alkophos to precipitate if the grain hardness is excessive. Reducing this hardness, as used in this example, was done in a water softening system using a phosphate. As the quantity of water is critical to the production of a homogenous product, this aspect of the addition must be controlled. The problems created if excess water is present will be layering when there is compression at the molding stage later on in the process; whereas, too little water will cause poor binding and low cohesive strength.

1.5 pounds of silicone were then added to the water (Union Carbide R-64, having a 60% solids content). Silicone should be added to the water before adding Alkophos in order to keep the materials that would cause precipitation as dilute as possible. About 47.5 pounds of Alkophos C (43% solids content) were added to the wet phase mix; and a small amount (1.50 pounds) of Aqua Black (carbon black, 50% solids content) were added for coloring only to give the final product a distinctive gray color and to mask smudges on the product from handling. The constituents of the wet phase were thoroughly mixed to form a homogenous mixture in 4 to 10 minutes. The wet phase mixture was then ready for mixing with the dry phase constituents.

The dry phase was prepared by adding 76 pounds of expanded perlite to a rotating tumbler type mixing drum which included a counterrotating whip. As the mixing drum rotated with the expanded perlite in it, 6.5 pounds of dry sodium tetraborate and 8.0 pounds of dry sodium silicate were then added to the dry mix. Then, 5.76 pounds of chopped glass fiber were added and the dry mixture was tumbled and agitated for a time sufficient to provide thorough mixing, but not long enough to break up the perlite particles before they were mixed with the liquid mixture. A suitable time for this tumbling and mixing may be about 3 to 7 minutes. The liquid mixture was then sprayed into the dry phase mixing drum to wet the dry mixture and form a damp product having the consistency of damp sand.

The product formed was then dropped to the bottom of the drum and carried out of the drum into a receiving hopper and carried out of the hopper by conveyor. The product was then collected and weighed to get the correct amount of damp product which when compressed to a specified volume and subsequently dried would provide a final product having a desired low density of 10.5 lb/ft^3.

The molded product was then placed in a drying oven and dried at 500°F for 12 to 48 hours until the product dried thoroughly and uniformly throughout its thickness. The important thing is that the drying procedure be such that the 500°F temperature is reached throughout the product so that the endothermic reaction of the Alkophos takes place. The dry, molded product had a composition of about 64% expanded perlite; 17% Alkophos; 5.5% sodium tetraborate; 6.8% sodium silicate; 5% chopped glass; and the balance about equally of silicone and Aqua Black. This product was characterized by superior properties when compared with product made from compositions at either end of the required ranges for this product, as shown by the following drop test results.

	Inches Dropped to Produce Failure
Water	
A 52%	14
B 35%	9
C 42%	20.5
Chopped glass	
A 35%	30
B 1.0%	28
C 2.8%	33
$Na_2B_4O_7$	
A 3.0%	22
B 1.75%	28
C 2.5%	31
Silicone	
A 2%	17
B 0.33%	15
C 1.25%	23
Perlite	
A 32%	11
B 20.50%	33
C 26%	22.5
Alkophos	
A 21%	31
B 13.5%	12
C 17.5%	33

The test employed was the ASTM Drop Test C-487-64. In the foregoing test results A is high end limit of chemicals and their results; B is low end limit of chemicals and their results; and C is the preferred range chemical composition and results. In all cases, the density is 10 lb/ft^3.

The product formed resisted spalling and thermal shock as illustrated by a test in which the product was heated to 1500° to 1800°F and then dipped in water at room temperature. In prior art products, spalling or hairline cracks and chips appear in the presence of such conditions.

The product also inhibits corrosion in stainless steel pipe. It also is nonleaching, as illustrated by observing the product for a period of 10 days in a controlled environment having a relative humidity of 50%. No whiskers or crystalline growth characteristic of leaching occurred on the surface of the product during the 10-day test period.

The product of the example also has low thermal conductivity (K factor) of about 0.57° at 600°F mean temperature and can function as an insulation product even at 1800°F without decomposition.

Perlite Mixtures for Underground Conduit

A process described by *J. Barrington; U.S. Patent 3,655,564; April 11, 1972; assigned to Insul-Fil Manufacturing Company* relates to a water-repellant thermal insulation for an underground conduit. When a conduit is buried underground, its life may be prolonged by shielding it from the moisture in the earth, thereby preventing, or at least retarding, corrosion of the surface of the conduit. When such a conduit is used to convey fluid (either liquid or a gas) in most instances it is necessary thermally to insulate it in order to impede the transfer of heat to or from the fluid in the conduit.

These functions have been performed by surrounding the underground conduit with a mechanical mixture of expanded perlite particles coated and mixed with a hydrophobic substance, and then sintering part of this mixture by exposing it to a temperature of at least 180°F, usually by passing through the conduit a fluid having such a temperature. However, the necessity of sintering this mixture presents a significant problem where the underground conduit is not to convey a fluid (for example, if it is to house electrical conductors), or is to convey a fluid which will not reach a temperature sufficient to cause sintering (for example, if it is to convey a fluid used in an air-conditioning system).

It has been discovered that, by mixing together particles of hydrophobically coated expanded perlite of a particular size with particles of a particular but different size having a hydrophobic surface, the dual functions of thermal insulation and moisture shielding can be performed without the need to sinter the mixture.

The preferred form of the process is a mechanical mixture of two sizes of particles: (1) Particles of a size that between 90 and 100% of them are retained on a Tyler Standard Sieve No. 100, and between 70 and 90% of them pass through a Tyler Standard Sieve No. 8. These particles, preferably, are of a size that substantially all will pass through a Tyler Standard Sieve No. 3, and substantially none will pass through a Tyler Standard Sieve No. 100; and 75% will pass through a Tyler Standard Sieve No. 8, but that 75% will not pass through a Tyler Standard

Sieve No. 50. Particles of this size make up 50% by volume of this mixture, but may make up 25 to 75% by volume, depending upon considerations treated hereinafter. (2) Particles of a size that between 80 and 100% of them are retained on a Tyler Standard Sieve No. 200 and between 80 and 100% of them pass through a Tyler Standard Sieve No. 20. These particles, preferably, are of a size that substantially all will pass through a Tyler Standard Sieve No. 20 and substantially none will pass through a Tyler Standard Sieve No. 210; and about 60% will pass through a Tyler Standard Sieve No. 30, but that 60% will not pass through a Tyler Standard Sieve No. 100.

It has been determined that, if carefully sized particles are not used, moisture will percolate between the interstices of adjacent particles. Thus, the mixture, without further processing, does not provide an effective moisture barrier. The larger particles of the mixture (1) above are expanded perlite coated with a high softening point petroleum asphalt hydrocarbon, a by-product of the refinement of petroleum asphalt (solvent precipitated asphalt resin).

In an application of the process, a loose-fill, water-repellant insulation is placed around the periphery of a conduit within a ditch. The thickness of the layer used depends principally upon the temperature to be maintained in the conduit. The mixture is then tamped. The nature of the mixture, as well as the tamping, creates an interlock of the interstices between adjacent particles and thereby prevents, or at least significantly retards, percolation of moisture around the several particles of the mixture. This in turn protects the outer surface of the conduit from exposure to moisture emanating from the surrounding earth, thereby preventing or retarding corrosion; and the presence of the expanded perlite particles effectively thermally insulates the conduit.

That such a result occurs was proven by the following test: A plurality of holes were made in the lower portion of the side of a cylindrical container, near where the side was joined to the bottom of the container. A mixture of the preferred form of the process, as described above, was placed in this container. After tamping this mixture, a layer of approximately one inch was formed resting on the bottom of the container. Three inches of water were placed onto this layer. After one day in this condition, there was no evidence of any moisture penetration of the tamped layer, let alone any discharge of water through the holes at the lower portion of the container. After five days in this condition, there was still no evidence of moisture penetration of the tamped layer or of discharge through the holes in the container.

On the other hand, a similar test was conducted on a prior art mixture of equal volumes of asphalt coated expanded perlite particles and asphalt in flake form, some of these flakes being up to one-fourth inch across. Immediately upon inserting the water onto the layer of tamped mixture, water was discharged out of the holes at the bottom of the container.

Frost Barrier

L.C. Rubens and D.J. Sundquist; U.S. Patent 3,839,518; October 1, 1974; assigned to The Dow Chemical Company describe a plastic foam structure which is substantially water impermeable and comprises particles of foamed thermocollapse resistant plastic material securely bonded together in a matrix of a solid binder. The particles have an initial unfoamed solid shape which may be either geometrically regular, such as a sphere and a cylinder, or may be entirely irregular. The unfoamed particles must have two or more opposing outer surface portions which have a distance therebetween of 0.6 to 4.0 millimeters with at least 90% of the thermocollapse resistant material forming the particles lying between the opposing outer surface portions. The solid binder substantially uniformly coats the outer surfaces of the foamed particles.

The method for making the plastic foam structure comprises mixing unfoamed particles of thermocollapse resistant material with heated liquefied binder, the heated binder supplying sufficient heat to cause the particles to foam. The particles are then foamed to a volume of at least 50% of their final volume during mixing to insure a substantially uniform coating of the binder on the outside surfaces of the particles.

After the foaming particles and liquid binder have been mixed, they are formed into any desired shape. Thereafter, the foaming particles and liquid binder are restrained thereby causing the foaming particles to be securely bonded together in a matrix or binder. The restrained plastic foam structure is then cooled to a temperature at which the particles will not continue to foam and at which the liquid binder is transformed into an essentially solid state.

The mixing step of the method described above for making a plastic foam structure can be accomplished in a wide variety of mixing apparatus, such as screw conveyors. It is usually necessary to include heating means in the apparatus to prevent cooling of the foaming particles to a temperature at which they will not continue to foam during the mixing step. The solid binder must be in a liquid state capable of substantially uniformly coating the outside surfaces of the particles, at least during the time, and at the temperatures, required to foam the particles. The foaming temperatures for the particles vary from 75° to 200°C.

It has been found that the plastic foam structure is useful in a frost barrier construction exposed to an environment experiencing freezing conditions, such as a pavement construction for vehicles or a runway construction for airplanes. Such frost barrier construction comprises a surface layer adapted as a load-bearing and wear-resistant surface and a second layer of the plastic foam structure located below the surface layer. There may or may not be additional layers of cohesionless non-frost-susceptible material located between the surface layer and the layer of plastic foam structure to distribute normal surface loads evenly over the layer of plastic foam structure.

The combined layers, including the layer of plastic foam structure, reduce the design thickness requirements for a frost barrier construction by substantially preventing frost penetration into a frost–susceptible soil on which the layers rest thereby providing a frost-free area in the soil below the layers. In permafrost regions where it is necessary to build a pavement or runway construction on a frozen frost-containing soil, the combined layers are useful in substantially preventing heat penetration into the frozen frost-containing soil, thereby preventing the soil on which the frost barrier construction rests from thawing during warm climatic seasonal periods.

The load-bearing and wear-resistant surface layer of a frost barrier construction can consist of any conventional materials, such as asphalt, concrete, or a mixture of glass fibers and polyester plastic material.

It has also been found that the plastic foam structure is useful in a thermo-insulation construction such as in a pipeline, both above and below ground, adapted to transport gaseous and liquid materials; and in a building, more particularly in the foundation and basement wall and floor areas. In a thermo-insulation construction, the article to be insulated has a substantially continuous layer of the plastic foam structure disposed adjacent to at least a portion of one or more of its outer surfaces. The layer will substantially reduce heat loss from the article.

Figure 6.2a is an isometric view of a layer of plastic foam structure showing particles of foamed thermocollapse resistant plastic material securely bonded together in a matrix of solid binders. Figure 6.2b is a side elevation representation of an apparatus used to form a layer of a plastic foam structure on graded frost-susceptible soil.

FIGURE 6.2: PLASTIC FOAM THERMAL INSULATION

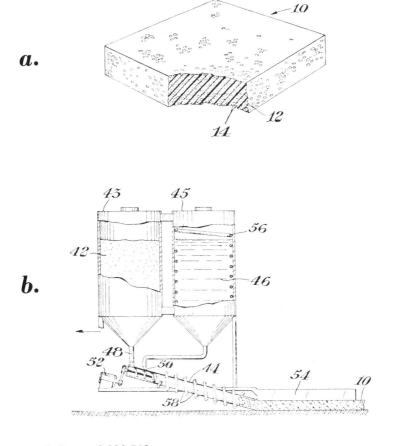

Referring to Figure 6.2a, a layer of plastic foam structure **10**, having partially crosslinked polystyrene foamed particles **14** securely bonded together in a matrix of solid asphalt binder **12** is illustrated. The asphalt binder is a commercially available roofing grade asphalt. The partially crosslinked polystyrene foamed particles are prepared from spherical expandable beads having an approximate diameter of 0.6 to 4.0 millimeters. The expandable beads were prepared by suspension polymerization from the following mixture:

	Weight Percent
Styrene monomer	93.16
Isopentane	6.7
Divinylbenzene	0.05
Peroxide catalysts	0.09

The layer of plastic foam structure was formed by placing a mixture of heated liquefied asphalt and foaming particles in a picture frame mold, foaming the same, and then cooling the mixture to a temperature at which the particles would not continue to foam, and the asphalt binder solidified.

Referring now to Figure 6.2b, an apparatus, with only the critical details shown, for making a continuous layer of the plastic foam structure **10** is illustrated. The apparatus has holding tanks **43** and **45** for expandable beads **42** and liquefied asphalt **46**, respectively. The asphalt is maintained in a heated condition by electric heating coils **56** at a temperature of 180° to 210°C. The expandable beads are fed into a screw conveyor **44** through a tube **48** by gravity flow, and the hot liquefied asphalt is sprayed into the screw conveyor as shown at **50**. One or more screw conveyors may be used, depending on the desired width of the layer **10**. The screw conveyor is driven by motor **52**.

The expandable beads and the hot liquefied asphalt are mixed in the screw conveyor and advanced in the exit of the conveyor where they are formed into a layer of plastic foam structure **10**. The expanding beads and asphalt are maintained at about 150°C while in the screw conveyor. Heating coils **58** surround the screw conveyor to help maintain the temperature of the mixture. The beads are expanded into particles having a volume of 80 to 90% of their final volume while in the screw conveyor to insure a substantially uniform coating with the asphalt.

After a layer of plastic foam structure is formed, it is restrained by a skid plate **54** to insure that the particles will be securely bonded together in a matrix of the asphalt. Cooling of the layer of plastic foam structure takes place while under the skid plate, thereby essentially terminating the foaming of the expanded particles and solidifying the asphalt. Representative physical properties of the plastic foam structure, including compressive strength, insulating factor and water pickup, as compared with other more conventional insulating materials, are shown in the following table.

Sample No.*	Foam Particle Density (pcf)	Total Density (pcf)	Compressive Strength**	K Factor***	. . . . .Water Pickup		
					10 days	6 weeks	12 weeks
1	5.5	14.5	75.9	—	—	Nil	0.2
2	5.9	13.9	44.8	—	—	0.35	0.55

(continued)

Sample No.*	Foam Particle Density (pcf)	Total Density (pcf)	Compressive Strength**	K Factor***	. . . . . Water Pickup†		
					10 days	6 weeks	12 weeks
3	5.1	11.8	45.5	—	—	0.19	0.16
4	5.2	11.3	42.9	—	—	0.44	0.28
5	3.5	12.5	—	0.306	—	—	—
6	4.0	10.8	—	0.280	—	—	—
7	4.5	14.7	—	0.305	—	—	—
8	5.5	12.8	—	0.288	—	—	—
9	6.0	14.2	—	0.318	—	—	—
10	4.0	4.0	—	—	1.1	1.43	—
11	3.5	3.5	—	—	0.37	0.69	—
12	3.0	3.0	—	—	0.50	1.27	—
13	2.5	2.5	—	—	0.33	1.64	—
14	—	—	50.0	0.208	0.27	0.34	—
15	2.6	2.6	81.2	—	—	0.46	—
16	2.5	2.5	—	0.190	—	—	—

*Samples 1 to 9 are lightly crosslinked polystyrene particles in an asphalt binder made according to the process. Samples 10 to 13 are polystyrene foam bead boards made by conventional steam molding of expandable beads. Samples 14 to 16 are polystyrene foam boards made by a conventional extrusion process.

**Compressive strength is in units of pounds per square inch at a 5% deflection.

***K factor is in units of Btu/hr/ft²/°F/in (K factors of less than 0.4 are considered very satisfactory for frost barrier construction requirements).

†Water pickup data are in units of volume percent pickup (samples tested by submerging them in ambient room temperature water).

Oscillation Damping Layer for Conduit

A process described by *P. Matthieu, O. Leuchs, F. Glander, H. Kuypers and D. Pelz; U.S. Patent 3,698,440; October 17, 1972; assigned to Kabel-und Metallwerke Gutehoffnungshutte AG, Germany* relates to thermally insulated conduit which is used to transport liquid or gaseous materials in a heated or cooled condition. Such conduit consists of at least two corrugated concentric tubings.

These conduits are particularly used for long distance heating and usually include a heat insulating layer, such as polyurethane foam, disposed between inner and outer tubings. The conduit can be made in long lengths and lends itself to shipment wound on drums in a manner similar to that used for electrical cables. However, a disadvantage resides in known conduit constructions where the tubings are corrugated to increase the flexibility. Thus, with very long lengths of conduit and at certain rates of flow of the media passing through the inner tubing, such inner tubing is stimulated and swings in its longitudinal axis because of the corrugation.

This process provides a thermally insulated conduit comprising a pair of concentric corrugated tubings, together with a layer of oscillation damping material between.

Referring to the following figures, Figure 6.3a illustrates a longitudinal sectional view of a thermally insulated conduit; Figure 6.3b is a transverse sectional view of the insulated conduit.

FIGURE 6.3: THERMALLY INSULATED CONDUIT

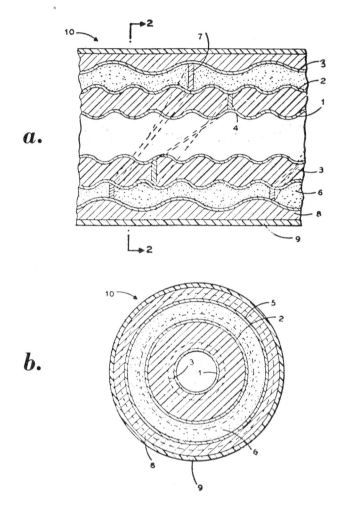

a.

b.

Source: U.S. Patent 3,698,440

As shown in the figure, **10** designates a thermally insulated conduit. The conduit comprises an inner tubing **1**, which may be formed of copper or other suitable material including synthetic resin foil. The tubing is transversely corrugated. Disposed about tubing **1** is a tubing **2**, concentrically related thereto and which may be formed of steel and derived from a metal tape converted to a tubular form about inner tubing **1**, with a longitudinal slit to provide a seam which is welded in a known manner.

A layer **3** of oscillation damping material is introduced between the tubings by way of the slit portion in tubing **2**. The tubings are held in spaced relation by a helical supporting strip **4** of plastic or other suitable material. The damping material **3** is of a thixotropic nature and may suitably consist of a hydrous dispersion of polyvinyl acetate, which is filled with vermiculite flake or particles. The material may be further loaded with powdered barium sulfate.

Conduit **10** may include an outer sheath **5** which is formed of steel from a tape folded about tubing **2** and longitudinally seamed in a known manner, as by welding. Disposed between the sheath and tubing is a layer of heat insulating material **6**, such as hard polyurethane foam. Spacer means **7** is also provided between the tubing and the sheath in the form of a helical strip of plastic, or the like, which is introduced by way of the slit portion of the sheath during its formation.

On the outer surface of the sheath there is disposed a corrosion inhibiting layer **8** followed by an extruded jacket **9** of polyvinyl chloride, or other suitable resin. It has been found that the thixotropic material **3** increases its plasticity when the conduit is subject to sudden deformation as incipient oscillations, and, thus, is effective to dampen such oscillations. Also, this layer may comprise alternating sections of hard and soft material along its length, which also acts as a damping means to break up any oscillation system. It is noted that a single layer of material may be used in the conduit which exhibits both oscillation damping characteristics and is also a thermal insulator.

Calcium Silicate Composition for Stainless Steel Pipe

R.F. Shannon; U.S. Patent 3,923,674; December 2, 1975; assigned to Owens-Corning Fiberglas Corporation describes a strong, light-weight insulation material having a calcium silicate binder which inhibits corrosion of stainless steel. The insulation material is substantially devoid of the usual soluble chloride impurities, and includes from 1 to 15% of sodium or potassium silicate, and preferably has no more than 1,000 parts of chloride ion in 1,000,000 parts of the insulation material.

By maintaining a relative weight ratio of sodium silicate to soluble chlorides of at least 50:1, it has been found that such soluble chlorides as may be present in the insulation material are effectively inhibited to an extent sufficient to permit the use of alkaline earth metal silicate insulation materials in direct contact with metallic members, such as iron and steel and particularly stainless steel, without the occurrence of damaging corrosion. Further, it has been found that by the use of sodium silicate in such amounts in the insulation material, stressed stainless steel objects are capable of being effectively protected against damaging chloride corrosion even when situated in localities where excessive atmospheric concentrations of chloride contamination prevail.

Example 1: An aqueous slurry having a water to solids ratio of 6.0 is prepared on a relative weight basis by mixing, together with 1,048 parts of water and 25.0 parts of spiculated asbestos fibers, such finely comminuted constituents, and in the relative parts by weight (dry weight basis) as follows: 46.0 quicklime, 25.0 quartz, 28.8 diatomaceous earth, 5.0 clay, 18.7 inert filler materials, and a solution of sodium silicate. Numerous different types of suitable inert filler materials may be used and many representative types are set forth in U.S. Patent 3,001,882. The sodium silicate may be a solution of 82 parts by weight of $Na_2O \cdot 3.25SiO_2$ in 123 parts by weight of water.

The slurry as thus prepared is poured into an open or cavity type mold and thereafter indurated in a steam autoclave operated at steam temperatures and pressures of approximately 208°C and 250 psig, respectively. The steam induration is continued for a sufficient length of time (ordinarily 4 to 6 hours) to convert the slurry into an indurated self-sustaining body of insulation material shaped to conform to the configuration and size of the mold in which it is contained during induration. Suitable methods of such induration are fully described in U.S. Patent 2,547,127.

Example 2: As another preferred example, a body of calcium silicate thermal insulation material may be prepared in the same manner set forth in Example 1, with the exception that instead of introducing the sodium silicate as a solution, the sodium silicate may be introduced into the slurry in finely divided solid or powdered form, preferably corresponding to a particle size such that by screen analysis 100% of the sodium silicate will pass through a 40 mesh screen and at least 75% will pass through a 100 mesh screen.

Example 3: As a comparative standard for determining the effectiveness of sodium silicate as a corrosion inhibitor for soluble chloride, another body of calcium silicate thermal insulation material was prepared in the same manner as described in Examples 1 and 2, except that the addition of sodium silicate was omitted from the processing of the material.

Example 4: As an example of an alkaline earth metal silicate material containing soluble chloride in amounts where the weight ratio of sodium silicate to soluble chloride is substantially less than 50 to 1, another exemplary calcium silicate thermal insulation material was prepared from the same slurry formulation as in Example 1, except that an excessive amount of chloride was purposely introduced into the slurry by the introduction of 5% vinyl chloride coated glass fibers as a fibrous replacement for a like amount of asbestos fibers.

Example 5: Another body of calcium silicate thermal insulation material was prepared which corresponded to Example 1 with the exception that phenolic coated glass fibers in the amount of 5% by weight of the slurry solids were substituted for a like amount of asbestos fibers.

The stress-corrosion effects of insulation materials containing soluble chloride, as well as the corrosion-inhibiting effects achieved by sodium silicate additions to such insulation materials, are clearly established by representative test data obtained from tests of insulation materials having formulations corresponding to Examples 1 through 5. Such tests also included the preparation of representative samples of insulation material corresponding to each of the foregoing examples, but having the sodium silicate omitted for purposes of determination of soluble chloride content of the insulation materials by conventional methods of quantitative analysis.

The corrosion tests were carried out on stainless steel specimens since stainless steel has been found to be particularly susceptible to corrosion by insulation materials containing soluble chloride. The testing of the corrosion effects involved placing specimens of the insulation materials upon stressed U-shaped sixteen gauge stainless steel members in accordance with standard testing procedures. One conventional method for determining soluble chloride content was employed for determining the soluble chloride contents represented in the following table. The method involves placing a 12.5 g pulverized sample of the heat insulation material in a 250 ml Erlenmeyer flask together with 125 ml of freshly distilled water. The contents of the flask are then placed under a reflux condenser and refluxed for 4 hours. The refluxed contents of the flask are then vacuum filtered, using a Buchner funnel, through a white ribbon filter paper. The resulting filtrate is then evaporated to dryness while being maintained in an alkaline condition as determined by the addition of methyl orange indicator.

The dry residue is then digested in 25 ml of distilled water on a hot plate and again filtered through white ribbon paper into a 100 ml volumetric flask, with the residue and filter paper being washed thoroughly with hot distilled water. The resulting filtrate is then acidified with nitric acid and diluted with sufficient distilled water to make up a 100 ml sample.

Aliquots of the sample are then taken according to expected amounts of chlorides present in the sample. Ordinarily, a 25 ml aliquot of the sample will suffice. To the aliquot there is added an equal volume of acetone. If smaller aliquots are taken, they may be diluted to 25 ml with distilled water followed by an addition of 25 ml of acetone. Then, utilizing a magnetic stirrer and a microburette, the aliquot is titrated with an 0.002 normal solution of silver nitrate until a potentiometric end point of 245 mV is obtained using a silver electrode and a calomel electrode with an ammonium nitrate salt bridge. The amount of soluble chloride is determined in conventional manner upon the basis of the amount of silver nitrate used in the titration.

The comparative results of the soluble chloride analysis and the stainless steel corrosion testing appear in the table below.

Example	Soluble Chloride (ppm)	Stress Corrosion
1	175–150	none
2	125–150	none
3	125–150	substantial stress corrosion
4	3,150–3,200	extreme stress corrosion
5	75–100	none

From the results appearing in the table, it is quite evident by comparison of the similar calcium silicate formulations of Examples 1 and 2, each of which contained equivalent amounts of sodium silicate amounting to approximately 5% (dry weight basis) of the slurry solids; and Example 3, which contained no sodium silicate in the formulation, that the addition of sodium silicate in the slurry effectively tends to inhibit stress-corrosion resulting from the presence of soluble chloride in the insulating material. Furthermore, the addition of 5% sodium silicate was ineffective to inhibit severe chloride stress-corrosion in Example 4, where the soluble chloride content was between 3,100 and 3,200 ppm, or the equivalent of a sodium silicate to soluble chloride weight ratio of between 15.6 and 16.1 to 1, i.e., substantially less than 50 to 1 minimum limitation of this process. Even lesser amounts of soluble chloride in the insulation material are to be preferred, however, to provide a substantial excess of sodium silicate in the material which is capable of providing built-in protection against stress-corrosion attending the use of the insulation material in installations which are frequently subject to external sources of high chloride contamination.

Pipe Lagging Material

A process described by *J.W. Echerd and W.K. Watters; U.S. Patent 3,769,072; October 30, 1973; assigned to H.K. Porter Company, Inc.* relates to a pipe lagging material such as for use in securing insulation on a pipe conduit. Such materials generally consist of a base, such as fabric, which is wrapped or lagged about the outside of insulation on a pipe to hold the insulation in place and thereby protect the contents of the pipe against thermal gain or loss. It is highly desirable that such pipe lagging materials be flame-resistant, safe to handle, and easily permanently bonded to pipe insulation at a work site to form a nonfriable protective coating about the insulation. It is also essential that the lagging material possess sufficient adhesive qualities to be retained on the pipe under various conditions of use.

The process involves pipe lagging materials providing a rewettable, nontoxic, drapable lagging material which is adapted to be wet with water at the work site and adhesively applied to a pipe surface to form, upon drying, a substantially rigid, nonfriable, flame-resistant covering. The lagging material can be prepared and stored indefinitely in a dry condition prior to use, and can be used to cover insulation or wrapped directly on the pipe itself as an insulator, as desired.

This is accomplished by impregnating a suitable base, such as a porous fabric, with a liquid slurry of a hydrophilic inorganic composition which dries to form, with the base, a drapable, nondusting lagging material. The composition can be hydrated with water to form a substantially rigid, nonfriable mass of interlaced needle-like crystals firmly bonding the lagging material to a desired surface. The lagging material possesses both excellent wet and dry adhesion, even to relatively smooth surfaces, so that subsequent coating with an adhesive is not required to permanently secure the material on the pipe.

In preparing the liquid slurry for impregnation of the base, it is necessary that the viscosity of the slurry, i.e., the degree of hydration of the hydrophilic components, be controlled so that the slurry readily penetrates the base and is thereto bonded, after drying, without greatly decreasing the drapability. In this respect, dispersion of the composition in a liquid containing approximately equal amounts of water and a water-soluble compound, such as a low molecular weight alcohol, ester, or ketone, will effectively limit the viscosity of the slurry to produce a flexible, drapable material upon drying.

The porous base used in forming the lagging material may be of various fiber types, e.g., cellulosic, synthetic, proteinaceous, mineral, glass, asbestos, ceramic, siliceous, or any combination of the foregoing reinforced with wire; and it should possess sufficient porosity to permit impregnation with the adhesive composition. Where a high degree of flame resistance is desirable, loosely woven asbestos-base fabrics, i.e., fabrics containing at least 50% by weight asbestos fibers, have been found to be quite satisfactory.

For the impregnant composition of the base it has been found that dusting or flaking of the dry impregnant from the base can be prevented and wet adhesion is enhanced by the use of hydrophilic siliceous materials, such as clays (e.g., bentonite and kaolin) and hydrous magnesium silicate. In addition to these siliceous materials, the inorganic composition may contain other inorganic compounds, such as plaster of Paris, which are capable of forming a crystalline network when hydrated with water.

To apply the lagging material to a pipe conduit or other suitable surface, the material is wet with water at the work site and then wrapped about the pipe, or it can be wet at some remote location and placed in a moisture-, vapor-proof barrier for storage and transportation to the work site. While wet the lagging material will readily adhere to the surface upon which it is placed and upon drying the material forms a rigid covering bonded to the surface. The lagging material will usually be sufficiently wet after about 30 seconds contact with water. However, it may be left in the water for extended periods without significant leaching out or losing its adhesive character.

Example 1: A plain weave asbestos-base fabric (approximately 80% by weight asbestos) of approximately 0.75 lb/yd^2 is impregnated by dipping it into a liquid slurry containing the following components:

	Parts by Weight
Water	25
Isopropyl alcohol	30
Bentonite (clay)	20
Plaster of Paris	20
X-12 (flame retardant)*	5

*A proprietary composition based upon ammonium salts sold by E.I. du Pont de Nemours & Co., Inc.

After impregnation, the dry weight of the fabric is approximately 1.5 lb/yd^2.

Example 2: A lagging material is prepared as in Example 1, utilizing the following slurry composition:

	Parts by Weight
Water	37
Isopropyl alcohol	38
Bentonite (clay)	24
NaBF$_4$ (flame retardant)	1

Example 3: A lagging material is prepared as in Example 1, utilizing the following slurry composition:

	Parts by Weight
Water	36
Isopropyl alcohol	38
Magnesium silicate	25
Antimony silicooxide (flame retardant)	1

Upon drying, one face of the material formed in accordance with Example 3 is then coated with a water dispersion of latex form of neoprene of sufficient thickness to form a continuous film on the coated face of the fabric.

The lagging materials prepared as in Examples 1, 2 and 3, upon wetting and application to a pipe conduit, exhibit excellent wet adhesion to the pipe and upon subsequent drying form substantially rigid, nonfriable protective coverings adhesively bonded thereto.

Sponge Rubber Core for Pipe Wrap

J.J. Berwanger; U.S. Patent 3,654,061; April 4, 1972; assigned to Hood Sponge Rubber Company describes a pipe wrap comprising a unitary strip of rubbery material having an enveloping skin and a sponge core or interior, the skin having a resistance to the passage of moisture that is great as compared to that of the sponge interior.

The relatively moisture impervious skin is quite thin and completely envelopes the highly porous sponge interior so that when the wrap is applied to a pipe, the skin serves as a moisture barrier for the porous sponge core to prevent deterioration of the insulating properties of the core. In many instances, this eliminates the need for applying a multiplicity of wraps of different materials to the pipe.

At the longitudinal margins of the wrap are outwardly extending flanges that are sized and are so offset from each other that when the strip is helically applied about a pipe, the flanges overlap and form a helical seam which does not produce a bulge or lump but, on the contrary, provides a wrapped sheath on the pipe in which the outer surface of the sheath is substantially smooth and continuous. Furthermore, a strip of a given size constituting the pipe wrap of the process is sufficiently flexible within its limits to be helically wrapped around pipes within a range of diameters and still produce a sheath with a smooth exterior surface.

AUTOMOTIVE—CERAMIC INSULATORS

Ceramic Mixture for Manifold Reactor

Y. Takeuchi; U.S. Patent 3,991,254; November 9, 1976; assigned to Nippondenso Co., Ltd., Japan describes a lightweight insulating structure having a vibration resisting property in which spaces in various insulating structures are filled uniformly with a slurry having a high fluidity, and the slurry is heated to expand and solidify the same so that strong insulating layers firmly fused to the insulating structures are formed; and which can be used, for example, in automobiles for their exhaust pipes, manifold reactors, catalyst convertors, and insulating covers.

In manufacturing the insulating structure in accordance with the process, a heat resisting material consisting of 20 to 70% by weight silica, 15 to 80% by weight alumina, and 0 to 30% by weight magnesia, such as cordierite, mullite, sillimanite and kaolin, is mixed in a water solution of a phosphate compound, for example aluminum primary phosphate, to prepare a base slurry. If necessary, to this base slurry is added an acid and a substance which gives a gas on reaction with the acid. Instead of the acid and the substance, a material which loses its volume at high temperatures may be dispersed in the base slurry. Further, instead of the acid, substance and material mentioned above, a light-weight aggregate may be mixed in the base slurry.

The space in the double structure is filled with the slurry thus prepared by pouring or filling under pressure, or filling under reduced pressure. The filled slurry is heated at a temperature of less than 500°C for the purpose of drying, and is then fired at an elevated temperature to form a ceramic insulating layer.

This ceramic insulating layer is a very superior one which has a low thermal conductivity, an excellent insulating property, a high maximum resistible temperature, a high strength, and a small coefficient of thermal expansion. It has been proven that this ceramic insulating layer is quite useful as a high temperature insulating structure.

Example 1: Heat resisting materials, each consisting of mullite, sillimanite and kaolin, were separately mixed in a water solution of aluminum primary phosphate to prepare slurries. The slurries produced had the following compositions: 50 to 75% by weight of a heat resisting material; 25 to 50% by weight of a 50% water solution of aluminum primary phosphate.

By simple pouring, the spaces between the inner walls and the outer walls of double structures were filled with the slurries thus prepared. The slurries were fired to form ceramic insulating layers. The following table shows the properties of the ceramic insulating layers produced in this manner. In case mullite was used as the heat resisting material, the maximum resistible temperature and the coefficient of thermal expansion of the insulating layer were respectively 1300°C and 4.7–5.3 x 10^{-6}. The insulating layer also had excellent strength. Thus, the ceramic insulating layer using mullite is particularly useful principally as an insulating structure for use with a manifold reactor which requires a superior vibration resisting property and a high strength because it is directly subjected to the vibrations of the engine. The ceramic insulating layers using sillimanite and kaolin as the heat resisting material showed the same excellent results as those obtained by the ceramic insulating layer using mullite.

Heat resisting material	Thermal conductivity (kcal/m.hr. °C)	Maximum resistible temperature (°C)	Coefficient of thermal expansion	Density g/m³
Mullite	0.5 – 0.7	1300	4.7 – 5.3 ($\times 10^{-6}$)	1.3 – 1.7
Sillimanite	0.5 – 0.7	1300	4.8 – 5.0 ($\times 10^{-6}$)	1.2 – 1.5
Kaolin	0.4 – 0.7	1000	4.8 – 5.0 ($\times 10^{-6}$)	1.3 – 1.6

Example 2: Heat resisting materials each consisting of mullite, sillimanite and kaolin were separately mixed in a water solution of aluminum primary phosphate using the same mixing ratios as in Example 1 to prepare slurries. To each slurry was added less than 10% by weight of one or more acids selected from the group consisting of such common acids as phosphoric acid, hydrochloric acid and sulfuric acid. In order to effect effervescence, in each slurry was uniformly dispersed less than 1% by weight of a powdery metal which yielded hydrogen on chemical reaction with the acids mentioned and which had been selected from the group consisting of such metals as aluminum, iron, calcium, magnesium, manganese, nickel, tin and cadmium. The spaces between the inner walls and the outer walls of double structures were filled with the slurries thus rendered effervescent, and the slurries were fired to form ceramic insulating layers. The insulating layers manufactured in this manner were porous.

The following table shows the properties of these insulating layers. In case mullite was used as the heat resisting material, the maximum resistible temperature and the coefficient of thermal expansion of the insulating layer were respectively 1300°C and 4.7–5.3 x 10^{-6}, the values being the same as those obtained in Example 1.

The tested insulating layer had a slightly lower strength than the insulating layer of Example 1, but had a thermal conductivity of 0.3–0.6 kcal/m-hr-°C which was lower than those obtained in Example 1. Thus, this ceramic insulating layer is particularly useful as an insulating structure for use with a double exhaust pipe, muffler and an insulating cover which are complicated in shape and construction and require a superior insulating property and medium degree of strength. The ceramic insulating layers using sillimanite and kaolin as the heat resisting material showed the same excellent results as those obtained by the ceramic insulating layer using mullite.

Ceramic	Thermal conductivity (kcal/m-hr-°C)	Maximum resistible temperature (°C)	Coefficient of thermal expansion	Density g/cm³
Mullite	0.3 – 0.6	1300	4.7 – 5.3 ($\times 10^{-6}$)	0.7 – 1.0
Sillimanite	0.3 – 0.6	1300	4.8 – 5.0 ($\times 10^{-6}$)	0.6 – 1.0
Kaolin	0.3 – 0.6	1000	4.8 – 5.0 ($\times 10^{-6}$)	0.7 – 1.0

Foamable Perlite Particles

F. Noda and Y. Takeuchi; U.S. Patent 3,958,582; May 25, 1976; assigned to Toyota Jidosha Kogyo KK and Nippondenso Co., Ltd., Japan describe a high-temperature heat-insulating structure having a ceramic heat-insulating layer fired at a high temperature. The ceramic layer is made from a paste prepared by mixing an aqueous aluminum phosphate solution with foamable perlite particles or by mixing an aqueous monoaluminum phosphate $Al_2O_3 \cdot 3P_2O_5 \cdot 6H_2O$ solution with foamable perlite particles and ceramic fiber.

The vibration-resistant light heat-insulating structure is made by introducing a highly fluid paste into various heat-insulating structures uniformly, and expanding and simultaneously solidifying the paste by heat, while causing the heat-insulating layer to adhere rigidly to the structure, which may be an exhaust pipe, manifold reactor, catalytic converter, or the like, of an automobile.

Example 1: After 210 g water has been mixed with 210 g monoaluminum phosphate to form a viscous liquid, 85 g of perlite and 92 g of ceramic fiber (grain size 200 mesh) were introduced into the liquid and stirred to produce homogeneity, thereby forming a thin paste.

The paste was then introduced into a metal shell which has a space having a diameter of 54 mm and a height of 50 mm. The cover of the metal shell was secured by bolts and the metal was heat solidified in a furnace at 350°C for an hour and fired in a furnace at an elevated temperature of 650°C for an hour.

In firing, a portion of the sample overflowed from the metal shell through small gaps therein due to expansion. The resultant heat insulating material (bulk density 0.72 g/cm³) was cut into test pieces 20 x 20 x 40 mm. Two types of test pieces were then produced from the above test pieces; one (having a rate of volumetric contraction of 2 to 3%) which was fired in a furnace for 50 consecutive hours at a temperature of 1250°C; and the other (having a rate of volumetric contraction less than 1%) which was fired in a furnace for 150 consecutive hours at a temperature of 1200°C.

Heat impact tests in which the specimens were suddenly heated from room temperature to 1000°C and forcibly air cooled from 1000°C to room temperature were repeated 1,000 times for each specimen. However, no cracks were generated. Also, vibration tests for a metal shell containing the fired heat insulating composition were carried at 90 Hz, at 45 g for 5 consecutive hours. No abnormalities resulted.

Example 2: After 210 g of water had been added to 210 g of monoaluminum phosphate to prepare a viscous solution, 2 g of an expansion restrainer Ebit (an acid corrosion inhibitor) was added to the solution in order to prevent spaces from being generated within the heat insulating member due to expansion. 85 g of perlite (as a foamable perlite), 92 g of ceramic fiber, 60 g of reaction accelerator Takibine (formula $Al_2(OH)_5 \cdot Cl \cdot 2.4H_2O$) and 60 g of alumina strengthener were also added to the solution and stirred to homogeneity, thus forming a thin paste.

The paste was introduced into the same metal shell as in Example 1 and the cover was secured by bolts. The metal shell was then heat solidified in a furnace at 300°C for 45 minutes and additionally fired for 2 hours at an elevated temperature of 600°C. The resultant heat-insulating member (bulk density 0.78 g/cm^3) had the same heat-resisting temperature and heat impact resistance as that of Example 1 and had no abnormal expansion, but the solidification time was shortened and the strength of the insulating member was increased.

Aluminum Foil Sandwich

A process described by *Y. Kaneko and F. Noda; U.S. Patent 3,963,547; June 15, 1976; assigned to Toyota Jidosha Kogyo KK, Japan* involves the manufacture of a composite heat insulating material by bonding a ceramic material which foams when heated to one or both sides of a piece of aluminum foil, filling a vessel or a predetermined volume to form a heat insulating layer with layers of such pieces and then heating them to cause foaming.

Such a composite heat insulating material, which represents layers of aluminum foil sandwiched at definite intervals between layers of a foaming ceramic material, does not pulverize as easily under vibration as does a conventional material composed of foaming ceramic material alone. At the same time, since the sandwiched aluminum foils mutually radiate the heat, a far greater heat insulating effect is achieved than in the case of single foaming ceramic material.

Example 1: Figure 6.4 shows a partially cutaway oblique view of a heat insulated pipe according to the process, where reference 1 indicates the outer pipe, 2 the aluminum foil, 3 the foaming ceramic material, 4 the inner pipe, 5 a bolt hole in the flange for fitting the heat insulated pipe, and 6 the flange itself.

To manufacture a heat insulated pipe according to the process, the inner pipe is first sprayed with liquid water glass (JIS-3) and an aluminum foil also coated with water glass is applied. After successive wrappings of such water glass coated aluminum foils, the inner pipe is inserted into the outer pipe, the flange is welded thereto, and the resulting double walled pipe is heated in a furnace at 250°C to cause the water glass to foam, producing a heat preserving pipe. The water glass to be used in the example consists essentially of 28 to 30% SiO_2, 9 to 10% Na_2O, less than 0.02% Fe, 0.2% insolubles and the balance water.

The heat insulated pipe thus produced consists of an inner iron pipe (outer diameter 40 mm) and an outer iron pipe (inner diameter 56 mm), the heat insulating layer being 8 mm thick, with five 20 μ aluminum foils interleaved therein. When a gas at a temperature of 450°C was passed through the inner pipe, the outer pipe registered a surface temperature of 115°C. When the foamed body consisted of water glass alone with no insertion of aluminum foil, the surface temperature of the outer pipe attained 190°C.

FIGURE 6.4: HEAT INSULATED PIPE

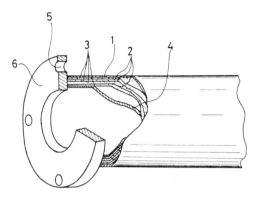

Source: U.S. Patent 3,963,547

Example 2: Instead of the water glass employed in Example 1, a mixture of liquid water glass (JIS-3) with 35% of aluminum hydroxide was used and the same process as in Example 1 was carried out to manufacture a heat insulated pipe.

When an exhaust gas at a temperature of 450°C was passed through the inner pipe, the outer pipe registered a surface temperature of 105°C. By contrast, when an exhaust gas at 450°C was passed through an inner pipe heat insulated with a foam layer alone, having no sandwiching of aluminum foil, the outer pipe registered a surface temperature of 185°C.

STEAM INJECTION WELLS

Gel Composition

A process described by *J.W. Howland and J.C. Rosso; U.S. Patent 3,642,624; February 15, 1972; assigned to Gulf Oil Corporation* relates to the thermal stimulation of wells and more particularly to an insulating liquid to be inserted in the annulus between steam injection tubing and casing of a steam injection well. In the thermal insulating fluid composition, a small amount of water is dispersed in a stable emulsion in a heavy mineral oil. The composition includes soap-forming ingredients, specifically lime and fatty acids, which react adjacent to the steam injection tubing to form a coating of soap on the tubing and to form a gel after the injection of steam begins. Bentonite organic base compounds and asbestos particles are dispersed through the composition to provide a thick stable thermal fluid that has high gel strength and yet retains mobility to facilitate removal of the fluid after the steam injection.

The essential characteristics of a thermal fluid for reducing the transfer of heat from the tubing of a steam injection well to the surrounding casing are a low thermal conductivity, a high viscosity or gel strength to reduce convection currents and thereby reduce transfer of heat by convection, and an opaqueness that will reduce radiation from the tubing to the casing. The thermal fluid should be stable both with respect to maintaining a mobility that will permit the fluid to be removed from the well and with respect to retention of its gel strength to minimize settling of solid particles in the composition. It is preferred that the fluid be thixotropic to make possible a low viscosity during pumping of the thermal fluid into the annulus and a high gel strength to prevent settling of solids after the liquid is in place.

A thermal insulating fluid was prepared in accordance with this process having the following composition:

	Parts per Barrel of Insulating Fluid
Bunker C	27.3 gal
Diesel fuel	4.2 gal
Water	1.68 gal
Bentonite organic compound*	6.0 lb
Tall oil or tallene (tall oil pitch)	42.0 lb
Calcium oxide (quicklime)	10.0 lb
Finely divided asbestos fibers	40.0 lb

*Dimethyl dioctadecyl ammonium bentonite

The above composition was used to fill the annulus between the tubing and casing of a steam injection well in which steam was injected into a formation approximately 4,000 feet below the surface. Steam was injected at a temperature of approximately 600°F for approximately 7 days. After completion of the steam injection, the well was produced at a rate of 300 barrels per day with a top hole temperature of 160°F for approximately 5 weeks. A workover rig pulled the pumps and rods and the thermal insulating fluid was pumped out of the well.

A cement bond log run on the well showed excellent bonding of the entire interval and showed no change from the cement bond log run prior to steam injection. A comparison of before and after cement bond logs on two other wells showed significant changes in the cement sheath. In both of the other wells, good bonding before the steam injection had become nonexistent or very poor after the steam injection. Neither of the other wells had fluid in the tubing casing annulus during the steam injection, but the tubing in the injection string of one of the wells was aluminum coated tubing.

Three samples of the thermal fluid were taken as the fluid was pumped from the well. One sample was taken at the top of the well, one from the middle (2,000 feet plus or minus) and one from the bottom (4,000 feet plus or minus). The samples were tested for density, concentration of solids, and shear strength. The results tabulated in Table 1 show the thermal insulating fluid was stable and there was no measurable settling of the solids.

TABLE 1

	Density (psf)	Solids (%)	Shear Strength (lb/100 ft^2)
Top	64.0	14.65	320
Middle	64.2	14.12	300
Bottom	64.0	14.00	300
Before injection	65.2	15.20	330

The effect of the bentonite organic base compound and water on the gel strength and thixotropic properties was tested by preparing compositions containing the Bunker C fuel oil, diesel fuel oil, tall oil and calcium oxide in the proportions set forth for the thermal insulating fluid used in the well and varying the concentration of water and bentonite organic base compound. The initial gel strength and gel strength after 10 minutes were measured on a Fann viscometer. The difference in the two gel strengths is a measure of the thixotropic effect of the composition. The results of the tests are set forth in Table 2.

TABLE 2

	. . . . .Sample Number				
	1	2	3	4	5
Water, %	0	0	4	2	4
Bentonite compound, lb/bbl	0	6	0	2	6
After aging 2 hr					
Initial gel	6	7	8	7	10
Gel after 10 min	7	8	9	10	19
Thixotropic effect	1	1	1	3	9
After aging 16 hr					
Initial gel	8	14	7	13	17
Gel after 10 min	10	16	8	15	25
Thixotropic effect	2	2	1	2	8

It will be noted from Table 2 that the gel strength of Sample Number 5, which contains both the water and bentonite, is higher than the gel strength of any of the other samples. Moreover, the thixotropic effect of Sample No. 5 is at least three times the thixotropic effect of any of the other samples.

The insulating properties of the composition described above that was placed in the well were compared with a similar composition, designated as the control sample, of mineral oil, soap-forming components and asbestos but not containing the bentonite compound and water, in a test cell consisting of a section of 2½ inch tubing mounted and sealed inside a section of 7 inch casing. A resistance wire heater inside the tubing supplied heat to the tubing. Temperature measurements were made by thermocouples located on the outside of the 2½ inch tubing and on the outside of the 7 inch casing. The results of the tests are set forth in Table 3.

The thermal insulating fluid of this process has excellent stability as well as excellent insulating properties. After an extended period in the tubing-casing annulus of a steam injection well, the fluid could be readily pumped from the annulus. Although the fluid was easily pumped, substantially no settling of the asbestos fibers in the fluid had occurred.

As shown by the data presented in Tables 2 and 3, the addition of dimethyl dioctadecyl ammonium bentonite to a gelled heavy mineral oil resulted in important increases in gel strength and reduction in heat transfer.

TABLE 3

Control	Input Wattage	Tubing Temperature (°F)	Casing Temperature (°F)	Temperature Difference (°F)
Fluid 1	216.6	445	145	300
Control	216.6	412	151	261
Fluid 1	253.2	472	158	314
Control	254.2	396	156	240
Fluid 1	306.0	592	178	379
Control	306.0	482	175	256

Magnesium Silicate Thickener

C.L. Jacocks; U.S. Patent 3,719,601; March 6, 1973; assigned to Continental Oil Company describes a process for insulating thermal injection wells where a liquid insulating medium comprising a liquid mineral oil plus a finely divided fibrous magnesium silicate or asbestos is employed as the liquid insulating medium.

The portion of the insulating medium which is made up of a finely divided fibrous magnesium silicate or asbestos can comprise in the range of 1 to 10 weight percent of the insulating medium. An example is Avibest-C, a microcrystalline silicate, which is a hydrated magnesium silicate having 37 weight percent Mg as MgO, 43 weight percent Si as SiO_2, 4 weight percent Fe as Fe_2O_3, and 12 weight percent H_2O. It is a light-grey microcrystalline, fibrous powder having a bulk density of 10 lb/ft³, a density of 2.2 g/cc, a pH of 7.5, a refractive index of 1.5, a surface area of 68 m²/g. It is obtained by treatment of naturally occurring chrysotile. Typical microcrystals have a diameter of 200 A and a length of 4000 A. An example of a preferred asbestos product is Calidria asbestos, a short fiber particulate chrysotile asbestos product.

The liquid insulating media are particularly advantageous for insulating thermal injection wells for secondary or tertiary hydrocarbon recovery where such liquid insulating media are used to fill the annulus between the casing and injection conduit. Temperatures of such environments often range from 150° to 650°F. More generally the liquid insulating media of this process can be used at any temperature below the decomposition point of the particular hydrocarbon used at which temperature the insulating media remain liquid. Pressure sufficient to maintain in the liquid state can be used. Often temperatures in the range of 450° to 650°F are particularly suitable.

Example 1: A series of runs were made where Calidria asbestos and Avibest-C, a microcrystalline fibrous magnesium silicate, were each mixed in a bright paraffinic mineral oil stock (having a viscosity index of 95 and produced by an extraction of a lube residuum) by blending the materials in the oil with a Waring Blender for 10 minutes in amounts shown in Table 1.

A Haake Rotovisco viscometer, a rotating bob instrument that can be used with fluids at elevated temperatures was used to measure the viscosity of the samples. Data are reported in Table 1, and constitute the viscosities measured for a 176 inverse second shear rate test.

TABLE 1

Temperature (°F)	Bright Stock	. . . . Calidria Asbestos			. . . . . Avibest-C		
		1.5 wt %	5.5 wt %	7 wt %	1.5 wt %	2 wt %	3 wt %
100	486	762	1,287.4	1,664.2	730	753.6	902.8
200	43	85	272.8	391.9	123	119.3	216.3
300	20	25	69.8	112.1	46	49.4	116.0
350	15	16.5	55.7	76.8	30	40.0	83.9
400	10	-	47.0	81.5	22	36.1	64.3
450	12	-	51.7	76.8	21	31.4	51.7
500	12	-	51.0	72.9	21	31.4	47.0

From these data it is readily seen that the insulating media of the process have greatly improved insulating properties as compared to mineral oils which have been previously used. Thus, 7 weight percent of Calidria asbestos product increased the viscosity six-fold at 500°F. This is calculated to result in a minimum reduction of 40% in the calculated heat transfer coefficient. All mixtures were sealed in pressure bombs and held at 550°F for one week. No evidence of phase separation or thermal degradation was noted in any of the process compounds at the end of this period. Cab-O-Sil, a finely divided silica product, was tested in a similar test and found to be unsatisfactory because of temperature instability of the resulting liquid insulating media.

This example demonstrates the superior insulating properties of the fluid insulating media of this process as compared to the hydrocarbon alone, which was previously employed as an insulating medium. Vastly improved viscosities are obtained. Calculation of the Grashof number and heat transfer coefficient for such media can readily be calculated; see the *Handbook of Fluid Dynamics,* Streeter, page 6-32 and 48.

Example 2: A series of thermal conductivity runs were made to compare the insulating properties of exemplary insulating media of the process with mineral oil, which had previously been employed as an insulating medium.

A conductivity cell consisting of a 1⅞ inch outside diameter pipe mounted concentrically inside a 4-foot length of 4½ inch outside diameter steel casing was fabricated. The annular space was approximately 1.06 inches wide. The annular volume was approximately 6,400 cc. 8 iron-constantan thermocouples were embedded in the inner wall of the pipe and the outer wall of the casing. Inner and outer thermocouples were positioned opposite each other at 10-inch linear intervals and 90° radial offset between adjacent positions. All thermocouples were attached to a Texas Instruments electronic recorder with a range of 0° to 600°F.

A boiler (a cylindrical steel vessel of approximately 2,000 cc capacity) was connected to the bottom of the column by a length of ¾ inch pipe. The condenser mounted on the top of the inner pipe was cooled with tap water. The system was vented to the atmosphere. Liquid from the condenser returned to the boiler through a ½ inch copper tube mounted outside the casing.

The main source of heat was a 3,990 watt, 220 volt heater placed around the boiler. Additional heating was required to maintain reflux of the Dowtherm, a chlorinated phenol heat transfer medium which was employed as a heating medium in the reflux system. This was supplied by three 2 foot long, 110 volt flexible heating tapes placed around the pipe at the top of the boiler, the bottom of the pipe, and the top of the pipe. Variable resistors were used to control voltage to all heaters. The casing was wrapped with asbestos insulation.

To maintain reflux through the system, the boiler was held at 1000°F and the flexible tapes were held at maximum voltage. The time required to stabilize the reflux from room temperature was about 3 hours. The temperature recorder was started when the boiler reached the boiling point of the heat transfer medium (approximately 450°F). The annulus was closed and the reflux was maintained under a maximum pressure of 400 psi.

The difference in temperatures between the inner pipe and the outer casing was determined by a pair of thermocouples approximately two-thirds of the way up the column. The difference in these temperatures at varying temperatures of the internal pipe were indicative of the insulating value of the various fluid insulating media evaluated. The essence of data from a series of runs is presented below in Table 2.

TABLE 2

	Internal Temperature (°F)	External Temperature (°F)	Temperature Difference (°F)
Run 1 (10 lb/bbl Avibest-C* in Bright Stock**)	460	170	290
Run 2 (10 lb/bbl Avibest-C in Bright Stock)	505	250	255
Run 3 (10 lb/bbl Avibest-C in Bright Stock)	460	250	210
Run 4 (10 lb/bbl Avibest-C in Bright Stock)	495	250	245
Run 5 Control (Bright Stock)	325	250	75

*Avibest-C product is a microcrystalline magnesium silicate.
**Bright Stock is a 95 V.I. paraffinic mineral oil. Bright Stock is produced by an extraction from a lube residuum.

These data clearly demonstrate the superior insulating properties of an exemplary liquid insulating medium of the process in comparison to a mineral oil liquid insulating medium of the prior art.

Example 3: Runs where exemplary liquid insulating media of the process were allowed to stand in a pressure bomb at 500°F demonstrated that no appreciable settling of the finely divided fibrous magnesium silicate or asbestos occurred upon examination after a period of two weeks.

This example demonstrates that the improved insulating media of this process have satisfactory stability at high temperatures.

LOW TEMPERATURE APPLICATIONS

Frozen Substrates and Roadways

According to a process described by *F.C. Gzemski; U.S. Patent 3,903,706; September 9, 1975; assigned to Atlantic Richfield Company* a layer of expanded perlite particles or granules, each particle coated with a bituminous coating composition, is applied to a frozen substrate whether it be permafrost or tundra. This layer of particles can range in thickness from as little as one inch up to as much as one foot, the thickness being varied in accordance with the insulation requirements. After this layer of insulating material has been applied, the gravel layer will be greatly reduced since it is no longer required to be the sole insulating material but merely furnishes the bearing surface for the wheeled vehicles or other structures, such as buildings, tanks or the like, which may be placed thereon.

A preferable polymer asphalt composition is shown in U.S. Patent 3,637,558 (1972). These compositions contain asphalt and a urethane which is the reaction product of a diisocyanate with an intermediate polyhydroxy polymer having an average of at least about 1.8 predominantly primary, terminal, allylic hydroxyl groups per molecule and being an addition polymer of 0 to 75% by weight of an alpha-olefinic monomer of 2 to 12 carbon atoms, and about 25 to 100% of a 1,3-diene hydrocarbon of 4 to about 12 carbon atoms, the intermediate polyhydroxy polymer having a viscosity at 30°C of about 5 to 20,000 poises and a number of average molecular weight of about 400 to 25,000. The urethane component is at least partially uncured when combined with the asphalt and the curing is completed when the asphalt is heated.

A typical coated perlite will have a thermal conductivity, K factor, of about 0.46 measured as Btu/hr/sq ft/°F/inch thickness at a mean temperature of 75°F. Thus the thickness required to provide sufficient insulation to maintain the substrate in a frozen condition can be calculated readily from the known history of the ambient temperature variations in a particular area and the duty dictated by the type of superstructure.

Rigid Fibrous Panel for Cold Storage Applications

D.A. Blewett and R.G. Adams; U.S. Patent 3,905,855; September 16, 1975; assigned to Owens-Corning Fiberglas Corporation describe a rigid panel which has particular adaptability for cold storage walls with the panel being self-supporting and washable and having fire-retardant and insulating characteristics with high impact resistance. The panel comprises a layer of highly densified glass fibers held in a rigid form by a binder with a washable plastic film adhered to the face of the layer. The plastic film can be adhered to the fibrous layer by an intermediate adhesive sheet which can also serve as an ultraviolet radiation barrier. A metal mesh or hardware cloth can also be used in or with the fibrous layer for rodent control.

The method of making the panel includes densifying or compressing a fibrous layer, curing a binder in the layer to form a rigid self-sustaining mass or board, and adhering a plastic sheet or film substantially at the same time to the layer in one operation. The binder is cured with the fibrous layer compressed by heat applied to both sides of the layer, with the heat applied to the side having

the plastic film being initially retarded or delayed which has been found to prevent wrinkling of the plastic film on the layer. The fibrous panel is highly densified and exhibits a high degree of rigidity with hard, high impact-resistant surfaces not previously associated with panels made of mineral fibers and, specifically, glass fibers.The panel exhibits rigidity and has impact-resistant surfaces
comparable to good quality wood panels, and when provided with a plastic film
on an exposed face thereof is readily washable. The fibrous panel also has reason-
ably good insulating properties and exhibits greater fire-retardant properties than
wood. In addition, the plastic film surface of the panel can be textured in a
variety of patterns for decorative purposes.

The particular characteristics of the fibrous panel make it especially suitable for
use in walls of cold storage chambers or freezers. Proper sanitary conditions can
be readily maintained through the washability of the plastic film while the fire-
retardant and insulating properties are especially suitable for cold storage applications.

Referring to Figure 6.5a, a rigid building panel is indicated at **10** and includes
a highly densified fibrous layer **12** with a plastic film or sheet **14** adhered to a
surface. The film **14** can be adhered to the fibrous layer **12** by adhesive shown
in the form of a phenolic-bonded paper **16**. The layer **12** consists of discontinuous fibers held in a permanent rigid shape by a cured binder. The layer has a
density in the order of 35 to 60 pcf and exhibits high impact resistance, for
example the resistance being similar to that of a good quality wood panel. The
layer also exhibits good insulating qualities, better than plywood, because of
its fibrous character, and is also highly fire retardant. The fibrous layer **12** itself, by way of example, has a flame spread rating in the order of 12 to 15,
while the plastic film **14** has a flame spread rating of less than 20. This provides the composite panel **10** with a rating under 25 and therefore is classified
as a Class I fire-retardant material.

The layer **12** is made from uncured pelts or pads of fibers which are highly densified. For example, to produce the layer **12** in a thickness of one-quarter inch,
five uncured pelts of fibers each having a thickness of three inches and an apparent density of three-quarters pcf are compacted to the one-quarter inch thickness under a pressure of about 180 psi. The fibers of the pelts can be made by
a conventional rotary process with diameters in the range of 0.00027 to 0.00032
inch. A phenol-formaldehyde binder can be employed, but the binder is uncured or only partially cured prior to the pelts being placed in a press and densified.

The plastic film or sheet **14** preferably is a polyvinyl fluoride commercially known
as Tedlar. The sheet has a thickness of 1 to 4 mils, preferably 2 mils, and is
substantially nonporous and washable, which is particularly necessary for cold
storage and freezer applications. The polyvinyl fluoride is preferred because it
will withstand a molding or curing temperature in the order of 350°F used
in the forming press. It also is fire resistant, as discussed above.

The sheet **16** is thicker and stiffer than the film **14** to help provide a smooth
base or backing surface. It employs a heat-softenable adhesive on both sides to
provide a bond or "glue-line" between the film **14** and the layer **12**. The sheet
16 is commercially available and is preferably black so as to effectively block
ultraviolet radiation.

FIGURE 6.5: RIGID FIBROUS PANEL

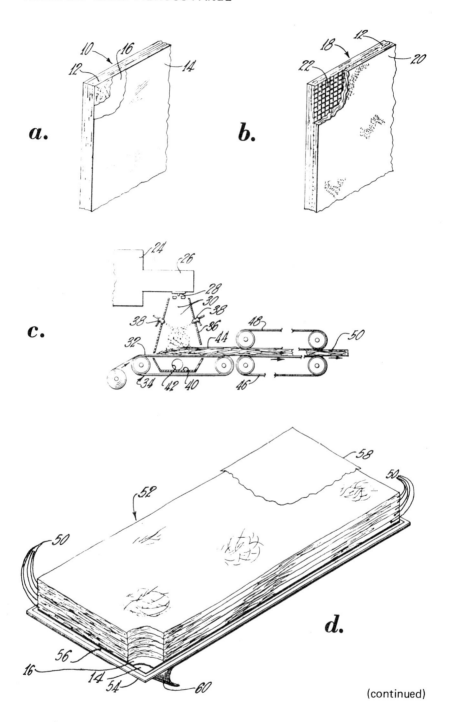

(continued)

FIGURE 6.5: (continued)

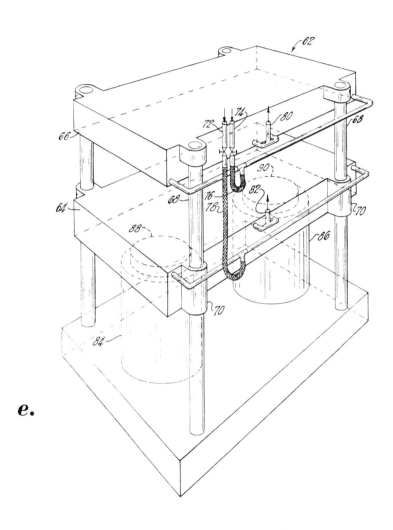

e.

(a)   Fragmentary view in perspective of a fibrous panel
(b)   View similar to Figure 6.5a of a modified fibrous panel
(c)   Schematic view in elevation, with parts in section, of a line from which
          layers or pelts of glass fibers for use in the panels of Figures 6.5a and
          6.5b can be produced
(d)   View in perspective, with parts broken away, of a caul plate and charge,
          ready to be placed in a press, for producing the panel
(e)   Schematic view in perspective of a press to receive the charge of Figure 6.5d

Source: U.S. Patent 3,905,855

A slightly modified building panel **18** of Figure 6.5b has the highly densified fibrous layer **12** but a modified plastic film **20** having a textured surface formed in the press by a suitable textured layer placed in the press. The textured patterns possible for the plastic film **20** are substantially unlimited and can be produced by employing sheets or layers of such materials as felt, leather, canvas, linen, etc. in the press contiguous with the plastic film.

As shown, the panel **18** has a metal mesh or hardware cloth **22** associated with the fibrous layer **12**, being placed within the pelts when the charge is made. This mesh prevents rodents and particularly rats from chewing through the panel **10** and gaining access to the interior of the cold storage chamber or freezer. Openings in the mesh **22** of one-quarter to one-half inch have been found to be suitable.

The pelts used to form the fibrous layer **12** can be produced on a fiber-forming line, schematically shown in Figure 6.5c. Glass is melted in a tank **24** and is supplied through a forehearth **26** to a fiber-forming unit **28**. Fibers indicated at **30** are attenuated from the unit **28** and move downwardly, being deposited on a belt **32** of a conveyor **34**. Binder is applied to the fibers in a forming hood **36** by suitable spray applicators **38**. The fibers are gathered on the belt with the aid of a vacuum chamber **40** located below the upper run of the belt **32**. A vacuum is maintained in the chamber **40** by an exhaust fan communicating with an exhaust duct **42**.

The fibers **30** with the binder applied are moved toward the right, as viewed in Figure 6.5c, when deposited on the belt **32**. A resulting layer **44** of the fibers **30** is then carried between two conveyors **46** and **48** which are spaced apart a predetermined distance to form the surfaces of the layer **44** in smooth, parallel relationship. A resulting pelt or pad **50** for the press charge is thereby produced. The pelt **50** can be carried away by an exit conveyor to a point where it can be cut to predetermined shape and size. Alternately, the pelt can be stored on a reel and transported closer to the press, at which time the pelt can be cut to size. In this state, the binder of the pelt is not cured, or at the most only partially cured, so that the pad is readily deformable; however, the binder still provides a degree of structural integrity for the pelt.

Since a plurality of the pelts or pads **50** are employed in contiguous, superimposed relationship in the press in order to make the fibrous layer **12**, additional binder is sometimes applied to the surfaces of the pelts to assure that they will adhere to one another and not form a weak plane at their junctures. To assure that sufficient binder is located at the pelt surfaces, without adding additional binder, binder can also be applied to the pelts by using a technique in which the fibrous layer from the fiber-forming unit is immersed in a pool of binder and subsequently passed through mechanical squeeze rolls to assure thorough distribution of the binder.

A press charge for producing the panel **10** is indicated at **52** in Figure 6.5d. The charge **52** is built up on a caul plate **54** which provides a smooth, planar, heat-conducting, supporting surface for the charge **52**. The plate **54** is, of course, larger in area than the pelts, in this instance measuring somewhat in excess of 4' x 8' in order to produce 4' x 8' building panels. By way of example, the caul plate can be made of 12 to 16 gauge steel.

A suitable release sheet **56** of paper is laid directly on the caul plate **54**. The release paper can be of various suitable materials, one being a commercially available brown 40-lb paper. The release paper does not affect the operation of the press but merely assures ease in separation of the final panel from the surface of the caul plate **54**. The plastic film **14** is next placed on the release paper **56** and the phenolic-bonded or adhesive sheet **16** is placed on the plastic film **14**. The five fibrous pelts **50** are next placed on the adhesive sheet **16** in superimposed, contiguous relationship therewith and also with the plastic film **14**.

With the final panel **10** having a thickness of about one-quarter inch, it will be seen that the five pelts **50**, each having a three-inch thickness, are reduced to a thickness of about one-sixtieth the original thickness, and preferably from one-fortieth to one-eightieth the original thickness. This substantial reduction is required to result in the high-density panel having the high impact and rigid characteristics achieved in the process. Of course, other combinations of pelts can be used, for example, fewer thicker pelts or more thinner pelts. The particular combination of the five pelts, however, assures more uniform density throughout the panel, which might not be as readily achieved with fewer thicker pelts, whereas more thinner pelts would require more labor and handling.

After the plastic film **14**, the adhesive sheet **16**, and the pelts **50** are in place on the caul plate **54**, a roller is preferably moved back and forth on the top surface of the upper pelt **50** to eliminate any excess air trapped between the pelts or within them. Subsequently, a top release sheet **58** of paper is laid over the upper surface of the uppermost one of the pelts **50**. This release sheet should be sufficiently pervious to permit air to escape as the composite is subjected to the closing and molding pressures. A commercially available brown 40-lb paper also has been found to be suitable for this release sheet.

In accordance with the process, a layer or sheet **60** of insulating material, such as canvas, is applied to the bottom of the caul plate **54**. In such an instance, after the charge **52** thereon has been pressed and molded, and the resulting panel removed, the caul plate **54** with the canvas **60** can be cooled prior to receiving another charge and again being placed on the press. Rather than having the canvas **60** adhered thereto, however, a separate canvas belt can be employed which extends through the press above the lower platen and carries thereon the caul plates in sequence. This is a particularly advantageous expedient for higher production rates. However, the principle is nevertheless the same, namely that the insulating layer delays, retards, or inhibits flow of heat from the lower platen to the caul plate and to the composite layer.

When the composite charge is ready, the composite along with the caul plate **54** and the insulating layer **60** are moved into a press **62** of Figure 6.5e. The press **62** includes a lower heated platen **64** and an upper heated platen **66**. The upper platen is stationary, in this instance being supported on posts or ways **68**, while the lower platen is movably guided by bearing blocks **70** on the ways **68**. Both of the platens **64** and **66** are heated by a fluid, such as oil, supplied by lines **72** and **74** through flexible lines **76** and **78** to end portions of the platens **64** and **66** with the oil being exhausted through suitable conduits **80** and **82** from central portions of the platens. The oil passages located internally of the platens **64** and **66** are of a conventional nature and are not shown.

The lower platen **64** is moved in vertical directions by fluid-operated cylinders

84 and 86 having piston rods 88 and 90 supporting and engaging the lower platen 64. The number and size of the fluid-operated cylinders 84 and 86 depend upon the size of the platens 64 and 66 as well as the cylinder diameters and pressures employed. In a typical 4' x 8' platen, six of the cylinders 84 or 86 are employed, by way of example. In any event, the pressure needed to achieve the required densification of the pelts 50 is substantial, being in the order of 180 psi, to reduce the five pelts 50 to a thickness of one-quarter inch.

When the charge 52 and the caul plate 54 with the canvas 60 are placed in the press, the platens 64 and 66 are at an operating temperature in the order of 350°F, and preferably in a range of 300° to 400°F. As the platens start to close, and contact the upper surface of the upper release paper 58, heat is substantially immediately transferred from the upper platen 66 to the upper pelt 50. On the other hand, heat from the lower platen 64 is delayed in being transferred because of the canvas 60, and the heat also must be applied to the caul plate 54, which is at a temperature substantially lower than the operating temperature of the platens, prior to heat being applied to the plastic film 14 and to the adhesive sheet 16 and the pelts 50.

The retarded transfer of heat upwardly from the lower platen 64, as compared to the direct transfer of heat from the upper platen 66, prevents the plastic film 14 from creeping and wrinkling as would otherwise occur. Apparently this is achieved because the adhesive sheet 16 becomes sufficiently tacky to prevent the plastic film 14 from creeping before the film 14 can become heated sufficiently to do so. In any event it has been found that by delaying or retarding heat transfer by virtue of the insulating sheet 60, the fibrous layer 12 can be produced with the film 14 adhered thereto all in one operation. Further, the bond achieved between the film and layer in the single operation is much stronger than that obtained by densifying and producing the fibrous layer 12 in one operation and subsequently applying the plastic film 14 to a surface thereof in a subsequent separate operation.

In order to produce the textured panel 18 of Figure 6.5b, a textured layer in the form of linen, leather, felt, canvas, etc. can be employed on the caul plate 54 in place of the lower release paper 56. The textures which can be achieved with this arrangement are substantially unlimited.

High Compressive Strength Foam for Cold Storage Plants

R.F. Shannon; U.S. Patent 3,726,755; April 10, 1973; assigned to Owens-Corning Fiberglas Corporation describes a foam product that has high compressive strength along at least one of its major axes. The foam body contains glass foam pellets and a matrix of a foamed binder. The pellets are arranged in an end to end abutting engagement along at least one major axis of the foamed body. The matrix of a foamed binder fills the interstices between the glass foam pellets. The insulation materials, which are capable of withstanding relatively high compressive loads, can be used in load-bearing structures, as for example in the floors of walk-in type refrigerators and cold storage plants.

Example 1: A 12" x 12" x 2" composite was made by tumbling 68 grams of a foamable resin mix, later to be described, with 515 grams of one-eighth inch nominal diameter glass foam pellets having a true density of 10 pounds per cubic foot and a bulk density of 6.8 pounds per cubic foot. These pellets have interstitial voids of 32%. The tumbling procedure coated the pellets with the resin and the resin was placed in a suitable apparatus with the platens positioned to provide a 12" x 12" box. The upper platen is lowered upon stops which spaced the upper platen two inches from the conveyor. The platens were heated to 350°F to cause the resin to foam and fill the interstitial voids. The final product weighed 575.5 grams and had an equivalent density of 7.6 pounds per cubic foot. The foamable resin mix was made by blending a mix A and a mix B immediately before application to the pellets. Mix A and mix B had the following compositions.

Ingredients	Amount, g
Mix A	
Resin A of U.S. Patent 2,979,469 (refrigerated)	250.0
Tween 40 (polyoxyethylene sorbitan monopalmitate)	12.5
Diatomaceous earth	13.0
Diazoaminobenzene	14.3
Methanol	15.0
Benzene	10.0
Mix B	
Urea	12.5
37% HCl (refrigerated)	38.7

The finished phenolic foamed glass composite had the structure wherein the glass pellets were in engagement with each other along the three major mutually perpendicular planes of the composite.

Example 2: A Styrofoam glass foam pellet composite is made by placing 394 grams of one-quarter inch nominal diameter glass foam pellets having a true density of 8.6 pounds per cubic foot and a bulk density of 5.2 pounds per cubic foot, in a mold all sides of which rest against the glass pellets. The voids between the glass pellets are filled with a polystyrene foam having a density of approximately 3 pounds per cubic foot by forcing the foamed polystyrene of U.S. Patent 2,669,751 into the mold in the manner indicated by the patent. The finished composite weighs approximately 485 grams and has a density of 6.4 pounds per cubic foot.

Inorganic foamed materials can also be forced into the interstitial voids between glass foamed pellets to produce similar results. The production of the glass foam pellets is also described in detail.

OTHER APPLICATIONS AND PROCESSES

Reinforced Plastic Storage Tanks for Chemicals

A process described by *D.H. Bartlow and M.E. Greenwood; U.S. Patent 3,860,478; January 14, 1975; assigned to Owens-Corning Fiberglas Corporation* relates generally to an expansion joint for composite wall constructions, especially those for large diameter plastic tanks.

A sandwich wall construction is provided having a slit in one of the reinforced plastic layers of the wall and a shaped expansion member overlaying the slit in bridging relationship. The member has a pair of base sections adhered to the layer on opposite sides of the slit and a central portion spaced from the wall and extending between the base sections. Although the expansion member is preferably of a substantially rigid material, such as a hardened resin reinforced with glass fibers, the member is shaped such that the base sections can move toward and away from each other in accord with the expansion or contraction of the wall.

The expansion member overlays the slit in the one layer of the tank wall and is bonded to the layer on both sides of the slit. Since the expansion member is of substantially rigid material, it is capable of carrying or transferring some of the load on the wall. Therefore the expansion joint of the process does not weaken the tank wall to the extent of the prior art expansion joints. Furthermore, since the expansion member is continuous and is integrally bonded to the wall on both sides of the slit, the joint is moistureproof and suited for use vertically as well as horizontally. This feature is especially significant on insulated tanks where moisture may cause deterioration of the foam insulating layer.

The insulated storage tank **10** shown in Figure 6.6a is typical of those having the sandwich type wall construction discussed previously. In general the tank comprises a cylindrical shell **12**, a generally flat base **11**, and a dished end cap **13**. Referring to the sectional view of the tank wall **15**, as shown in Figure 6.6b, the inner fluid-containing shell **20** is usually of glass fiber reinforced plastic. In many cases the reinforcement is provided by continuous windings of glass rovings providing sufficient tensile strength to withstand the pressures of the contained liquids.

The inner shell is typically between 0.150 to 0.500 inch thick depending on the application. The insulating core **22** is preferably a urethane foam. The thickness of the insulating layer will vary according to the application and typically is between 1 to 4 inches. The foam core is applied on a mandrel and will bond to the inner shell. The outer shell **24** again preferably consists of a plastic resin reinforced with either continuous or discontinuous glass fibers. The outer shell is primarily intended as a protective covering for the insulating layer **22**, and is generally about one-tenth inch thick. Through the integral bonding of the insulating layer to both the inner and outer layers **20** and **24**, a sandwich structure is created providing a stronger tank.

In the normal process for making the tank **10**, the inner layer **20** is first formed on a rotating mandrel. While the resin of the inner layer **20** is still uncured and tacky, the insulating layer **22** is foamed in place. Finally, before the foam has fully set the outer layer **24** is formed. After the resin of layers **20** and **24** and the foam layer **22** are fully cured an integral bond between the three layers is effected.

A temperature differential is established across the sandwich wall when in use since the fluid contained therein is usually at a temperature different than that of the environment. Since the inner shell **20** and the outer shell **24** are frequently of different composite materials, they will have different coefficients of expansion. Consequently the expansion or contraction of one shell will likely result in tensile or compressive stresses imparted to the other shell. In some cases these stresses are so severe as to cause a failure in the tank structure either in the form of a buckling of the inner shell, or a fracturing of the outer shell.

FIGURE 6.6: FLUID STORAGE TANK

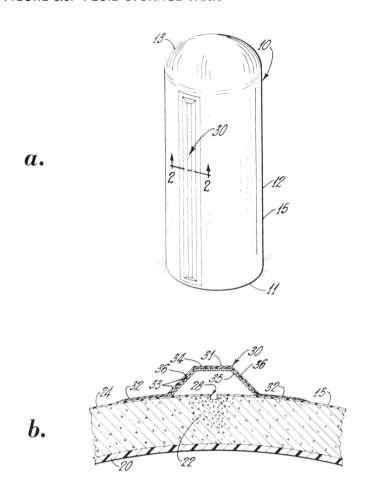

a.

b.

(a)   View in perspective of a vertical insulated storage tank
(b)   View in cross section through the wall of the tank shown in
          Figure 6.6a as taken along the line **2—2** in Figure 6.6a

Source: U.S. Patent 3,860,478

In order to provide for the relative expansion or contraction between the inner and outer layers of the wall **15** it has become necessary to incorporate an expansion joint such as that indicated generally at **30** in Figures 6.6a and 6.6b. Referring to Figure 6.6b, the joint comprises a slit **28** provided in at least one of the tank shells; in this case the outer shell **24**. The slit may or may not extend through the foam layer **22**. The width of the slit will vary in the particular application, but a width of one-eighth to one-quarter inch has been found suitable in most cases. Furthermore, as used here, the word slit is meant to cover any type of separation which will allow the outer shell 24 to expand or contract in accordance with the expansion or contraction of the inner shell, including cuts, fissures, grooves, splits, etc. Unless otherwise provided for, the slit 28 will expose the foam insulating core to the effects of the environment. Since most urethane foams are susceptible to deterioration due to moisture, it is necessary to prevent moisture from entering into the slit. In addition, the slit 28 weakens the outer shell causing an increase in the loading on the foam layer.

To compensate for these prejudicial effects of the slit, the shaped expansion member 31, shown in Figure 6.6b, is incorporated in the wall structure. The member 31 overlaps the slit and extends the length thereof, as shown in Figure 6.6a. The expansion member is made by applying a cardboard form 35 in bridging relation over the slit 28 and taping it to the outer layer 24. One or more layers of reinforcement mat 33 saturated with hardenable resin are then laid up over the cardboard form. Although many types of reinforcements could be used, mat composed of glass fibers, including bonded mat, chopped strand mat, continuous strand mat, scrim, and woven rovings are preferred. The resin impregnated mats extend onto the tank wall on both sides of the cardboard form, and the resin of the member 31 is integrally cured with the resin of the outer layer 24.

Although the expansion member is made of a substantially rigid material, i.e., reinforced plastic, it is shaped such that it can flex in accordance with the expansion or contraction of the outer layer 24. To this end the member comprises a pair of spaced base sections 32 adhered respectively to the outer layer 24 on opposite sides of the slit 28, and a central portion abridging the slit and extending between the base sections. In the particular embodiment shown, the central portion is trapezoidally shaped and consists of a plateau region 34 spaced from the outer layers 24 above the slit, and two oppositely inclined leg portions 36 connecting the plateau to the base sections.

Due to the shape of the member 31, when expansion or contraction of the outer layer 24 takes place, the base sections 36 can move toward or away from each other in response to the movement of the layer 24. Although the trapezoidal shape shown is particularly suitable for this purpose, it will be readily understood that other configurations including semicylindrical, square, rectangular, triangular, etc. could be used to perform in a similar manner.

Lehr Roll Composition for Glass Manufacture

A process described by *F.L. Jackson and J.W. Axelson; U.S. Patent 3,954,556; May 4, 1976; assigned to Johns-Manville Corporation* relates to asbestos/talc compositions intended for high temperature service in glass and steel, aluminum, and other metal manufacturing processes. More particularly, it relates to millboard compositions from which can be produced Lehr rolls which come into contact with the high temperature surfaces of glass and metal sheets.

Lehr rolls also find extensive use in the processes for manufacturing plates of steel, aluminum and other metals where high temperature formation of the plates is involved. In some of these processes, the metal plates are annealed in annealing ovens in which temperatures as high as 1200°C are maintained. Lehr rolls are mounted inside these ovens to support and convey the plates through the ovens. In addition to the obvious requirement of high temperature stability under such conditions, surface smoothness of the Lehr rolls is also necessary, for with many of the metal sheets being produced, such as those of stainless steel, good surface finish is a requirement.

The crux of this process lies in the combination of the chrysotile asbestos and the talc component. Thus it has been found that a combination of chrysotile asbestos and talc is not only highly satisfactory for use in such high temperature environments, but in fact has better properties under equivalent conditions than do the conventional chrysotile/amphibole materials.

Two separate sets of millboard sheets were made to determine the properties of the product. One set, whose properties are listed in Table 1 below, were made as laboratory hand sheets, using standard laboratory techniques. A second set, whose properties are listed in Table 2, were made on a wet process papermaking machine. The chrysotile fibers were first beaten for 10 minutes at a load of 220 newtons and then transferred to the wet machine mixer along with the talc, cement, and clay. Sufficient water was provided to prepare a slurry containing 1 to 2% solids. The wet machine was set to run at 8.7 meters per minute with maximum vacuum. The flow rate was adjusted to achieve minimum overflow from the vat. In addition, about 0.125% (based on dry solids) of a commercial synthetic nonionic water-soluble polymer of acrylamide (Seraparan NP10) was added to the water slurry.

In addition, three control samples containing large amounts of amosite asbestos and representative of prior art compositions were tested. One was produced as a laboratory handsheet, one on the aforementioned wet machine, and a third in a commercial millboard plant. The properties of all three were sufficiently similar that they can all be accurately represented by a composite average of the three. These data are shown in Table 3. In the examples below, percent moisture was calculated according to the formula:

$$\frac{W_w - W_d}{W_w} \times 100\%$$

W_w = wet weight; sample weight exposed to the atmosphere
W_d = dry weight; sample dried at 300°F for 24 hours; cooled in a desiccator and weighed

Percent ignition loss was determined by the following formula:

$$\frac{W_d - W_i}{W_d} \times 100\%$$

W_i = ignition weight; sample held at 1200°F for 24 hours; cooled in a desiccator and weighed

Compression and recovery were determined according to ASTM Method F-36. "Normal" compression and recovery are the properties of the dried millboard. "650°C" compression and recovery are properties of the millboard after it has been subjected to 650°C for 16 hours while under a compression of 5,860 kg/m^2. For comparison purposes, the higher the value of modulus of rupture, the stronger the millboard; the lower the compression value, the more resistant the millboard is to compression; and the higher the recovery value, the more resilient the millboard.

TABLE 1

Formulation, wt. %	Samples			
	1C	2C	3	4
Chrysotile fiber	25*	35*	25**	35**
Talc	60	50	60	50
Portland Cement	10	10	10	10
Bentonite Clay	5	5	5	5
Properties				
Density, gm/cm³	1.10	1.04	1.14	1.15
Ignition loss, %	6.2	7.6	6.2	7.1

*1:1 mixture of 5K and 6D fibers
**6D fiber

TABLE 2

Formulation (wt. %)	Samples				Average
	1A	1B	2A	2B	
Chrysotile fiber*	25	25	35	35	—
Talc	60	60	50	50	—
Portland Cement	10	10	10	10	—
Bentonite Clay	5	5	5	5	—
Properties					
Density, gm/cm³	1.06	1.06	0.99	0.99	
Ignition loss, %	5.8	5.9	6.0	6.1	5.9
Compression, %					
normal	21.4	21.8	22.5	21.1	21.7
650°C	26.5	26.1	25.9	26.6	26.3
Recovery, %					
normal	54.4	55.8	55.5	50.0	53.9
650°C	25.7	25.6	22.8	22.8	24.2

*1:1 mixture of 5K and 6D fibers

TABLE 3

Formulation, wt. %	
Asbestos fiber	
Chrysotile	25
Amphibole (Amosite)	25
Talc	35
Portland Cement	10
Bentonite Clay	5
Properties	
Density*	0.99 gm/cm³
Ignition loss*	6.7%
Compression**	
normal	27.3%
650°C	26.9%
Recovery**	
normal	43.4%
650°C	13.4%

*Average of three samples
**Average of two samples

It will be apparent from the preceding data that the compositions of this process produce a product which has a density and ignition loss comparable to that of the prior art materials, despite the fact that the amphibole asbestos component previously believed to be critical is entirely absent from the composition. The products provide a superiority in compression and recovery, both under normal conditions and after heating at 650°C. Since ignition loss and compression and recovery values comprise extremely important criteria for assessing the quality of Lehr rolls, it will be apparent that the products of this process can be used as Lehr roll materials quite superior to those materials previously available.

Jewelry Repair

A process described by *D.D. Banks; U.S. Patent 3,975,142; August 17, 1976* relates to a composition for absorbing and dispersing rather than transmitting heat energy which has particular utility as a coolant in the repair of articles such as jewelry and dental appliances. The mixture is preferably about 100 parts by weight of asbestos powder, 5 parts by weight of boric acid, and 0.75 to 1 part by weight of glycerol. Water is then added to provide a proper paste-like consistency, though the composition is effective without particular regard to consistency. Thus consistency can be provided with regard to convenience of use rather than operability with regard to thermal protection. For the preferred mixture, 25 to 30 parts by weight of water provides a convenient consistency.

The optimum ratio between the asbestos powder and boric acid has been found to be 20 parts by weight of the asbestos powder to 1 part of boric acid, although this ratio may be varied within limits. It has been found that increasing amounts of boric acid with respect to the asbestos powder present will tend to cause crystallization and formation of deposits around the material being heated. However, reduction of the boric acid beneath the optimum ratio will tend to inhibit the fire retardant properties of the composition.

When used, the portion of the article to be protected is pressed or inserted into the composition. Alternatively, the composition is merely applied to a portion of the article which is to be protected, or at a position between the portion to be protected and the area to which heat is to be applied. When heat is applied to bring an unprotected portion of the article to an elevated temperature, for instance in the range of 1000° to 5000°F, the heat flux which flows towards the protected portion of the article is absorbed and dissipated by the composition.

An example of the use to which the composition can be advantageously put is to be found in the sizing of rings. Sizing of rings is accomplished by cutting the rear portion of the ring to the proper length and, in the instance of silver rings, soldering the ends of the ring together at a temperature of about 2500°F. The composition of the process is disposed in a container and is employed conveniently as a support with the portion of a ring not being soldered pressed into the composition in order to minimize fire scale and to protect, for instance, gems or other portions of the ring. A torch or other soldering implement may then be brought into contact with the portion of the ring to be soldered in a conventional and well-known manner. It has been found that utilization of the composition of the process does indeed minimize scaling of the ring and effectively protects heat-sensitive portions of the rings. After soldering is completed, the protective composition can be readily removed from the ring by rinsing with water. No objectionable residue and very minimal heat scaling results.

Machine Parts for Nuclear Reactor Installation

F. Schweigert and P. Schroder; U.S. Patent 3,740,247; June 19, 1973; assigned to Siemens AG, Germany describe a method of producing heat insulation for machine parts. The heat insulation is wear resistant and able to withstand heat and pressure variations and is adaptable for use in a nuclear reactor installation subject to mechanical vibrations. The process comprises the steps of preparing insulating bodies of glass wool or rock wool, adjusting the fibers of insulating bodies of glass wool or rock wool to the shape of the machine parts, coating the fibers of the insulating bodies with a heat-resistant and permanently elastic binder, connecting the fibers of the insulating bodies to each other with the binder, coating the entire mass of insulating bodies with a layer of the binder, and providing openings in the binder layer for pressure balance.

The insulating bodies are saturated with the binder and are subsequently dried. The binder is applied to the fibers of the insulating bodies during the preparation of the insulating bodies. The fibers of the insulating bodies are coated with a fiber of thinned silicon rubber.

Since a silicon rubber mass is normally relatively tough, such a mass is preferably thinned prior to its processing, with a relatively volatile agent or agents. A suitable thinning agent may comprise, for example, a chlorinated hydrocarbon such as, for example, toluol, xylol, benzol and ligroin. The most practical thinning ratio of volume shares, depending upon the structure and strength of the insulating bodies, is 1:3 to 1:5.

In order to permit the solvent to escape as rapidly as possible from insulating bodies saturated therewith, it is preferable to dry the insulating bodies on a wire net having rough meshes, at a temperature of 40° to 50°C. Higher temperatures are to be avoided, since they may lead to a formation of bubbles, due to the evaporation of the solvent which occurs too rapidly. Moreover, it is advantageous to provide the dried insulating bodies which have been saturated with silicon rubber with a silicon rubber jacket or layer, so that a direct contact between the fibers and the housing of the machine parts is prevented.

Furthermore, sectional edges should be immediately covered, sheathed or encased. The covering of the sectional edges is best provided by coating with an unthinned silicon rubber such as, for example, a siloprene paste, or by injection with silicon rubber having approximately 10% thinning. After the silicon rubber has hardened, the casing is provided with fine holes at various locations so that there may be pressure balance during a pressure load without damage to the insulating bodies.

In tests undertaken under conditions approaching actual operating conditions, it has been observed that insulating bodies produced in accordance with the process meet the requirements completely and adequately. Furthermore, tests for overheating have been conducted at 300°C for up to 50 hours, and have caused no damage.

Stainless Steel Processing Equipment

A.P. Mueller; U.S. Patent 3,639,276; February 1, 1972; assigned to Panacon Corporation describes molded thermal insulation material for use in contact with austenitic stainless steel chemical-processing equipment to inhibit stress corrosion cracking. The insulation consists of at least 60% by weight of cellular expanded perlite, bonded by a dried, inorganic, water-soluble binder. The binder consists of a major proportion of a mixture of sodium silicate and potassium silicate and a minor proportion of sodium hexametaphosphate, and the molecular ratio of the silica to the alkaline oxide is at least 3.6 to 1.0.

The potassium silicate consists of at least about 40%, by dry weight, of the total silicate in the dried binder, and the total binder constitutes at least about 20% by weight of the total insulation composition. In the manufacture of the molded insulation, the mixture of perlite and the binder, in slightly damp condition, is placed in the mold cavity and subjected to light molding pressure, sufficient to compact the material into the exact configuration of the mold cavity without expelling water from the mixture. The molded insulation is removed from the mold and oven dried at a temperature of about 250°F.

Heat Barrier for Explosives

C. Ré, E.O. Conrad and J.R. Conrad; U.S. Patent 3,692,682; September 19, 1972; assigned to Dyna-Shield, Inc. describe material which serves as a protective heat barrier for explosives, or other combustible products, and which may be applied as a paste or a liquid to the external or internal surface of the casing containing the explosive, to create a hard coating for the casing of extremely low heat transfer characteristics. The material is composed of a mixture of water-extended polyester (WEP) and particles of a heatproofing material such as powdered or fibrous asbestos, or powdered glass.

Example: A coating, according to this process, may be prepared from the following ingredients, in parts by weight:

  (a)   1 part of a water-extended polyester resin. This material is available commercially as WEP-27 (Ashland Chemical Co.).
  (b)   0.0075 to 0.25 part cobalt.
  (c)   0.005 to 0.01 part dimethylaniline (DMA).
  (d)   0.8 to 1.2 part water.
  (e)   Trace (2×10^{-4} part) of an inhibiter such as a mixture of methyl Cellosolve (80%) and hydroquinone (20%).
  (f)   Trace ($\frac{1}{16}$ part) of powdered or fibrous asbestos or powdered glass (determined by specific requirements).
  (g)   0.004 to 0.02 part of a catalyst such as methyl ethyl ketone peroxide (determined by desired gel time).

The polyester resin, cobalt and the DMA are mixed together as one group which is designated as the carrier. The asbestos (or powdered glass), water inhibitor and catalyst are mixed as another group which is designated as the heat retarder. The two components form the heat barrier coating of the process when mixed together.

Rescue Suits

A process described by *H. Görlach, W. Stroot and A. Kreckl; U.S. Patent 3,801,422; April 2, 1974; assigned to Akzona Inc.* involves a multilayer fibrous insulating article containing fibers composed of a polyacyloxalamidrazone in chemically combined form with at least one of the following metals: zinc, tin, cadmium, barium, strontium, calcium, antimony or tantulum. Especially good results have been achieved with the metal-containing polyacyloxalamidrazone when combined with a cellulosic polymer, the two polymers being spun in common from an aqueous alkaline solution and the metal subsequently introduced into the resulting filaments or fibers.

Polyacyloxalamidrazones are known polymers which have recurring units of the formula:

$$\left[-NH-N=C-C=N-NH-\underset{\parallel}{C}-R-\underset{\parallel}{C}- \right]$$

in which R represents a straight chain or branched saturated or unsaturated aliphatic radical having 2 to 12 carbon atoms or a cycloaliphatic, araliphatic, aromatic or heterocyclic radical, preferably a substantially divalent hydrocarbon radical of 2 to 6 carbon atoms, i.e., a structure in which an occasional hetero atom such as oxygen, nitrogen or sulfur has no material effect. These polymers may be prepared according to conventional procedures, for example, by reacting oxalic acid bisamidrazone with one or more dicarboxylic acid dihalides by the process described in Belgian Patent 705,592.

Insulating materials having a weight of up to 1,000 grams per square meter are especially suitable for use as protective clothing and rescue suits. Such insulating materials provide no difficulties in fabricating such clothes or suits and the articles made up from these materials provide the wearer with a high degree of protection against heat and flames as well as sufficient freedom of movement. In insulating materials used for this purpose, at least the fibrous outer surface layer which in the suit constitutes the layer furthest away from the body advantageously consists of a woven or knitted fabric of one or more of the aforementioned metal compounds of polyacyloxalamidrazones.

If the flame-resistant layer is used as the innermost layer, preferably in the form of a fibrous web or flecce, the surface layer may be in the form of a woven or knitted fabric of polyamides (nylons), polyesters (e.g., polyethylene terephthalate) or mixtures of these polymers with cotton or with woven or knitted rayon fabrics which have been treated with the usual commercial flame-retarding agents.

A few practical examples of articles made with the multilayer insulating material are given below. The sheets of insulating material are composed of several and at least two dissimilar layers. If the sheet is used for protective clothing or rescue suits, then the individual layers are so arranged that the first layer (1) as set forth in each example constitutes the layer which will be farthest away from the body and the last layer constitutes the layer closest to the body. The permeability to air was determined according to DIN 53 887 (German Industrial Standards), using a Frank-Hauser test apparatus at a vacuum measured as a 20 mm column of water. The thermal effect was determined by the method of the *Handbuch der Werkstoffprufung*, 2nd edition, Vol. 5 (1960), p. 1135.

All weights of the multilayer or composite articles in the form of sheets, webs, or the like, are given as weight per unit of covering area, i.e., grams per square meter or g/m^2 in abbreviated form. The permeability of the multilayered article is measured in units of liters of air per 100 cm^2 per minute, abbreviated as l/100 cm²/min. While the thermal efficiency is a measure of the flame resistance, each individual layer consisting of or containing the polyacyloxalamidrazone-metal fibers was also tested and found to be flame resistant in terms of being difficultly combustible up to a substantially flameproof layer.

Example 1:
   (1) Knitted polyamide fabric; weight = 132 g/m².
   (2) Continuous fibrous fleece of a mixture of 40% of polyterephthaloyl oxalamidrazone and 60% of cellulose having a tin content of 30% by weight; weight = 69 g/m².
   (3) Crimped fleece of polyethylene terephthalate fibers; weight = 150 g/m².
   (4) Continuous fibrous fleece of a mixture of 40% of polyterephthaloyl oxalamidrazone and 60% of cellulose having a tin content of 30% by weight; weight = 69 g/m².
   (5) Knitted rayon fabric dressed with 30% of tris(2,3-dibromopropyl) phosphate; weight = 196 g/m².

The weight of this composite sheet of five layers of insulating material is 616 g/m², the permeability to air is 313 l/100 cm²/min and the thermal efficiency is 61.4%.

Example 2:
   (1) Woven fabric of a mixture of 67% of polyester and 33% of cotton; weight = 200 g/m².
   (2) Continuous fibrous fleece of a mixture of 40% of polyterephthaloyl oxalamidrazone and 60% of cellulose having a zinc content of 20% by weight; weight = 69 g/m².
   (3) Crimped fleece of polyester; weight = 150 g/m².
   (4) Fleece of a mixture of 40% of polyterephthaloyl oxalamidrazone and 60% of cellulose having a zinc content of 15% by weight; weight = 69 g/m².
   (5) Knitted fabric of a mixture of 40% of polyterephthaloyl oxalamidrazone and 60% of cellulose having a calcium content of 20% by weight; weight = 286 g/m².

The weight of this multilayer sheet of insulating material is 774 g/m², the permeability to air is 103 l/100 cm²/min and the thermal efficiency is 68.1%.

Example 3:
   (1) Knitted polyamide fabric; weight = 132 g/m².
   (2) Continuous fibrous fleece of a mixture of 40% of polyadipinoyl oxalamidrazone and 60% of cellulose having a zinc content of 15% by weight; weight = 286 g/m².
   (3) Continuous fibrous fleece of a mixture of 40% of polyadipinoyl oxalamidrazone and 60% of cellulose having a tin content of 15% by weight; weight = 138 g/m².
   (4) Knitted rayon fabric finished with 30% by weight of tris(2,3-dibromopropyl) phosphate as a flame retardant; weight = 196 g/m².

The weight of this composite sheet of insulating material is 752 g/m², the permeability to air is 285 l/100 cm²/min and the thermal efficiency is 49.9%.

Hot Hydrogen Environments

R.E. Riley and J.M. Taub; U.S. Patent 3,740,340; June 19, 1973; assigned to the U.S. Atomic Energy Commission discovered that composite metal carbide-metal oxide materials of the composition $MC-M'O_2$ where M and M' may be Ti, Zr, Hf, V, Nb, Ta, Th, or U, and the metal carbide content may range from about 25 to 75 volume percent, have low thermal conductivities and adequate structural properties in a flowing hydrogen environment heated to 2000°C and higher. The designation $MC-M'O_2$ is a general one and does not necessarily imply a stoichiometric monocarbide and/or dioxide. This is true also of any reference to a specific metal carbide or metal oxide as, e.g., ZrC and ZrO_2. These materials are thus highly suited for use as high temperature insulators in a hot hydrogen atmosphere. In particular, it has been found that a nominal 25 volume percent ZrC-75 volume percent ZrO_2 composite has essentially the same thermal conductivity after 16 hours at 2300°C in flowing hydrogen as it possesses after its initial exposure to that temperature.

Airtight Housing Construction

A process described by *E. Frei; U.S. Patent 3,854,261; December 17, 1974; assigned to Hermann Pieren AG, Switzerland* provides for the construction of a housing for apparatus working in sound-absorbing, heat-insulating or airtight conditions. The walls, bottoms and ceiling are assembled of elementary self-supporting insulation plates, each plate comprising two parallel metal sheets arranged at a distance from each other, the rims of which are pressed together and bent to form a shank and a channel, the metal sheets defining an interior space between. Such housings are used, e.g., with air conditioning and ventilation plants, or as an encasing for machines and apparatus.

Referring to Figure 6.7, the housing will be preferably made of elementary insulation plates 1 which are connected together to form a housing. The plates 1 have the same dimensions and are therefore interchangeable. The plate rims are bent towards the outside to form shanks 2, as it can be seen from Figures 6.7a, 6.7b, and 6.7c. The shanks 2 of the two adjacent plates 1 which lie in one plane are so arranged that they lie close to each other across their whole length; whereas the shanks 2 of the two adjacent plates, which are arranged perpendicularly to each other, do not contact at all. The two adjacent plates 1 are screwed together by means of bolts 3 and nuts 4, which bolts pass the shanks 2 and reinforcement angles 5 and 9 on both sides of the shanks 2. In this way the walls, bottom and ceiling and finally the whole housing will be put together.

Each reinforcement angle 5 placed on the inside of the shank 2 has a tongue 5a, which is adapted to be inserted in a channel 14 formed by the shank 2 and the bent rim of the sheet 1a of the plate 1, whereby its free end lies approximately 0.0788 inch from the bottom of the channel. There will be put on the shank 2 with the screwed-on reinforcement angles 5 and possibly 9 an insulating means 6, 6a. Depending on the arrangement of the plates 1 to one another (perpendicularly or in one plane) the insulating means is covered from outside either by a cover lid 7 or a cover lid 8.

The cover lids according to Figure 6.7d are of two different types, one of which is denoted with the reference number 7, and the other one with the reference number 8. Each cover lid 7 and 8 has rims 7a and 8a which are bent inwardly,

FIGURE 6.7: AIRTIGHT HOUSING SHELL

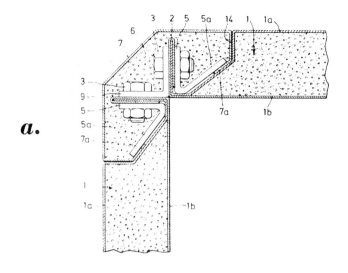

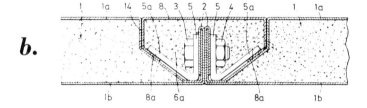

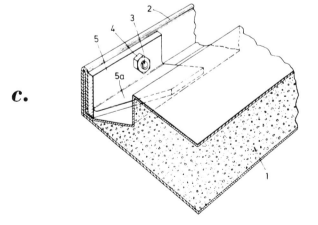

(continued)

FIGURE 6.7: (continued)

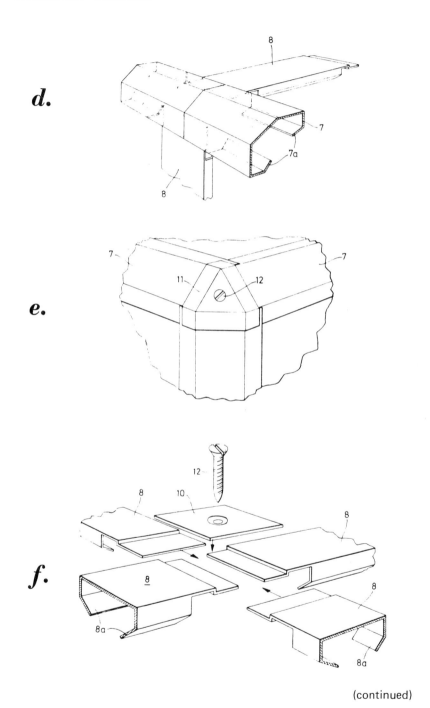

d.

e.

f.

(continued)

FIGURE 6.7: (continued)

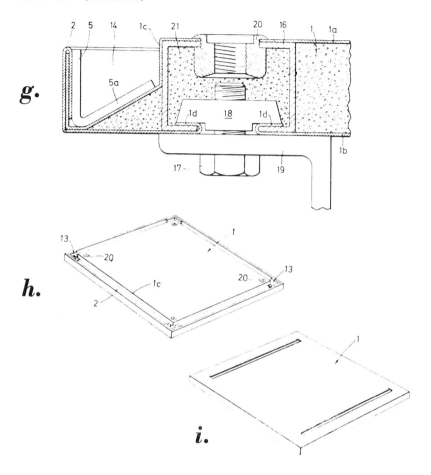

g.

h.

i.

(a)   Longitudinal sectional view of two plates arranged perpendicular to each
         other with an insulation means and a cover lid
(b)   Longitudinal sectional view of two plates arranged to each other in one
         horizontal plane with an insulation means and a cover lid
(c)   Diagrammatic sectional view of a single plate with a reinforcement angle
(d)   Diagrammatic view of three cover lids which are put together
(e)   Diagrammatic view of an intersection of three cover lids with a cover cap
(f)   Diagrammatic view of the disassembled intersection point of four flat
         cover lids with the cover cap
(g)   Enlarged sectional view of the rim portions of an elementary insulation
         plate with a reinforcing rail
(h)   Diagrammatic view of the plate according to Figure 6.7g
(i)   Diagrammatic view of the elementary insulation plate from another side
         as in Figure 6.7h

Source: U.S. Patent 3,854,261

so that by pushing the cover lids **7**, **8** on the shanks **2**, the rims **7a**, **8a** will snap behind the tongues **5a** of the reinforcement angle **5**, whereby the insulating means **6** will be force-pushed against the shank **2**. This will result in an airtight seal between the cover lids **7** and **8** and the shank **2**. The reinforcement angles **5** are preferably placed at the corners formed by the shanks **2**, so that there are at least two reinforcement angles **5** at each side of the plate **1**. The cover lids **7** and **8** are aligned with the upper surfaces of the plates **1** when they are assembled.

On their shorter sides, the flat cover lids **8** have end portions **8b** (see Figure 6.7f) which are inwardly reset and which abut against each other when they are assembled. These end portions **8b** are covered by a cover cap **10** which is in alignment with the outer surfaces of the cover lids **8** and so with the outer surfaces of the plates **1**. This has an aesthetic effect. The cap **10** is screwed on the cover lids **8** by means of a countersunk bolt **12** which enters a small gap between the assembled cover lids **8**. Cover caps **11**, which are screwed on the cover lids **7** by means of a countersunk bolt **12**, are used to cover the ends of three cover lids **7**, converging in a corner. This can be seen in Figure 6.7e. The cover lids **7** and **8** as well as the insulating means **6** have the same length as the respective sides of the plates **1**.

The elementary insulation plate **1** according to Figure 6.7g consists of two parallel zinc metal sheets **1a**, **1b** arranged at a distance from each other. The four rims of the plate **1** are pressed together and bent in such a way as to form a shank **2** which is perpendicular to the surface of the plate **1**. A channel **14** is made between each shank **2** and the edge **1c** of the metal sheet **1a** by bending the rims. The shanks **2** close an interior space between the metal sheets **1a**, **1b** which space is filled with an insulation sheet **15**. The plate **1** can be made more rigid by inserting pieces **13** in its corners.

The insulation sheet **15** consists of a hard, porous material which is during the production of the plate consistently connected with the inside surfaces of both the metal sheets **1a**, **1b** and forms a lasting hard insulation reinforcement of the plate **1**. In order to enable the use of apparatus (not shown) and in order to reinforce the whole plate **1**, a reinforcing U-rail **16** is arranged alongside the small sides thereof. The rail **16** is inserted between the metal sheets **1a**, **1b** at the proximity of the rim of the plate **1** in such a way that it is held on its one side by the edge **1c** of the metal sheet **1a** and on its other side by the hard insulation sheet **15**. The free ends of the rail **16** are held by the flanged ends **1d** of the metal sheet **1b** on that side of the plate **1** which faces the channel bottom **14**. The so-formed slot (see Figure 6.7i) serves to receive a bolt **17** which is adjustable. The interior space of the rail **16** is filled with a soft, porous material **21**. The washer **18** with a threaded hole is disposed on the free flanged ends of the rail **16** in which the bolt **17** carrying, e.g., a supporting arm **19**, can be screwed.

On the opposite side to the side of the free flanged ends the plate and the rail **16** are pierced by a riveting socket **20**. This riveting socket serves for holding apparatus or for securing cross and longitudinal bracings for large apparatus.

Intumescent Composition and Foamed Product

A process described by *S.R. Riccitiello and J.A. Parker; U.S. Patent 3,819,550;*

June 25, 1974; assigned to NASA; and U.S. Patent 3,730,891; May 1, 1973
provides an intumescent composition or agent which comprises the reaction
product of a para-benzoquinone dioxime with a mineral acid such as sulfuric
acid, phosphoric acid, and polyphosphoric acid. A preferred method for preparing the intumescent agents includes the slow addition of the mineral acid
to the selected amount of dioxime. During the addition of the acid, the temperature of the reaction is kept below 50°C in order to prevent the development
of an active exothermic reaction.

It was found that the proportions of the reactants could vary as long as the mol
ratio of the dioxime to the acid is maintained between 0.25 and 2.00. In physical appearance, the reaction products using different acids and different proportions of such acids ranged from powders for low acid ratio to a paste as the
ratio of acid was increased.

The intumescent agents prepared in the manner described may be used per se as
intumescent agents or they may be combined with or incorporated into other
compositions in the preparation of intumescent composites. For example, the
intumescent agent, per se, either as the dry powder or in the form of dry pellets,
may be blown into or otherwise placed in the space between building walls, void
spaces in ship and airplane structures, etc., as fire protective agents. Exposure of
such structure to intumescent or higher temperatures will cause the intumescent
agents to foam and swell to many times (60 to 200) their original volume and
fill the void space. Since the products formed by intumescing or foaming the
intumescent agents are both highly fire-resistant and heat-insulating, filling the
void space of a structure with the intumescing agents of the process will tend to
suppress and prevent spreading of the fire across the walls of the structure. The
following examples illustrate the process.

Example 1: This example illustrates the preparation of the p-benzoquinone
dioxime-sulfuric acid intumescent composition using a mol ratio of dioxime to
acid of 0.5. To 1.38 grams of p-benzoquinone dioxime was added, while stirring, 1.96 grams of concentrated sulfuric acid (96%; density 1.84) within about
5 minutes. The mixture produced an exothermic reaction. It was cooled while
stirring and maintained at a temperature of about 50°C. After cooling, the
product was found to be a greenish-black free-flowing powder.

Example 2: This process illustrates the preparation of the fire-resistant heat-
insulating foam by heating the product of Example 1. One gram of the p-benzoquinone diamine-sulfuric acid reaction product of Example 1 was placed in a
container and then heated on a hot plate maintained at a temperature of about
150°C. The reaction product melted to a dark liquid and produced a thick foam
as soon as the material temperature reached 150°C (about one minute). 0.72
gram of foam was produced for a yield of 72%. The foamed product was cooled,
dried, and broken up to a fine particle size. The foamed product had a density
of 0.6 lb/ft^3, was insoluble in such powerful solvents as dimethyl formamide,
dimethyl sulfoxide, and hot sulfuric acid, and withstood temperatures of over
600°C without combustion or destruction when exposed to a JP-4 aviation fuel
fire.

A fire of this type exposes the product to a total flux of 10 Btu/ft^2/sec. The
product foam density varies from 0.6 to 15 lb/ft^3, depending on the acid used
and the weight fraction of acid used. This is particularly true in the case when

polyphosphoric acid is used. At a weight ratio of 1.380 grams dioxime to 1.96 grams of acid, the resulting foam is approximately 16 lb/ft^3. The foams for all the reactions are chemically inert to most common solvents such as acetone, petroleum ether and benzene as well as the more exotic solvents like dimethyl sulfoxide, dimethyl formamide, and hot sulfuric acid. The foams produced in the manner described are black, coherent and mainly closed cell. The foams have compressive strengths varying from $\frac{1}{2}$ to about 10 psi, depending on the density and condition of preparation.

Other intumescent compositions were prepared by the procedure of Example 1 with p-benzoquinone dioxime and different proportions of sulfuric acid, phosphoric acid, and polyphosphoric acid. Tables 1, 2 and 3 below set forth typical examples of such compositions and the corresponding intumescing temperature and char (foam) yield obtained by heating the respective samples in a thermogravimetric analysis (TGA) apparatus at 3°C/min under nitrogen.

TABLE 1: p-BENZOQUINONE DIOXIME-SULFURIC ACID INTUMESCENT COMPOSITION

Sample	Weight (g)	Composition	Mol Ratio	Temperature* (°C)	Foam Yield* (%)
1	1.38	D**	0.5	138	72
	1.96	H_2SO_4			
2	1.38	D**	1.0	111	52
	0.98	H_2SO_4			
3	1.38	D**	1.5	108	49
	0.66	H_2SO_4			
4	1.38	D**	2.0	110	50
	0.51	H_2SO_4			
5	1.38	D**	2.5	111	50
	0.39	H_2SO_4			

*Intumescence temperature and foam yield as determined by TGA at 3°C/min under N_2 atmosphere.
**p-Benzoquinone dioxime.

TABLE 2: p-BENZOQUINONE DIOXIME-PHOSPHORIC ACID INTUMESCENT COMPOSITION

Sample	Weight (g)	Composition	Mol Ratio	Temperature* (°C)	Foam Yield* (%)
1	1.38	D**	0.25	110	80
	4.62	85% phosphoric acid			
2	1.38	D**	0.50	105	76
	2.31	85% phosphoric acid			
3	1.38	D**	1.0	115	67
	1.15	85% phosphoric acid			

*,**Same as Table 1.

TABLE 3: p-BENZOQUINONE DIOXIME-POLYPHOSPHORIC ACID
INTUMESCENT COMPOSITION

Sample	Weight (g)	Composition	Temperature* (°C)	Foam Yield* (%)
1	1.38	D**	112	78
	0.98	Polyphosphoric acid		
2	1.38	D**	115	75
	1.96	Polyphosphoric acid		
3	1.38	D**	115	75
	0.51	Polyphosphoric acid		

*,**Same as Table 1.

Note: Polyphosphoric acid is commercially available and has the following general formula: $H_{n+2}P_nO_{3n+1}$ where n ranges from 2 to 4.

It is believed that the fire-resistant, heat-insulating foamed compositions obtained by heating the intumescent compositions are stable heteropolymers of the class commonly known as semiladder and ladder polymers. Such polymers are formed by condensation and hydrolysis reactions of ring or ring-forming reactants. The end product may be a heat-resistant polyquinoxaline-type polymer or a polyphenoxazine-type polymer.

Coated Hollow Polystyrene Granules

R. Heller; U.S. Patent 3,640,787; February 8, 1972 describes a method of producing shaped bodies of low specific gravity which comprises the steps of wetting a mass of discrete roundish hollow granules with a hardenable, preferably heat-hardenable, liquid binder material so as to cause substantially complete wetting of the outer surfaces of the granules and the formation of a coating of the liquid binder material. This is followed by mixing the thus-formed mass of wetted granules with a pulverulent solid material so as to adhere particles of the solid pulverulent material to the liquid binder coating on the hollow granules, in such a manner as to form on each of the hollow granules an outer shell which will consist of the coating of hardenable binder material and particles of the solid pulverulent material adhering thereto and at least partially embedded therein, and which outer shell encloses a nucleus constituted by the respective hollow granule.

Upon subsequent hardening of the hardenable binder material of the outer shell, granules are obtained which form a mass of shaped bodies of low specific gravity consisting of a plurality of roundish hollow bodies or granules, each comprising a hard outer shell consisting of the hardened binder material and the pulverulent material adhering thereto, and also including the plastic material of the initial hollow granule, for instance a foamed or blown polystyrene granule. The material of the polystyrene granule, or the like, may adhere as the thin inner layer to the inner surface of the outer shell, or may be disposed in the interior of the outer shell in some other manner, however, in any event only partially filling such interior space.

The cell walls of such cellular structure preferably are relatively thin down to about 0.05 mm, and the pulverulent material incorporated in the cell walls preferably has a maximum dimension somewhat smaller than the thickness of the cell wall and may be as small as about 0.01 mm. The individual hollow granules, such as swelled polystyrene granules, which are initially subjected to being coated with the liquid hardenable binder material and the pulverulent material, may be of any desired size, but very good results are achieved with hollow granules having diameters between 2 and 8 mm.

The flowable granular mass of the process as described above has the following characteristics: (a) low weight per unit of volume which generally ranges from a maximum of 300 kg/m³ to a preferred weight of 100 kg/m³; (b) considerable pressure resistance and shape retention of the individual granules which will prevent destruction of the same upon compression, transportation and mixing of the granular mass, for instance when incorporating the same into a concrete-forming mass; (c) resistance against unfavorable climatic conditions such as high humidity, extensive temperature changes and contaminations of the air; and (d) high resistance against microbes, insects and other biological pests.

By way of example, for producing one cubic meter of the completed unitary cellular structure according to the process, there may be required: 12 kg ±30% swelled Styropor granules of 2 to 8 mm diameter; 60 kg ±30% quartz powder of about 0.1 mm particle size; and 12 kg ±30% epoxy resin with required hardener addition. Thus the total weight will be 84 kg/m³ ±30%. The unitary cellular structure may be formed by compressing the binder-powder covered granules at a gauge pressure of about 2 atmospheres.

Carpet Scraps

A process described by *K. Lesti; U.S. Patent 4,029,839; June 14, 1977* involves comminuting the carpet scraps, then mixing them with a binder and pressing them into rigid construction elements. It is possible to use these elements for both sound and thermal insulation, since the carpet scrap filling is ideally suited for these purposes. The difficult problem of disposing of the carpet scraps in an environmentally acceptable manner is completely solved and at the same time a useful and valuable product is obtained.

The carpet scraps are comminuted so that in no direction do they have a dimension greater than 10 mm. The comminution may be such as to reduce them into strips whose length is between 2 and 10 mm. The binder is a foamable synthetic resin. A thermosetting resin may be employed or one which simply sets with time once its constituents are mixed together. In both cases the foaming of the resin will serve to form a compact and rigid element when the mass is hardened in a particular shape, as in a mold or between a pair of belts.

In the process the carpet is first comminuted and then mixed together so that a mass is produced of generally uniform characteristics. The comminuted carpet scraps are then mixed with one component of a foamable synthetic resin, such as a polyol. Thereafter the soaked mass is extruded on a belt and the other component, for instance an isocyanate, is added to the mixture and the mass is pressed between a pair of belts as the resin foams and hardens. The mass may be extruded onto a lower foil and an upper foil may be laid on top of the mass after the second synthetic resin component is added so that the construction element has a pair of decorative or uniform skins. Thereafter the continuously extruded mass is cut up transversely into individual construction elements.

GLASS FIBER BLANKETS AND MATS

INSULATION BLANKETS

Manufacturing Process

A process described by *R.L. Troyer; U.S. Patent 3,979,537; September 7, 1976; assigned to Johns-Manville Corporation* relates to thermal insulating materials and more particularly to a blanket of glass fiber for insulation and the methods of forming the blanket and fabrication of the blanket and its facing.

A thermally insulating body is provided as a blanket of resin bonded glass fibers having major flat faces with longitudinal margins. The blanket is slit longitudinally along one side just a short distance inward from the edge to define an edge strip which is temporarily held in position by the slit being an incomplete or perforated slit. The edge strip is adapted for physical removal as a later step. The edge strip is at least equal in width to the tab to protect the tab from being crushed or displaced relative to that portion of the facing secured to the primary blanket. The edge strip remains temporarily in position to provide a protective body adjacent the tab.

The blanket is formed by bonding fibers which are thermally insulating into a blanket, longitudinally slitting the blanket in a perforating manner to form a primary blanket and an edge strip, preferably with the strip paralleling a margin and securing a tab of sheet material to the primary length of blanket in registry with the edge strip of blanket to allow the edge strip of blanket to protect the tab during handling.

The tab may be reinforced by means such as folding a portion of the facing sheet upon itself or by use of a supplemental tape. When the latter method is used, the blanket with a thin flexible facing sheet and a supplemental tape is to be bonded to the blanket. The facing and tape are simultaneously applied to the blanket with the tape lying between the facing and the blanket in registry with the edge strip already formed by slitting the blanket along its longitudinal margin. The combination is cured in an oven to set the adhesive and then wound on a mandrel of sufficient diameter to prevent the blanket from obtaining a permanent set to enable it to unroll to a flat form. Since the tape is impervious to the adhesive, it masks the facing from the edge strip making the edge strip removable along the slit when the tab is to be exposed. At the same time, the tape beneath the tab overlaps to the primary blanket by a short distance. The tape also reinforces the tab.

The insulating blanket eliminates the need for stapling or gluing of tabs to form a vapor barrier as required in the prior art. Use of the tape and protective, disposable, edge strip-like protective portion eliminates wrinkles of the tab to give a better appearance. The rigidity of the blanket by reason of relative high density eliminates sagging and makes it easier to handle. Appearance is also enhanced because the flat overlapping tab is hardly discernible giving a smooth finish impression.

Referring to Figure 7.1a a manufacturing line in schematic form is illustrated for the manufacture of blanket insulation of fiber glass wool where filaments 10 are formed into elongated glass fibers 12, softened by a flame attenuator 14, and coated with binder by the binder applicator 16 to form a blanket 24 on the chain belt 18. The thickness of the blanket is exaggerated for illustration purposes. The blanket passes between the hot rolls 20 where the surfaces are cured and onto platens 22 where the insulation is cured to the desired thickness.

Spaced-apart rotary knives 26 trim the edges of the blanket while a slotted disc 27, turning against a back-up roll at a peripheral speed equal to the linear speed of the blanket, slits the blanket in a perforated manner to form an easily disposable strip-like blanket portion 54 along one longitudinal margin 62 of the blanket. Any other means may be used just so the edge strip is at least partially severed from the blanket for subsequent physical removal. Immersed in an adhesive bath 30 is a roll 32 having a doctor roll 34 for applying adhesive to facing 36 being fed from facing roll 38.

As the facing with adhesive thereon is advanced toward the blanket, it is joined by tape 40 from roll 42 and both are brought in contact with the blanket, at the point of tangency between the blanket and directional roll 44. The combination is passed through an oven 58 where the adhesive is cured to form a bond between the facing and the blanket and the tape respectively. What has been formed is a rigid blanket of thermal insulation having a thin flexible facing 36. A longitudinal portion thereof overhangs the blanket as a cantilevered tab when portion 54 of the blanket formed by the perforated slit is removed. That tab, reinforced by the tape, can be used to seal butted joints when lengths of the blanket are installed side by side.

Finally, the blanket, including edge strip 54, is wrapped or packaged in rolls 28 about a mandrel 46 to form a convenient handling package. While the line outlined above produces a glass fiber blanket 24 it is to be noted that other materials can be fiberized by the usual means of spinning, drawing, attenuating and blowing for all or part of the insulating blanket. The glass fibers 12 are formed from filaments softened by the flame attenuator 14 to attenuate the filaments into lengths of individual elongated fibers 12. Collection of the fibers on the chain belt 18 results in a build-up to form a blanket on the belt as it is continuously moved around the supporting rolls 48.

FIGURE 7.1: GLASS FIBER INSULATION BLANKET

a.

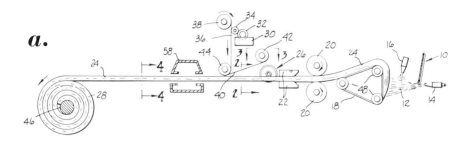

b.

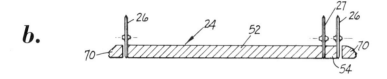

c.

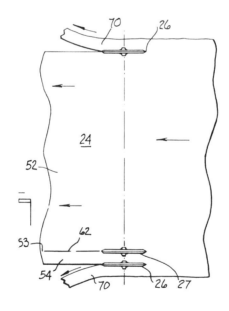

d.

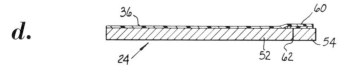

(continued)

FIGURE 7.1: (continued)

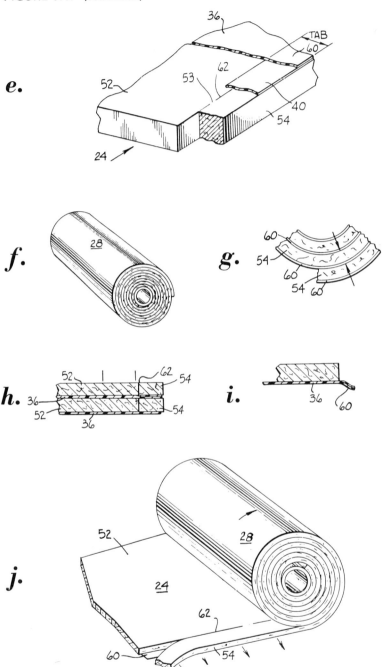

(continued)

FIGURE 7.1: (continued)

k.

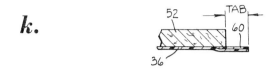

l.

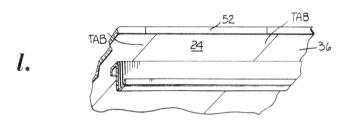

(a) Simplified schematic of an elevational view of a line for manufacturing blanket insulation
(b) Cross-sectional view taken generally along line 2–2 of Figure 7.1a and showing a slitting operation
(c) Plan view of the slitting operation of Figure 7.1b taken generally along line 3–3 of Figure 7.1a
(d) Cross-sectional view taken generally along line 4–4 in Figure 7.1a
(e) Fragmentary perspective view of the blanket insulation with cut-away portions to illustrate the disposable edge strip, tape and sheet
(f) Roll or package of insulating material
(g) Fragmentary end-view of rolled insulating blanket for illustrating relative positions of edge strip and tab
(h) Fragmentary view illustrating relative position of edge strip and tab
(i) Deformed tab
(j) Roll of insulation partially unrolled and showing partial removal of the edge strip to expose the tab
(k) Cross-sectional view of a portion of the blanket and tab
(l) Perspective view of several strips of insulating material over a roof purlin in side-by-side relationship

Source: U.S. Patent 3,979,537

Sufficient adhesion in the binder, applied by the binder applicators as the fibers **12** are attenuated, exists to maintain the fibers in a loose blanket **24**. The blanket is then passed through the hot rolls **20**, which are rotated, to cure the surfaces of the major faces of the blanket by virtue of the heat released from the hot rolls.

In passing between the platens **22** the surface cured blanket **24** is pulled through an opening formed by parallel inner faces of the opposing platens **22** which is smaller than the thickness of the entering blanket. By compressing the blanket as it is pulled past the platens and applying heat from the platens the blanket is given the desired thickness and cured to retain its shape and give it rigidity. In the utilizations contemplated here, densities of the order of 1 to 2 pcf have been employed in thicknesses of from 1 to 3 inches with sufficient binder to avoid any substantial sag of faced material when supported on five-foot centers. Division of blanket **24** into the primary portion **52** and secondary edge strip portion **54**, which becomes a readily detachable strip, is accomplished with a slotted circular saw **27** against a backing roll to leave readily severable bridges **53** of insulating wool between **52** and **54**, see Figure 7.1e.

A thin flexible facing **36** is placed atop the blanket or board over its full width fed from the facing roll **38** as illustrated in Figure 7.1a. Typical of the facings **36** used are unplasticized films such as chlorinated polyethylene and polyvinyl chloride. Other moisture impervious films may be used. An adhesive bath **30** with a roll **32** rotating in a supply of adhesive in the bath is provided to apply a controlled amount of adhesive to a doctor roll **34** which in turn applies the adhesive to the inner face of the facing **36** before it is joined on the directional roll **44** by the supplemental tape **40**, which may be a more flexible vinyl.

The tape reinforces tab **60** and prevents adherence of strip portion **54** to facing **36**. The tape roll **42** feeds the tape onto the directional roll which is rotated to apply the facing and the tape to the blanket at the point of tangency between the roll and the surface of the blanket. Curing of the adhesive to form a bond between the facing and the blanket and between the facing and the tape is accomplished by heating the combination in an oven **58**. The strip portion is not adhered to the facing or the tape. The blanket insulation, now complete with the facing and the tape, is wrapped around a mandrel **46** which is driven by conventional means, not illustrated, by attaching the leading end of the blanket to the mandrel.

The smallest radius of wrap is large enough to prevent a set in the blanket so that the blanket retains its flat shape upon being unrolled. Substantial lengths, for example, in excess of one hundred feet, are wound in rolls **28** and then the blanket is transversely severed by a conventional means, not illustrated, e.g., a shear. The rolled blanket is removed from the mandrel and the new leading edge attached to the mandrel to continue the process. Side strip **54** is temporarily held in position under tab **60** by friction therebetween or by it incomplete severance from the primary blanket, or by both. If severance is complete throughout the margin, friction against the tab alone is normally sufficient to hold the strip in temporary position.

The side strip temporarily protects the tab from deformation by being adjacent to it for providing stiffening or backing. It is effective during packaging, such as placing in a roll and subsequent handling and unrolling for use. The strip is manually removable to expose the tab in cantilevered projection from the primary blanket.

The fabrication and purpose of the blanket **24** when completed with its facing **36** and tape **40** can be better understood by viewing the product as illustrated in Figure 7.1e. A fragmentary portion of the blanket is illustrated with its facing, tape and secondary edge strip portion **54** of the blanket all cut back for illustrative purposes. The primary portion of the blanket **52** in 1½ inch thickness and one pound density having a facing sheet of 4 mil unplasticized chlorinated polyethylene and 3 mil plasticized vinyl tape, will sag only a ¼ inch when supported on five-foot centers because of the rigidity given to the blanket by the curing of the platens **22**. The rigidity enables the primary portion **52** of the blanket to support a tab **60** created by the overhang of tape **40** and facing **36** when the secondary portion **54** or detachable strip of blanket **24** is simply broken away along the longitudinal margin **62**. The tape offers the advantage that it is impervious to the adhesive on the facing and masks the detachable edge strip **54** from the adhesive to maintain it free for removal. At the same time, the tape overlaps the primary portion and is attached by adhesion to the facing lending reinforcement to the tab in the transverse direction as well as the longitudinal direction of the blanket.

The tab **60** is protected from deformation relative to the blanket, e.g., bending or wrinkling, by the detachable strip remaining adjacent thereto up until the time of use. By overlapping the tabs **60**, as illustrated in Figure 7.1l, a moisture barrier between blankets is obtained simply by butting the blankets side by side. It is, therefore, important that the tab be maintained flat prior to assembly in order to form the seal. In addition, the flat tab **60** gives a neat appearance being hardly discernible.

The reinforcing tape **40** in the preferred embodiment is plasticized vinyl which conforms with the expansion or stretching of the facing to prevent wrinkling of the facing and the tab. Other flexible materials could be used. It should be understood that the blanket can be manufactured in widths other than five-foot increments and that, while 1½ inch thickness is preferred, thickness of 1 to 3 inches can be manufactured. The rigidity and, therefore, the sag of the blanket varies with thickness and density of the blanket. A typical density is 1 pcf. Since rigidity increases with an increase in density, the density can be varied to obtain a desired rigidity.

Additional facing second tabs on the margins of the blanket could be produced if desired. A blanket with two tabs could be used in a similar manner to the blankets of the prior art by pulling adjacent tabs of butted blankets together along the edges of the longitudinal margin and stapling the tabs. While the widths of the tab **60** and tape **40** are variable, the preferred widths are 2½ inches for the tab with the tape extending another ½ inch to lap the face of primary portion **52** of the blanket **24** giving it a width of 3 inches. Figure 7.1g illustrates a rolled package with the edge strips rolled. Figure 7.1g is a fragmentary end view showing edge strips **54** rolled in position against tabs **60** for temporarily protecting them by providing a backing or packing mass.

Preferably, edge strips are not completely severed from the main portion **24**, but as indicated the wound arrangement would be sufficient to temporarily retain them in position by friction even if severed. Preferably, strips **54** are slightly wider than the tab for providing more protection for the tab. As illustrated in Figures 7.1b and 7.1c, the very side extremity **70** is trimmed away by rotating saws **26** and is immediately discarded as scrap. Slotted circular saw **27** pressing against a back-up roll severs edge strip **54** in a manner previously discussed. Figure 7.1d illustrates the substantially completed product further down the forming line as illustrated in Figure 7.1a generally along line **4–4** prior to being rolled into a package.

Figure 7.1j illustrates a package of packaged insulation in the process of being unrolled and strip **54** being pulled from its position to expose tab **60**. Without the benefit of strip **54** during handling, it is obvious that endwise force on the tab of the insulation, whether in package form or not, would cause physical deformation as indicated in Figure 7.1i. This, of course, would interfere with its appearance and sealing efficiency. Sagging of a panel of insulation supported on five-foot centers is minor, in the range of ¼ inch. With tape **40** reinforcing tab **60**, sag at the tab will be even less, thereby resulting in the tab being pulled closer onto the adjacent panel.

Advantages of the insulating blanket include its board-like rigidity, its convenient roll packaging and its protected integral tab **60**. There is an elimination of the stapling requirement of the prior art blankets or a separate taping of joints by the combination of a rigid blanket having characteristics of board which holds its shape without stretching. The smooth flat tab can form a substantial moisture barrier by simply overlapping an adjacent butted blanket **24** . The need of additional supports to overcome the sag of blankets of the prior art is also eliminated. The overlap of the tab **60** conceals the butted joint between blankets and gives the impression that a series of blankets joined together are a single smooth surface. Handling of the blanket is also enhanced by the rigidity of the blanket making it easier to handle and position in place.

Hinge-Like Bent Blanket

J. Bondra, Jr., and W.R. Ruhling; U.S. Patent 3,958,385; May 25, 1976; assigned to Metal Buildings Insulation, Inc. describe a blanket structure in which the lamination of the facing to a semirigid fiber mat is performed with an adhesive which does not set until after the laminate is rolled for shipment or the like. In accordance with the process, the facing material is coated with a liquid adhesive; for example, a water-based latex adhesive is pressed into contact with a semirigid mat and is rolled before the adhesive dries or sets. The blanket structure is arranged, however, so that when the blanket is unrolled and assumes its flat uncompressed position, the facing is smooth.

In an illustrated example of the process, the semirigid mat is laterally cut from the face of the mat opposite the face on which the facing is laminated at regular intervals. The cuts preferably extend toward the laminating face at least two-thirds of the thickness of the mat so that the mat tends to bend in a hinge-like manner about a center of bending substantially adjacent to the facing. The blanket assembly including the laterally cut mat and the facing are rolled for handling and shipment before the adhesive at the interface between is allowed to set.

However, because of the hinge-like bending of the blanket during the rolling operation, there is little or no sliding action tending to cause an uneven surface on the facing when the blanket returns to its normal flat condition. Upon being released from the roll, the blanket springs back to its flat condition and the cuts are closed. The assembled blanket provides sufficient rigidity to assure that it remains in a flat and nonsagging position when it is installed even though stretching of the blanket is not required.

Decorative Facing

A process described by *G.A. Hoffmann, Jr.; U.S. Patent 3,835,604; September 17, 1974; assigned to Certain-Teed Products Corporation* relates to building insulation of the type embodying a strip of fibrous blanket and a facing sheet having edge or lip portions projecting beyond the blanket edges. This general type of insulation material is extensively used in the insulation of houses and buildings. Thus, the process provides building insulation of the general kind referred to with a facing sheet having a decorative pattern, so that the installed appearance of the insulation is aesthetic or attractive. Indeed it is contemplated that the facing sheet have a decorative pattern of such character that it will itself provide a wall or ceiling decoration, in view of which, in at least many installations it is not necessary to apply any additional wall covering material, such as wallboard or paper or other finishes commonly employed over building insulation. Complete manufacturing details are provided.

GLASS FIBER MATS AND OTHER PROCESSES

Fibrous Mat Production Apparatus

A process described by *G.N. Bolen, S.C. Dunbar and G.E. Smock; U.S. Patent 3,936,558; February 3, 1976; assigned to Owens-Corning Fiberglas Corporation* relates to fibrous bodies, to the manufacture of fibrous bodies and to applying binders to fibrous products. More specifically, the method relates to a process for bonding multifilament strands in a body of desired form, in which the filaments are of fibrous glass and are gathered into strands immediately subsequent to the filament forming operation, the strands then being collected and arranged in a body.

The fibrous mat production apparatus of Figure 7.2a includes a portion of a series of molten glass feeding bushings 21, depending from conventional glass melting tanks which are not illustrated. Continuous filaments 23 are drawn from the minute orifices of the bushings. For the purpose of being specific, it will be considered that there are 438 orifices in each bushing and the filaments drawn therefrom have an average diameter of 0.00068 to 0.00070 inch. A forming size is applied to the filaments as they pass over carbon applicator rolls 24 of the conventional size applicators 25. If the fibrous mat is to be used as reinforcement in an electrical grade laminate, a preferred composition of the forming size is shown in the following Example 1.

Example 1:

Ingredient	Percent by Weight of Total Size
Weak acid	0.01-0.10
Lubricant	0.01-0.10
Coupling agent	0.10-2.0
Water	remainder

Example 2: A specific size formulation found to give good results when a polyester resin emulsion is used in the binder is as follows.

Ingredient	Percent by Weight of Total Size
Glacial acetic acid	0.03
Pelargonic acid-tetraethylene-pentamine condensate	0.030

(continued)

Ingredient	Percent by Weight of Total Size
Caprylic acid-tetraethylene- pentamine condensate	0.01
γ-methacryloxypropyltrimethoxysilane	0.20
Water	remainder

The pH is kept 4.0-4.5

The filaments from each bushing, after the forming size is applied are grouped together to form a number of strands which are individually segregated as they travel within grooves over the respective gathering shoes 27a, 27b illustrated in Figure 7.2a. The primary division of the filaments into strand groups may be accomplished manually at the start of production.

The showing of six bushing positions is representative of a twelve bushing production line apparatus to form the preferred example of the product of this process. The first two and the last two bushings of the series of twelve bushings deposit the bottommost layers and topmost layers, respectively, of a mat 60 on the collecting surface of a conveyor 61. The filaments of each of the first two and the last two bushings are preferably split or divided into twenty-two strands with approximately twenty filaments per strand. The filaments of each of the intermediate or middle eight bushings are preferably split into fourteen strands with twenty-nine to thirty-one filaments per strand.

The sets of strands 28 from the first two and the last two bushings in the series pass down around grooved guide or aligning shoes and idler wheels before being drawn over pulling wheels 35 and projected downwardly therefrom. Similarly, the sets of strands 29 from the remaining eight intermediate bushings reach pull wheels 36 and are projected therefrom. The pull wheels are driven by motors 38 arranged in pairs between adjacent pull wheels longitudinally of the production line.

The traction between the strands and the surface of a pull wheel furnishes the pulling force that attenuates the glass filaments formed from the minute molten glass streams issuing from the orifices of the furnace bushing. This adherence of the strands to the pull wheel is evidently due to the cohesive effect of the forming size carried by the strands supplemented by other air and surface forces of attraction.

The strands projected from the pull wheels are deposited upon the foraminous conveyor 61 and accumulate to form a continuous mat 60. The conveyor, or subsequent conveyors to which the mat is transferred, carries the mat through a selective binder application area to an oven 65 for curing of the binder components on the filaments and strands of the mat. The cured mat issuing from the oven may have its edges trimmed, be inspected and then packaged at a suitable packaging station if immediate use of the product is not desired.

In the vertical cross section of Figure 7.2b there is shown in elevation the first pair of bushings 21 of Figure 7.2a and the apparatus associated therewith including the pull wheels 35 for depositing the strand on a conveyor. The pull wheels with elements of the apparatus cooperating therewith are shown in enlarged form in Figures 7.2c and 7.2d.

FIGURE 7.2: FIBROUS MAT PRODUCTION APPARATUS

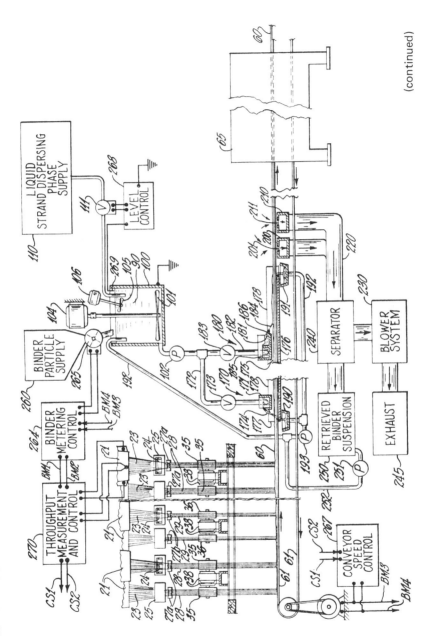

(continued)

a.

FIGURE 7.2: (continued)

b.

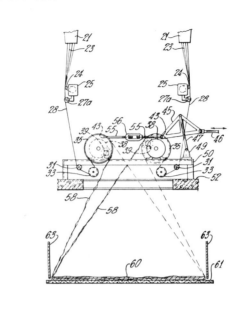

c.

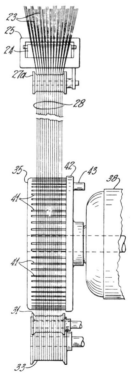

(continued)

FIGURE 7.2: (continued)

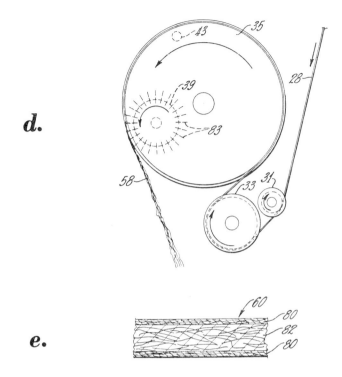

d.

e.

(a) Side elevational view of a portion of the production line
(b) Enlarged vertical cross section of the production line of Fig-
       ure 7.2a showing two pull wheels and associated apparatus
       for gathering filaments into strands and projecting the strands
       upon a conveyor
(c) Enlarged side view of one of the wheels of Figure 7.2b receiv-
       ing strands from a gathering shoe
(d) Front view on the same scale as Figure 7.2c of the pulling
       wheel
(e) Section of fibrous body

Source: U.S. Patent 3,936,558

As the wheels **35** and their associated apparatus are quite identical in structure
and function, the description of the wheels will generally apply to the apparatus
including wheels **36** for the intermediate bushings. From the guide or aligning
shoe **31**, which is grooved similar to the gathering shoe **27a** for maintaining the
strands separate and in spaced relation, the strands **28** are led around the idler
wheel **33** and over and around pull wheel **35**. The strands are released from the
pull wheel at a moving point reciprocating along an arc of the peripheral path
of the wheel. The release is effected by the successive projection of fingers **83**
of the oscillating spoke wheel **39** located within the pull wheel, through slots

41 in the cylindrical surface of the wheel. The strands are thus kinetically projected downwardly tangentially from the wheel and in a path moving back and forth across the conveyor **61**. The rear side of each pull wheel is covered by an independently mounted, oscillatable back plate **42** (see Figure 7.2c) on which the associated spoke wheel is carried. The back plate of the assembly including pull wheel **35** is arcuately oscillated through a rearwardly projection post **43**. The latter is driven by functioning of a fluid cylinder or other mechanism (not illustrated) which acts through the triangular link **45**, which pivots upon bar **47** mounted on the base **49** as shown in Figure 7.2b.

The rod extending from the cylinder or other activating mechanism is joined to the triangular link **45** by linking rod **46**. The base is positioned on the platform **50** which also supports the other wheel **35**, motors **38** and other equipment. The platform rests upon an operator floor **52**. Through the connecting assembly **55**, including the turnbuckle **56**, the transverse movement of the triangular link **45** is transmitted to a post **43** to also arcuately oscillate the back plate and spoke wheel **39** within the other pull wheel **35**. With a single means effecting the oscillation of both spoke wheels their action may be closely synchronized.

The group of strands **58** thrown down by the pull wheels **35,35** and the strands from the pull wheels following this pair are accumulated in mat form upon the traveling conveyor **61**, which may be a carbon steel chain construction. Side shields **63** define the edges of the mat **60** and prevent undesirable lateral overreaching of the strands.

The width of the conveyor covered by the mat in the case may be varied through a wide range of changing the oscillating arc length of the spoke wheels and the distance of the pull wheels above the conveyor. The side shields are adjustably mounted so that their spacing may be altered to match the desired width of the deposited material. For example, the width may be varied between limits of I4 to 84 inches.

The width of the pull wheels may be varied to accept a higher or a lower number of strands, the slots being made proportionately longer and the exterior portions of the fingers also made proportionately wider. The movement of the fingers **83** into the slots **81** and their momentary projection through the slots to release the strands is synchronized through a timing drive between the pull wheel and the spoke wheel. This may include a toothed pulley fixed upon the hub of the pull wheel, a cog timing belt running between this pulley and a pulley on a shaft upon which the spoke wheel is journaled.

The distance of the pull wheels above the conveyor, and the rotational speed of the wheels are so selected, in relation to the specifications of the strands being deposited, that the strands are projected with sufficient force to carry them as a band of generally constant form and in substantially regular paths to the surface of the conveyor or other collection surface.

Each group of strands is thus deposited in a reciprocating strip across the conveyor in a constantly repeating pattern and with substantially stable dimensions. A large range of relationships can be established between the strips laid by various pull wheels in the system shown, but any one product may be reproduced uniformly by locking the system into the dynamic relationship which has been found to produce the particular mat construction desired.

Referring to the sectional view of the mat produced as shown in Figure 7.2e it can be seen that the mat **60** incorporates upper and lower surface layers **80** of light strands and a central body portion **82** of heavier strands. Since each bushing supplying 438 filaments, the upper and lower layers **80** comprise two layers of 22 strands each from the first and second bushings and two layers of 22 strands each from the last two bushings in the series of twelve bushings. The intermediate body portion **82** includes eight layers of 14 strands each. As noted above, the strands in the upper and lower surface layers contain approximately 20 filaments per strand, whereas the strands in the central body portion **82** contain 29 to 31, or approximately 50% more filaments per strand.

However, there are more strands per unit area in each of the surface layers than in the central body portion. Thus, the smaller strands in the surface layers not only are able to lie more closely together to form smaller interstices because they are smaller in diameter, but there are more strands deposited per unit area so that a great many more interstices are formed than in the central body portion. This enables better particle entrapment capabilities when the mesh size of the surface layers, particularly the upper surface layer **80**, is changed by the process.

As the mat **60** leaves the strand deposition position area, it is conducted through a binder application area. At a first liquid impingement or flooding station **170**, a liquid suspension material **174** (preferably comprising solid binder particles suspended in a liquid strand dispersing medium) is distributed evenly across the mat-like collection of strands by a weir means **171**. A supply line **172** supplies liquid to the weir means from a suspension mixing tank or sump **100** via a pump **102**. A valve **173** may be utilized to control the flow of the liquid suspension to the weir and thus the amount of liquid impinging upon and flooding the strands in the mat-like mass **60**.

The liquid suspension **174** collects on a liquid-retaining means, such as a plate **176**, in a flood condition as noted at **175** to inundate the mat **60** completely. An end **177** of the liquid-retaining means **176** may be opened to enable the flow of excess suspension material **174** into a collecting trough **190** at the left side of the liquid-retaining plate **176**. Upwardly extending side walls or plates **178** are provided to prevent a flow transverse to the direction of travel of the mat and the stream formed by the flooded area **175**. The side plates or walls advantageously are spaced closely to the edges of the mats **60** to prevent flow of the suspension material **174** out through the edges of the mat.

The prevention of edge flow will maintain the dimensional qualities of the mat and will prevent a nonuniform binder distribution in the edge areas. The side walls **178** may be extended past the termination of the plate **176**, as noted at the right of plate **176** in Figure 7.2a, so that draining of the flood stream **175** from the mat will occur only down through the mat rather than out through the edges.

A flood stream flow will occur in the flooded area if the liquid suspension is provided at a rate to flood the mat completely and to move along therewith. A sufficient flow is advantageously provided by regulating the valve **173** so that the flooded area will become a stream moving at substantially the same rate and in the same direction as the mat. The strands in the mat will, therefore, not be disturbed from the orientation provided by the deposition apparatus. A second trough or catch basin **191** at the right of the liquid-retaining plate **176** catches the flood stream **175** as it drains down through the mat **60** as the conveyor **61** and the mat pass the right end of the plate **176**. Conduits **192** and the pump **193** cooperate to return the suspension materials collected in troughs **190** and **191** to the suspension mixing tank or sump **100**.

The length of the liquid retaining plate **176** and thus the interval during which the mat is maintained in the flooded condition is determined in response to the speed that the mat is moving through the flooded area and the soaking time necessary for the strand dispersing medium in the suspension materials to overcome or substantially overcome the forces holding the filaments together in strand form. The soaking interval is calculated or based upon the forces holding the smaller strands together that are lying in the upper surface of the mat since, as will be noted hereinafter, the dispersion of the strands in the upper layer **60** is a critical factor in changing the mesh size to provide a higher binder percentage content in the upper surface layer.

With some strands it may be possible through soaking alone to disperse the strands in the upper layer before the strands disperse in the immediate or central body portion of the mat **82**, depending upon the number of filaments in the strands and the relative treatments provided when the strands of the upper layer and central body portion are formed and deposited. But, the preferred and the most effective dispersal of filaments of the strands in the upper layer in the mat **60** may be accomplished by using a second liquid impingement station **180**.

The second liquid impingement station is spaced from the first impingement and flooding station **170** a distance, depending upon the speed of the conveyor **61**, which is adapted to provide a predetermined soaking or bond weakening interval. The second impingement station includes a weir **181** supplied via a supply line **182**, the amount supplied being controlled by a valve **183** connected between the pump **102** and the weir **181**. If desired a second weir **185** may be interposed between the supply line **182** and the weir **181** to provide a double weir construction which will reduce turbulence when high flow rates are utilized.

Apparatus for preventing or removing foam, such as bars placed across the surface of the liquid suspension in the weirs **185, 181,** may be utilized to prevent flow of foam into the flooded upper surface of the mat **60** and an interference with the strand dispersing and binder distribution operation being performed. Similarly, foam prevention or removal apparatus may be used with the weir **171**.

The control valve **183** in combination with the construction of the forward lip **186** of the weir **181** combine to provide a regulated predetermined forward velocity of the impinging stream **184** with respect to the mat **60** and the flood stream **175**. It is desirable to provide the impinging stream with a slightly higher forward velocity vector than that of the flood stream **175** and the mat for most effective strand dispersal in the upper layer.

A major portion of the excess suspension materials in the mat is removed by gravity drainage from the mat into the collecting trough **191** at the right of the liquid retaining plate **176** as the conveyor **61** and the mat clear the end of the plate **176**. Further excess liquid suspension materials are removed from the mat by passing the foraminous conveyor over a plurality of suction chambers **200, 210**, having suction slots **201, 211** formed in the upper side. A blower system

230 connected to the suction chamber 200, 210 by a conduit 220 and via a separator unit 240 causes a reduced pressure in the chambers and a positive flow of air downwardly through the mat 60 to extract excess liquid suspension materials therefrom. The separator unit separates liquid and particles that may be entrained in the air stream and diverts the suspension materials in the air stream to the retrieved suspension storage tank 250. The cleaned air exits from the separator 240 through an exhaust system 245. A pump 251 and a conduit 262 returns the retrieved suspension materials to the suspension mixing tank 100.

The powdered binder or binder particles are added to the liquid carrier at a rate equal to the rate of deposition of the binder particles from the suspension onto the mat. This may be accomplished by sensing the throughput of the twelve bushings by a throughput measurement and control unit 270. The control unit may provide a signal of leads BM1 and BM2 to a binder metering control unit 264 to regulate a binder feeding device 265 to control the amount of powdered binder put into the tank 100 from a particle binder supply 260 to maintain the percentage of binder particles supplied to weirs 171 and 181 substantially constant.

The throughput measurement performed by the unit 270 is accomplished by sensing the amount of heat supplied to the bushings 21 to maintain the molten glass therein at a desired attenuating temperature. If the amount of fibers being attenuated is greater, then the heat requirements are greater and a measurement of the throughput of the bushings is obtained.

It is also desirable to control the line speed of the conveyor 61 so that a uniform mat thickness is collected thereon. This may be accomplished by providing signals on leads CS1 and CS2 from the throughput measurement unit 270 to a conveyor speed control 267. This signal may be the same as or proportional to the signal applied on leads BM1 and BM2, since the production of a greater amount of fibers should cause the conveyor speed to be greater to maintain a uniform thickness of mat collection.

Since the rate retention of binder particles by the mat-like collection will depend upon the rate of fiber production and/or the rate of movement of the collection, signals representing the line speed control 267 are applied to the binder metering control 264 via leads BM3 and BM4. The dispersing medium 90 in tank 100 is constantly agitated by a stirrer 101 driven by a motor 104 to prevent the binder particles from settling out. In addition, an agitator 105 driven by a motor 106 is placed adjacent the point of initial contact of the binder particles with the liquid to promote a fast wetting of the solid binder particles.

The dispersing medium is provided from strand dispersing medium supply 110 via a valve 111 to the binder and agitation tank 100. To prevent settling out of the powdered binder or binder particles anywhere in the system, agitation is provided throughout the system by pumping large amounts of the liquid suspension from the tank through the binder application area in the closed loop system shown so that high rate of flow in all of the conduits and associated apparatus will prevent settling or separation.

The flow of suspension material from the supply 110 is regulated by a level control unit 268 which is responsive to the level in tank 100 as detected by a probe 269, to regulate the opening and closing of the valve 111. The unit 268 thus maintains a desired quantity or level in the tank.

As described, the mixing and application system is of the closed loop nature so that all materials are completely and efficiently used. There is no loss as a result of drainage, spillage, inefficient application, powder fall-out, dry powder fall-through, or air currents.

In summary, the binder application area includes a flooding weir **170** for applying binder (solid binder particles in suspension) to the mat **60**. The suspension stream **174** from the weir **171** preferably is flowed substantially vertically downwardly from the weir to completely inundate the mat. The fluid suspension strikes with sufficient force to flatten out all mole hills, tunnels and like irregularities which cause nonuniformities and which may have occurred in the deposition of the strands on the conveyor, but does not disturb the fiber orientation and does not affect mat uniformity. The no-load thickness of the mat is reduced by the flooding weir and the positive air flow therethrough during suspension extraction.

A soaking section is provided after the flooding weir **170** where the completely submerged mat is allowed to move with a pool of binder and strand dispersing medium at the same velocity so that all of the strands soak for a predetermined interval. The soaking interval may vary with the number of filaments in the strand and other conditions. It has been found that a soaking interval of as little as seven seconds may be sufficient.

The purpose of the soaking section is to break down any bonds or other forces that may be holding filaments together in strand form in the layer or layers that are to be filamentized. The excess binder and strand dispersing medium flowing from both ends of the soaking section over the ends of the plate **176** is collected in the catch basins **190, 191** and sent back to the sump mixing tank **100**.

A dispersing weir applies and impinges more binder and strand dispersing medium onto the top of the mat and the flood stream, after the soaking interval has expired. The fluid preferably flows from the weir **181** so that the horizontal vector of the impinging stream is substantially greater than the vertical vector thereof so that the impinging stream has a slightly higher velocity than that of the flood stream **175** and the mat **60** for most effective dispersing capability. The strands in the upper layer are therefore at least partially dispersed into their individual filaments to change the mesh size to a value which will mechanically entrap binder particles in the upper portion or surface layers.

In the preferred case, this is accomplished by the combination of providing strands in the upper layer which have fewer filaments per strand and thus are more readily reopened and by providing more of the strands per unit area in the upper layer so that the interstices of the strands lie closer together. This combination encourages a rapid change in mesh size to retain or entrap the binder in the upper layer before a mesh size change occurs in the central body portion that would retain more binder therein or substantially inhibit the flow of the binder particles out of the central body portion.

Some binder is, of course, desirably retained in the central body portion of the mat, whether liquid and/or particulate binder, by surface tackiness of strands and filaments in the central body portion and the lodging of the particles of the crossovers or intersections of the large strands, even though there are fewer strands per unit area in the central body section.

Since the bottom layer is also composed of strands having fewer filaments per strand, and since there are more strands per unit area, the soaking interval and the contact with the strand dispersal medium may be adjusted to overcome the forces holding the filaments together in strand form before a similar reaction occurs in the central body portion. Then, as the water or dispersant is draining out of the mat **60** into the catch basin **191** through the bottom layer, it is believed that a hole seeking flow may occur through the smaller interstices of the bottom layer which will cause at least a partial dispersement of the filaments of the strands in the bottom layer by filaments being carried toward the holes by the flow.

If a positive mechanical dispersing action is desired, an array of nozzles or a slit nozzle may be provided across the bottom of the plate **176** and connected to the sump **100** via a suitable flow regulator, to provide a flow which will impinge the lower surface of the mat **60** through the foraminous conveyor in the same fashion as the upper layer is dispersed.

There may be thus created a mesh size change which may not, depending upon the number of filaments per strand and the forces holding the strands together, change the mesh size of the bottom layer as much as the mesh size of the top layer is changed. All of the binder particles therein or going through because of the draining action may not be entrapped in the bottom layer.

Whether filamentized or not, the bottom layer at least provides substantially more intersections to filter some binder out of the draining flow from the central body portion and the bottom layer itself. Thus, there is a higher percentage binder content in the bottom layer than there is in the central body portion, although the binder content of the bottom layer may not be as high as that of the top layer. The higher binder content in the bottom layer will provide the mat with excellent surface properties and enhance the handleability of the mat.

It can be seen then, that one of the major principles of the process lies in circulating binder uniformly throughout a mat-like mass and then selectively changing the binder retention capabilities of one or more areas in the mat to enable the areas having a changed retention capability to have a higher binder content than the remaining areas of the mat. This is most effectively accomplished by insuring at least a partial positive dispersal of the strands with respect to the top layer of the mat.

In the preferred case, the overall mat has approximately an 80 mesh porosity so that binder particles of 100 mesh size or higher will tend to circulate or be distributed freely throughout the mat of the mass. When the mesh size of the mat is changed, as by the application of the filament dispersing impingement stream **184** from the weir **181**, the upper layer of the mat surface changes mesh size so that its porosity is at least 100 mesh, thereby entrapping 100 mesh size particles in the upper layer.

When using the product of this process as a reinforcement for products, the following improvements over other commercially used mats have been noted. The wet-through of the impregnating material is much faster. The forming press may be closed at a faster rate. No special pour pattern from the impregnating matrix is required. The higher porosity prevents dry spots or wash. The even horizontal binder distribution avoids the occurrence of soft edges. The wet-out by the impregnating matrix is better. There is less surface fiber prominence as a result of the finer strands. Heavily filled resins may be used. When the reinforced product is used as circuit mounting boards, the laminates have better punchability properties with less crazing and cracking in the circuit board surfaces and interior. The lower overall binder content, as compared to past reinforcing mats similarly used, enables the provision of more reinforcing glass fibers per unit area in the final product. The lower binder content and the coarser strands in the central body portion improve the moldability of the mat.

Colloidal Silica and Silicone Polymer as Binder

A process described by *H.A. Clark; U.S. Patent 3,944,702; March 16, 1976; assigned to Dow Corning Corporation* provides a coherent three-dimensional structure comprising a plurality of contiguous fibers, the said fibers being bonded at their points of contact with a composition consisting essentially of 10 to 75 weight percent colloidal silica uniformly dispersed in a matrix of 25 to 90 weight percent $RSiO_{3/2}$ in which R is selected from the group consisting of alkyl radicals of one to three inclusive carbon atoms, the vinyl radical, the 3,3,3-trifluoropropyl radical, γ-glycidoxypropyl radical and the γ-methacryloxypropyl radical, at least 70 weight percent of the matrix being $CH_3SiO_{3/2}$.

Regarding materials to be utilized as thermal insulation, a preferred structure is that of fiber glass nonwoven mats (batts) bonded with a composition comprising 30 to 50 weight percent SiO_2 (colloidal silica) and 50 to 70 weight percent $CH_3SiO_{3/2}$, the binder loading of such a structure being in the range of 3 to 25 weight percent. When utilized with glass fibers, this specific binder composition gives optimum retention of strength at elevated temperatures while being essentially nonflammable.

Example 1: One gram of glacial acetic acid was combined with 50 g of an aqueous dispersion of 13 to 14 mμ colloidal silica (30% solids and 0.32% Na_2O) after which 30.4 g of methyltrimethoxysilane were added. After mixing for one hour, the binder composition was filtered. The composition contained 50 weight percent SiO_2 and 50 weight percent the partial condensate (calculated on the weight $CH_3SiO_{3/2}$) in a methanol-water medium. After aging for two days to ensure formation of the partial condensate, portions of the composition were diluted with isopropanol to produce a series of binders containing from 0.4 to 3.2 weight percent solids having a pH in the range of 4.5 to 5.2.

Six specimens, approximately 100 mm wide x 305 mm long x 27.7 mm thick were cut from glass fiber mat which had been heat cleaned for 72 hr at 350°C to remove the starch sizing. The mats had a density of about 0.25 pcf. Five of the fiber glass mats were coated with the isopropanol diluted compositions by spraying until the mat appeared thoroughly wet. The sixth portion was used as an untreated control. The above-described fiber glass mats were placed between metal plates separated by 27.7 mm spacers and cured for 16 hr at 85°C. After removal from the curing oven, each specimen was weighed.

To determine the degree of bonding, the thickness of each mat was measured under a compressive load of 0.62 g/cm^2. A metal plate weighing 190 g and measuring 100 mm by 305 mm was placed on the cured mats and six measurements were made around the perimeter to obtain an average thickness. To determine if the cured binder deteriorated at high temperatures, the coated specimens were placed in a 305°C oven under a load of 0.62 g/cm^2. After 100 hr, the compressed mats were removed from the oven, weighed and the thickness was again measured. Results of this testing are tabulated below.

Fiberglass Mat	% Solids in Coating Composition	Mat Weight (g)		Mat Thickness (mm)	
		After 16 hrs. at 85°C.	After 100 hrs. at 350°C.	After 16 hrs. at 85°C.	After 100 hrs. at 350°C.
No. 1	uncoated control	3.3	—	2.0	—
No. 2	0.4	3.3	3.3	3.0	1.3
No. 3	0.8	3.0	3.2	5.7	1.8
No. 4	1.6	4.5	4.5	12.8	7.7
No. 5	3.2	5.4	5.3	19.2	14.2

One heat-aged mat No. (5) was compressed to 50% of its thickness and held for about two minutes at room temperature. When the compressive force was removed the mat returned to greater than 95% of its previous thickness. These data demonstrate that strong coherent glass fiber structures resistant to high temperature can be obtained by this process.

Example 2: Four different binder compositions containing from 25 to 50 weight percent partial condensate (calculated as % $CH_3SiO_{3/2}$) and 50 to 75 weight percent SiO_2, based on the solids content, were prepared by adding the appropriate amounts of methyltrimethoxysilane to portions of an aqueous dispersion of colloidal silica containing 34% SiO_2, less than 0.01 Na_2O having an average particle size of 16 to 22 mμ diameter to which acetic acid had been previously added. The compositions were shaken for one hour after which the pH of each was determined to be in the range of from 4.0 to 4.5. After aging for four days, the compositions were diluted to 4 weight percent solids by addition of isopropanol.

Heat-cleaned glass fiber mats, similar to those described above, except 56 mm thick, were weighed, then sprayed until thoroughly wet with one of the binder compositions. A fifth mat was uncoated and used as a control. The mat specimens were placed between metal plates separated by 56 mm spacers and cured for 15 minutes at 232°C. After curing the specimens were weighed, binder content calculated and thickness was measured under a 0.62 g/cm^2 load as described in Example 1. The coated mats were heat-aged for 100 hr at 350°C under compression as previously described. Results of this evaluation are tabulated below.

Mat	% $CH_3SiO_{3/2}$ in Binder	Binder Loading - %	Mat Weight (g)		Mat Thickness (mm)	
			After Curing	After Heat Aging	After Curing	After Heat Aging
No. 1	control	none	6.1	—	4.3	—
No. 2	25	18.0	7.2	7.1	15.5	8.7
No. 3	30	10.6	7.3	7.4	30.0	21.3
No. 4	40	12.9	7.0	6.8	32.5	21.3
No. 5	50	15.6	7.4	7.3	41.8	28.7

These data demonstrate that while the lower monomethylsilsesquioxane content (25%) gives some retention of strength at elevated temperatures, a monomethyl content of 30 weight percent and above give substantial improvements in retained strength. Two of the above specimens (No. 4 and No. 5) were held over the direct heat of a laboratory burner. Although the glass fibers melted away from the heat and a silica deposit formed on the fibers due to the decomposition of the binder, there was no smoke nor was a flame emitted. When the specimens were removed from the burner flame, no residual flame was observed.

Curing Oven for Mineral Wool

According to a process described by *B. Lundström; U.S. Patent 4,028,051; June 7, 1977; assigned to Jungers Verkstads AB of Goteborg, Sweden* a curing oven for impregnated mineral wool in the shape of mats comprises a pair of endless cooperating conveyors facing each other and being arranged in a housing and on either side of the cooperating parts of conveyors several pressure and evacuation chambers are located for feeding hot air through the mat.

The pressure and evacuation chambers are arranged in such a way that the air can be brought in mutually reverse directions through the mat and they are designed with a decreasing cross-sectional area in the direction away from the induction and evacuation ports. The interior spaces of the housing located outside the pressure and evacuation chambers are kept at a lower pressure than the ambient pressure by a pump unit.

Overall the process provides a curing oven in which an uniform aerodynamic air distribution is obtained along the whole mat width and length. The air can be brought in opposite directions, up-down, down-up through the mat. The heat leakage from the oven is minimized by maintaining a negative pressure and furthermore the oven is insulated. Since a pressure-supporting girder construction is used between the conveyor chains, the construction is so sealed, that there will be no hot air flow towards the conveyor chains and these will be kept at a moderate temperature.

The conveyor chains are led outside the hot zones and in this way less wear and an efficient lubrication is achieved, besides which the energy consumption will be lower. It is also possible to blow cold air along the chains. Furthermore, the process provides a curing oven with a greater capacity and a lower power consumption since a smaller quantity of hot gas is required in comparison with conventional curing ovens. Complete construction and operational details of the oven are provided.

Hard Granules and Resin-Coated Fibers

A process described by *C. Jumentier and A. Bonnet; U.S. Patent 3,745,060; July 10, 1973; assigned to Compagnie de Saint-Gobain, France* involves the homogeneous distribution of hard granules or particles throughout a mass of resin-coated mineral fibers to produce structural units in the form of sheets or slabs.

The products are composed of the mass of mineral fibers in lattice work form, particularly glass fibers, agglomerated with the dried and cured resin binder and having interspersed in the meshes of the mass, the separate hard and indeformable particles, either in solid form, such as sand, or in porous form, such as perlite or vermiculite, which render the structural units strongly resistant to physical deformation while enhancing the heat insulating characteristics. It has been determined that while the products of the process present a very low tendency to deformation, particularly compression, they retain strongly the high heat insulating capacity inherent in the porous structure of a mass of mineral fibers. This preservation of the high insulating quality is due to the fact that the hard particles or granules are in contact with the fibers of the meshes which encompass them only along points or lines of slight length and thus there is practically no formation of thermal conducting paths or bridges between the particles and the fibers.

The high degree of indeformability of the products arises from the fact that each particle impedes the local deformation of the network in which it is enclosed and that by reason of the homogeneous distribution of the particles in the entire mass, the deformation of the whole of the mass is prevented by the presence of all of these particles.

Figure 7.3a shows a part of a mass of mineral fibers which are joined together at cross points by a binder. Four of these cross points are marked **A, B, C, D.** If this mass is subjected to a mechanical stress, such as, for example, compression (Figure 7.3b), it is seen that the thickness of the mesh or lattice work of fibers decreases, and that the quadrilateral **A, B, C, D** is reduced to form quadrilateral **A', B', C', D'.**

Figure 7.3c shows the same fibrous structure as that shown in Figures 7.3a and 7.3b, but one in which hard, whole and indeformable particles or granules are introduced and interlocked between the meshes of the network of fibers. The preceding cross points are marked **A", B", C", D"** and occupy substantially the same relative positions as the cross points indicated in Figure 7.3a. If the mass is subjected to the same compressive stress as that imposed on the unit shown in Figure 7.3b, the resulting product is illustrated in Figure 7.3d. It is seen that the presence of each particle prevents deformation of the mesh in which it is enclosed, the points **A''', B''', C''', D'''** remaining in the same positions as points **A", B", C", D"** and that the assembly itself undergoes a decrease in thickness of much less extent than that in the case illustrated in Figure 7.3b.

FIGURE 7.3: HEAT INSULATING FIBROUS MASS

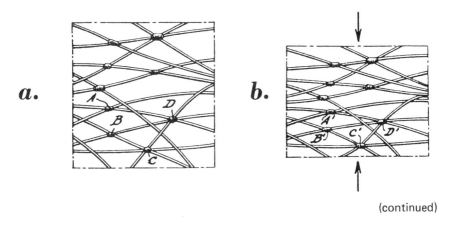

(continued)

FIGURE 7.3: (continued)

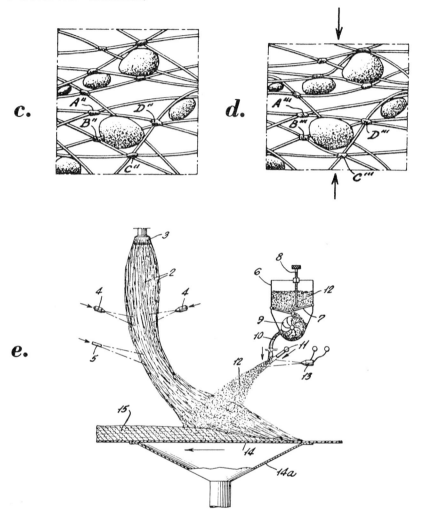

(a) View of a mass of interlocked mineral fibers, on a greatly enlarged scale, with binding agents incorporated

(b) View similar to Figure 7.3a, following the compression of the mass of fibers in a vertical direction

(c) View similar to Figure 7.3a with the inclusion of separate hard and indeformable particles in the network of the mineral fibers

(d) View similar to Figure 7.3c following the deformation of the mass of fibers by a compressive force of the same intensity as that employed on the mass shown in Figure 7.3b

(e) Front elevation, with certain parts in section of apparatus

Source: U.S. Patent 3,745,060

Figure 7.3e illustrates one form of an apparatus for obtaining a fibrous mass. Fibers 2, for example glass fibers, are produced by a machine 3, which may be a centrifuge body rotating at high speed and having a peripheral wall provided with orifices through which are projected by centrifugal force threads of material which are attenuated into fibers in a manner well known in the art. Spray guns 4 project a binding agent onto the mass of fibers and a nozzle 5 directs a jet of air onto the mass, to direct it toward the zone where hard granules or particles 12 are introduced.

The particles are contained in a receptacle 6, whose bottom is provided with ledges or movable shutters 7 with a feed regulator device 8. The particles issuing from the receptacle pass into a rotating drum 9 which assures a regular outflow of the particles wherefrom they flow by gravity through conduit 10. One or several nozzles 11 project a jet of air under pressure onto the particles in order to direct them toward the mass of fibers. A homogeneous spatial distribution of all the particles within the mass of fibers is assured by controlling the strength and direction of the air jet.

Spray guns 13 may project a binder onto the surfaces of particles before they are introduced into the mass of fibers. The mass of fibers with the particles incorporated therein then passes onto an endless cloth band or other air-permeable conveyor 14, under which is arranged a suction casing 14a to form a pad or mat 15 of the desired thickness. The passage of this pad into an oven results in polymerization and hardening of the binder and cohesion of the interengaging fibers of the mat at their points of crossing contacts.

Below are given examples of products of glass fibers according to the process as well as comparative data between these products and the same products which do not include hard and separate unitary solid or foamed indeformable particles, from the point of view of heat-insulating capability and resistance to deformation.

Example 1:

	(a)	Composition of glass, %	
		SiO	66.3
		Al_2O_3	3.0
		F_2O_3	0.4
		CaO	7.6
		MgO	3.4
		Na_2O	14.0
		K_2O	1.1
		B_2O_3	1.5
		BaO	2.0
		F_2	0.8

   (b)   Mean diameter of fibers: 6 microns
   (c)   Nature of binder: Phenol formaldehyde resin
   (d)   Nature of particles: Sand
   (e)   Mean diameter of grains: 0.2 mm

It is to be noted in the following table, that while the products have substantially the same insulating power, the load necessary to obtain the same reduction in thickness in the product in accordance with the process is nearly double.

Product	Composition of Product		Coefficient of Thermal Conductivity (kcal/mλ/°C)	Load Required to Reduce Thickness of Product by 25% (kg/m³)
Without sand	Fibers	38 kg/m³	28.7	470
	Resin	2 kg/m³		
With sand	Fibers	38 kg/m³	30.2	880
	Resin	2 kg/m³		
	Sand	60 kg/m³		

Example 2: Identical with those of Example 1:

    (a)   Composition of the glass
    (b)   Mean diameter of the fibers
    (c)   Nature of the binder
    (d)   Nature of the granules or particles
    (e)   Mean diameter of the granules

Product	Composition of Product		Coefficient of Thermal Conductivity (kcal/mλ/°C)	Load Required to Reduce Thickness of Product by 25% (kg/m³)
Without sand	Fibers	54.5 kg/m³	28.0	1,510
	Resin	5.5 kg/m³		
With sand	Fibers	54.5 kg/m³	31.0	2,300
	Resin	5.5 kg/m³		
	Sand	90 kg/m³		

Example 3:

    (a)   Composition of glass, %

SiO_2	69.0
Al_2O_3	2.3
F_2O_3	0.4
CoO	9.0
MgO	2.9
Na_2O	13.9
K_2O	0.2
B_2O_3	1.7
F_2	0.5

    (b)   Mean diameter of fibers: 6 microns
    (c)   Nature of binder: Phenol formaldehyde resin
    (d)   Nature of grains: Perlite
    (e)   Diameter of grains: 0.1 mm to 2 mm

It is noted in the following table, that while both products have substantially the same insulating capacity, the load required to attain the same reduction in thickness is nearly tripled for the product in accordance with the process.

Product	Composition of Product		Coefficient of Thermal Conductivity (kcal/mλ/°C)	Load Required to Reduce Thickness of Product by 25% (kg/m^3)
Without perlite	Fibers	36 kg/m^3	28.7	800
	Resin	4 kg/m^3		
With perlite	Fibers	36 kg/m^3	30.0	2,200
	Resin	4 kg/m^3		
	Perlite	18 kg/m^3		

ALUMINA-SILICA FIBERS

Polycrystalline Fibers of Alumina

S.P. Hepburn; U.S. Patent 3,996,145; December 7, 1976; assigned to Imperial Chemical Industries Limited, England describes a fibrous structure comprising staple fibers comprising polycrystalline alumina, alumina/silica or zirconia, the fibers having an average diameter of 0.5 to 5 microns, not greater than 20% by number of the fibers having a diameter greater than 5 microns and the fibers having a shot content of less than 5% by weight.

The fibrous structure may be in any form suitable for the application to which it is put, for example, a batt, mat, blanket, felt, cloth, paper or board. By shot is meant material of a nonfibrous nature. The fibrous structures are useful for many technological applications, especially for acoustic and thermal insulation. The process is illustrated by the following examples.

Example 1: 75 ml of aluminum oxychloride solution containing 11.2% by weight of aluminum and 8.1% by weight of chloride were mixed with 45 ml of a 2.1% by weight solution of a high molecular weight polyvinyl pyrrolidone. The mixture was concentrated by evaporation under reduced pressure to a viscosity of 45 poises. The concentrated solution was introduced into fiber spinning apparatus in which two high-velocity streams of air impinged from either side at an angle of 30° to a stream of the solution emerging from a 200-micron wide aperture under pressure. The air streams were at a temperature of 30°C and a relative humidity of 50%.

A mat of very fine fibers having lengths up to 10 cm and diameters in the range of 2 to 7 microns was collected on a gauze screen. The mat was heated up to 850°C over a 2 hr period to give fibers which were silky and flexible.

By this process sufficient fiber was obtained to form it into a blanket 24 inches long, 12 inches wide and 1 inch deep. The blanket density was measured to be 6 lb/ft^3, samples of fiber extracted from the blanket were tested for shot content (according to the method described in British Ceramic Research Association Technical Note 200), and a value of 0.9% by weight was found. The diameter distribution was examined (by optical microscopy, 625 counts, according to BS 3,406 Part IV, 1963) and the results are listed on the following page.

Diameter (μ)	Percent
7	<99
6	<97
5	<84
4	<58
3	<27
2	<6
1	0

Example 2: A solution suitable for spinning into a high-temperature-resistant yttria-stabilized zirconia fiber was prepared from 750 g zirconium acetate solution (22% w/w ZrO_2), 330 ml 1% w/w polyethylene oxide solution, 19.2 yttrium chloride hydrate and 3 ml concentrated hydrochloric acid.

The solution was concentrated by evaporation under reduced pressure to a viscosity of 15 poises at 20°C and placed in spinning apparatus with an orifice of diameter 0.004 inch. High velocity jets of air converging at an angle of 30° from either side of this hole drew down the jet of liquid from the hole to give fibers with a mean diameter of 3.0 microns. The fibers were fired at 850°C for one-half hour to give yttria-stabilized ZrO_2 fibers.

250 g of these fibers were dispersed, using a Silverson mixer, for 2 minutes in 15 liters of water. To this dispersion was added 100 ml of a silica sol (30% solids w/v) and 50 ml of a flocculating agent (0.02% polyacrylamide), the whole then being beaten for a further 1 minute. After addition of 100 ml of saturated aluminum sulphate solution, the mixture was beaten for a further one-half minute and then poured through a metal gauze filter. The water rapidly drained away, leaving a board 18 x 18 inches, which was dried for 10 hours at 60° to 80°C. It was subsequently fired at 800°C for one-half hour, the final thickness of the board being ¼ inch and the density being 14 lb/ft^3.

The fiber which was used to make the board was characterized in terms of its fiber diameter (determined by optical microscopy according to BS 3406 Part IV, 1963), and shot content (determined according to British Ceramic Research Association Technical Note 200). Shot content: 0.1% by weight.

Fiber Diameter (μ)	Percent
6	<100
5	<99
4	<94
3	<65
2	<17
1	0

Paper and Felt

L.V. Gagin; U.S. Patent 3,687,850; August 29, 1972; assigned to Johns-Manville Corporation describes a fiber composition and structure which is useful at temperatures of about 2000° to 2300°F. In particular, long staple fibers having silica and alumina as their major constituents can be produced in fine diameters so that they have high tensile strength. These fibers when in a shot-free form offer improved insulating qualities and are particularly suitable for manufacture of paper without the need of additional binders.

Fine fibers can first be formed from a glass composition by weight of 45 to 60% SiO_2, 12 to 18% Al_2O_3, 4 to 10% B_2O_3, up to 2% Na_2O and/or K_2O, and 16 to 26% CaO and/or MgO. These fibers have been produced in long staple form with average diameters of two thirds of a micron and average diameters of one and one-quarter micron. The long staple fibers are treated in sulfuric acid of about 0.75 pH at a temperature of about 200° to 210°F for about four hours, then washed free of the acid immediately. The resultant product is of a composition by weight of 76.0 to 90.0% SiO_2, 4.0 to 8.0% Al_2O_3, 4.0 to 10.0% CaO, 1.0 to 4.0% MgO, up to 0.5% Na_2O and/or K_2O, and 2.0 to 4.0% B_2O_3.

The product exhibits a resistance to temperature up to 800°F greater than the original material without degradation, thereby permitting its use to 2000° to 2300°F.

SILICATE FOAMS

COMPOSITIONS—ADDITIVES

Phosphoric Acid in Premix Composition

There has been a long and continued interest in expanded silicate foam materials. Such materials are noncombustible and fire resistant so as to find ready application in construction operations and related activities. It is interesting to note that the first U.S. patent for expanded silicates was issued in 1883 for the preparation of foam silicate structures for use as thermal insulation in fireproof safes.

J.C. Horai and C.W. Sheeler, Jr.; U.S. Patent 3,844,804; October 29, 1974; assigned to GAF Corporation describe a foamed silicate structure which has good dimensional stability and water insolubility characteristics, together with a lightweight, low density nature. The desirable strength and water insolubility are achieved by reacting the silicate during the foaming operation with a liquid silicate reactant, e.g., phosphoric acid. The reaction is controlled, however, so as to avoid inhibiting expansion of the structure. For this purpose, the liquid reactant is first mixed with an amorphous silica carrier, sufficient carrier being employed to absorb essentially all of the liquid reactant, thus forming a dry premix composition.

This premix is then blended with aqueous alkali metal silicate to form a moldable silicate mix that is heated rapidly to foam the mix and form the desired lightweight cellular foam structure. The silicate employed is preferably a combination of sodium silicate and potassium silicate, with the latter comprising from about 5 to 10% by weight based on the total weight of alkali metal silicate employed.

Example 1: A dry premix was prepared by mixing 8.0 grams of Hi-Sil-404 with 3.0 grams of 85.5% phosphoric acid. After thorough mixing, the silica carrier had completely absorbed the acid to produce a dry, freely flowable premix. Ten grams of this premix were mixed with liquid silicates for about 10 minutes at ambient temperature to achieve thorough blending and produce a moldable silicate mix. The liquid silicates employed were 85 grams of sodium silicate D having a weight ratio of Na_2O to SiO_2 of 1:2.00 and 5 grams of potassium silicate having a K_2O to SiO_2 weight ratio of 1:2.50. Both silicates were manufactured by Philadelphia Quartz Company. The moldable silicate mix was sufficiently fluid that it flattened out when put into a pile.

The foamed structure was produced by weighing 98 grams of the silicate mix into an iron mold having dimensions of 3" x 5" x 2", equalling 30 in^3 or 492 cm^3. The mold was covered with a perforated lid and placed in a muffle furnace maintained at an air temperature of 800°F. After heating for 1½ hours, the mold was removed and cooled to room temperature.

The intumesced foam structure had completely filled the mold. After being removed therefrom, it was precisely trimmed to exact dimensions of 3" x 5" x 2" and weighed. It had a density of 3.75 lb/ft^3 and possessed a uniform cell structure. The compressive strength of the structure was 9.8 lb/in^2. When exposed in a humidity chamber regulated to 90°F and 94% relative humidity, samples as above continued to be nonhygroscopic and maintained dimensional stability after 41 days. Equivalent foams formulated with 100% sodium silicate D, however, collapsed in the atmosphere after 7 days. After soaking in water at room temperature for 46 days, the silicate foam structure containing potassium silicate remained insoluble after 46 days, as compared to a 7 hour period of insolubility for samples made with 100% sodium silicate.

Example 2: 220 grams of the moldable silicate mix of Example 1 above were weighed into a Sheetrock mold having dimensions of 6" x 8" x ¾". The filled mold, after covering with a perforated lid, was placed in a microwave oven of 2,450 MHz and the oven set for 10 minutes automatic operation at about 220°F. After this heating period, the mold was removed from the oven and allowed to cool to room temperature. The foam silicate was removed from the mold and found to have a good regular shape of 6" x 8" x ¾". The foam silicate also had a uniform cell structure and a hard glassy surface or skin. The density of the product was found to be 9.9 lb/ft^3. It had a compressive strength of 12.2 lb/in^2.

Silane-Stabilized Foams

S.A. Gerow; U.S. Patent 3,661,602; May 9, 1972; assigned to E.I. du Pont de Nemours and Company describes stable aqueous foams which are prepared from aqueous solutions of alkaline ionic silicates containing a cationic surface-active onium compound and an alkyl silane, and optionally containing colloidal amorphous silica and a latent acid or salt gelling agent. Rigid foams are obtained by setting the aqueous foams with a salt or acid, then drying the set foam. Inclusion of the silanes increases foamability of the solution, stability of the wet foam, and water resistance and strength of the dried foam.

The foams, both in the wet stage and after setting, are characterized by very fine, uniform pore structure and low density. Pore diameter is generally no more than 1 mm, and average diameter is usually much less. Densities of both the wet and dried foams range from 3 to 40 lb/ft^3. The lower densities are preferred for thermal insulation applications, but the higher densities are preferred for applications in which substantial loads must be borne by the foam. The density of filled bodies bonded by these foams will of course vary with the density of filler material, but the density of the foams themselves falls within the range of 3 to 40 lb/ft^3. The wet foams are stable, even without added filler. They do not collapse and drain on standing in air at ambient temperatures for at least 24 hours. The dried foams are also characterized by high surface area, in the range 300 to 1,000 m^2/g as determined by nitrogen adsorption. This makes them desirable as catalyst supports. In the following examples, all parts are by weight unless indicated otherwise.

Example 1: A 5-quart Hobart mixer kettle is charged with 87.5 parts of sodium silicate (8.9% Na$_2$O, 29.0% SiO$_2$, 62.1% H$_2$O; of specific gravity 41.6°Bé at 60°F and having an approximate viscosity of 250 cp at 78°F), 12.5 parts of distilled water, 2.5 parts of 30% sodium methyl siliconate and one part of 50% hexadecyltrimethylammonium chloride. Upon adding the surfactant a white gel is produced. The mixture is blended for 10 seconds with a wire-ship beater at low speed and then beat at high speed for 10 minutes to produce a stable foam. The foam occupies a space 14 times greater than that of the starting mixture. It has a density of 9.8 lb/ft^3 in this wet state.

To the above foam is added 100 parts of ultrafine expanded perlite having a density of 2.5 lb/ft^3. The mixture is blended for 20 seconds at low speed to produce a low density, viscous mass. Specimens are molded and immediately subjected to an atmosphere of carbon dioxide gas for a period of 5 minutes. Specimens are air dried for 16 hours and then oven dried at a temperature of 50°C for an additional 8 hours. A low density insulation composition is obtained by this procedure and is characterized as follows: Density, 140 lb/ft^3; compressive strength, 114 psi at 5% deformation and lineal shrinkage on drying, <1%. After an 8-hour boil followed by 16-hour immersion, it had a water resistance such that it showed <0.25% loss in volume, a density of 12 lb/ft^3 and a compressive strength of 50 psi at 5% deformation.

Example 2: A 5-quart Hobart mixer kettle is charged with 175 parts of sodium silicate (8.9% Na$_2$O, 29.0% SiO$_2$, 62.1% H$_2$O; of specific gravity 41.6°Bé at 60°F and having an approximate viscosity of 250 cp at 78°F), 25 parts of distilled water, 5 parts of 30% sodium methyl siliconate and 2 parts of 50% hexadecyltrimethylammonium chloride.

Upon adding the surfactant a white gel is produced. The mixture is blended for 10 seconds with a wire-ship beater at low speed and then beat at high speed for 10 minutes to produce the stable foam. The foam occupies a space 14 times greater than that of the starting mixture. It has a density of 9.8 lb/ft^3 in this wet state. 10 parts of mineral wool is added and the mixture blended for 1 minute at high speed. 182 parts of ultrafine expanded perlite having a density of 2.5 lb/ft^3 was added and the mixture is blended for 20 seconds at low speed to produce a low density, viscous mass. Specimens are molded and immediately subjected to an atmosphere of carbon dioxide gas for a period of 30 seconds.

Specimens are air-dried for 16 hours and then oven dried at a temperature of 200°F for an additional 8 hours. A low density insulation composition is obtained by this procedure, and is characterized as follows: Density, 12.2 lb/ft^3, compressive strength, 100 psi at 5% deformation and lineal shrinkage on drying, <1%.

Pot Life Time, min	Density, lb/ft^3	Compressive Strength at 5% Deformation, psi
10	12.2	99.0
15	12.2	98.2
20	12.2	100.0
25	12.2	98.6

Example 3: The procedure of Example 2 is repeated except that 5 parts by weight of sodium methyl siliconate is omitted. The product is characterized as follows: Density, 12.2 lb/ft^3; compressive strength, 85 psi at 5% deformation and lineal shrinkage on drying, <1%.

Pot Life Time, min	Density, lb/ft^3	Compressive Strength at 5% Deformation, psi
10	12.2	85.3
15	12.2	60.5
20	12.2	58.1
25	12.2	55.5

Stabilizing Additives

A process described by *V.W. Weidman and P.C. Yates; U.S. Patent 3,725,095; April 3, 1973; assigned to E.I. du Pont de Nemours and Company* relates to stable, shapeable, aqueous siliceous foams, with a pH of at least 9, stabilized by the product of a reaction between a source of reactive silica and a surface active cationic, nitrogen-containing onium compound having at least 1 but no more than 2 alkyl hydrocarbon chains of 8 to 24 carbon atoms. The source of reactive silica is preferably in the form of dissolved silicate anions, but may also be present, in part, as reactive silanol (—SiOH) groups on the surface of colloidal silica having an average particle size of 5 to 200 millimicrons.

The total concentration of reactive silica plus nonreactive silica in the foams is at least 8% by weight calculated as SiO_2. The dissolved silicates, which are considered to be part of the reactive silica, comprise at least one dissolved alkaline ionic silicate selected from the group of lithium silicate, sodium silicate, potassium silicate and silicates of monovalent organic bases, the base having a basic dissociation constant at 25°C greater than 10^{-2}. The amount of reactive silica present on the surface of colloidal silica is determined by the formula

$$S = 0.08A = 218/D$$

where S is the percent of the total colloidal silica which is on the surface and available for reaction, A is the specific surface area of the colloidal silica in square meters per gram, and D is the average diameter of the colloidal silica particles in millimicrons. Also included in the total silica content of the foam is the nonreactive inner portion of the colloidal silica particles. The foam composition has a mol ratio of colloidal silica to silicate ion of 0:1 to 99:1.

The compositions of this process are useful in a variety of applications in which it is desired to obtain stable, shapeable foams which can, if desired, be converted into strong, very lightweight, refractory, water-insoluble structural materials. Thus the compositions can be employed either with or without fillers to prepare wallboard for structural applications, thermal insulation, insulation to decrease noise and ceiling tile. Since the compositions can be prepared continuously from a foam generator, and since they can be set at either room temperature or at elevated temperatures, they can be applied in place, such as by spraying the surfaces to be covered with a blanket of the foam, followed by an appropriate setting reaction.

Example 1: A 5-quart Hobart mixer kettle is charged with 200 parts of sodium silicate (8.9% Na_2O, 29.0% SiO_2, 62.1% H_2O; of specific gravity 41.6°Bé at 60°F and having an approximate viscosity of 250 cp at 78°F) and 2 parts of 50% hexadecyltrimethylammonium chloride (Arquad 16-50). Upon adding the surfactant, a white gel is produced. The two-phase system is blended for 15 seconds with a wire-ship beater at low speed and then beat at high speed for 6 minutes to produce a stable, shapeable foam. The foam occupies a space three times greater than that of the starting sodium silicate. It has a density of 30 lb/ft³ in this wet state. This foam is stable, without the addition of setting agents, for over 8 hours and may be shaped at any time during this period provided it is not allowed to dry. The foam does not collapse even upon drying.

To the above foam was added 90 parts of ultrafine expanded perlite having a density of 2.5 lb/ft³. The mixture is blended for 30 seconds at low speed to produce a low density, viscous mass. This foam is also stable, without setting, for over 8 hours. The foam may be used in this condition or it may be set by exposure to carbon dioxide gas. Specimens are molded and subjected to an atmosphere of carbon dioxide gas for a period of about 5 minutes. Specimens are air dried for 16 hours and then oven dried at a temperature of 50°C for an additional 8 hours. A low density insulation composition is obtained by this procedure, and is characterized as follows: Density, 20 lb/ft³; compressive strength, 110 psi at 5% deformation; lineal shrinkage on drying, <1%; water resistance, on 8-hour boil followed by 16-hour immersion: Loss in volume <0.5%; density, 15 lb/ft³ and compressive strength, 40 psi at 5% deformation.

Example 2: The procedure of Example 1 is repeated except that 65 parts of water are added to the sodium silicate prior to its foaming. The foam expands about five times the volume of that of the substituents and has a density of 16 lb/ft³. The foam is blended with 150 parts of perlite as above. This foam is stable and shapeable for over 4 hours. The filled foam is processed as in Example 1 to give a shape which has the following properties: Density, 12.5 lb/ft³; lineal shrinkage, <1%; compressive strength, 45 psi at 5% deformation; water resistance on boiling 8 hours followed by 16 hours immersion: Volume <0.5% loss; density, 11 lb/ft³ and compressive strength, 25 psi at 5% deformation.

Acid-Liberating Hardeners

W. Von Bonin, U. Nehen and U. Von Gizycki; U.S. Patent 3,850,650; November 26, 1974; assigned to Bayer AG, Germany have found that aqueous silicate solutions can be effectively foamed and that expansion of the silicate formed can also be obtained simultaneously with the hardening process by mixing the silicate solution together with an expanding or foaming agent and a hardener and allowing the resulting mixture to foam and harden, optionally in molds. Particularly suitable hardeners are substances which liberate carbon dioxide, carboxylic acids, sulfonic acids or mineral acids in aqueous media. It is preferred to use acid-liberating hardeners which, when added in the form of a 10% mixture to a soda waterglass solution containing 0.5% by weight of Na-C_{14}-alkyl sulfonate with a density of approximately 1.36 (Na_2O content approximately 8.6% by weight,

SiO_2 content approximately 25.4% by weight), cause this solution to gel in less than 15 minutes at 25°C. The process is illustrated by the following examples in which parts are parts by weight unless otherwise stated.

Example 1: Determination of the gelling time produced by the various hardeners or hardener mixtures, is described by way of example in the following with reference to some prototypes: all the substances and apparatus were tempered to 25°C. 100 grams of soda waterglass solution (density 1.36, Na_2O content approximately 8.6% by weight, SiO_2 content approximately 25.4% by weight), in which 0.5% by weight of sodium C_{14}-alkyl sulfonate had previously been dissolved, were then introduced into a 5 cm diameter beaker equipped with a stirring mechanism.

An L-shaped glass rod extends to the middle of the solution, the distance of the bent arm from the wall being about 5 mm. The glass rod is rotated at 250 rpm and acts as a stirrer, the beaker being stationary. The hardener substance is then introduced into the stirred waterglass solution and the time which elapses from this moment until the initially readily stirrable mixture solidifies, was measured. This time is designated the gelling time. The following gelling times were measured by way of example on the following substances active as hardeners.

Hardener	Minutes	Seconds	Hardener No.
Butyric acid chloride	–	7	1
Tetrapropenyl succinic acid anhydride	–	21	2
Hexahydrophthalic acid anhydride	–	30	3
Chloroformic acid butyl ester	–	48	4
Chloroformic acid isooctyl ester	1	18	5
Oleic acid chloride	1	31	6
Benzoyl chloride	2	1	7
2-(Methoxyethoxy)-4,6'-dichloro-s-triazine	2	21	8
3-Methyl benzoyl chloride	3	17	9
Oleic acid-acetic acid anhydride	4	26	10
Chloroformic acid phenyl ester	4	27	11
2-Chlorobenzoyl chloride	6	24	12
Tolylene diisocyanate	9	30	13
Benzene sulfonic acid chloride	10	19	14
Hexamethylene diisocyanate	12	57	15
Dimethyl sulfate	13	55	16
Phthalyl chloride, sym.	4	12	17
Isophthalic chloride	3	28	18
Chloroformic acid cresyl ester	3	30	19
5-Chloroisophthalyl chloride	10	10	20
Diphenol methane dichloroformate	2	50	21

Although the first of each pair of substances listed below generally has what is by definition too slow a gelling effect on its own, it can be a constituent of a mixture which produces the requisite gelling time through a suitable content of quick-acting hardener.

	Ratio by Weight	Minutes	Seconds
o-Toluene sulfonic acid chloride/benzoyl chloride	2:1	4	19

(continued)

	Ratio by Weight	Minutes	Seconds
2-Isopropyl-4,6-dichloro-s-triazine/chloroacetic acid ethyl ester	1:1	5	54
3,4-Dichlorobenzyl chloride/butyroylchloride	4:1	6	16
Phthalyl chloride/benzoyl chloride	1:1	1	49

The numbering of the hardeners in Example 1 is retained for the following examples.

Example 2: The production of foams which can be produced by the hand foaming process under different conditions is described by way of example with reference to hardener 7.

800 parts of 40% soda waterglass solution containing 4 parts of sodium C_{14}-alkyl sulfonate in solution, were heated to 45°C and poured into an open beaker equipped with a stirring mechanism. 90 parts of benzoyl chloride in admixture with 65 parts of trichlorofluoromethane were then added all at once while stirring with a propeller stirrer. Despite the increased temperature, the mixture could be effectively stirred for a few seconds before it begins to rise in the beaker. The foaming mixture was then immediately poured into a box mold of paper in which the mixture immediately began to rise, the rising process continuing at a greater rate after a few seconds because the reaction of hardener 7 produced a further increase in temperature.

After about 4 minutes, the foam had solidified in the box mold and could then be removed from the paper. Through continued reaction, some of the water present in the foam (approximately 140 parts) was automatically forced out of the foam. The finished, moist foam could then readily be cut into any shape. It had a density of approximately 0.3 and was uniformly fine-pored. It dried very quickly on standing in a current of air or in a recirculating air cabinet, after which it had a density of about 0.06. Foams of this kind can be used as an insulating material.

Hydrogen Peroxide and Acid-Liberating Hardener

W. Van Bonin, U. Nehen and U. Von Gizycki; U.S. Patent 3,864,137; February 4, 1975; assigned to Bayer AG, Germany have found that it is particularly advantageous to use hydrogen peroxide as the blowing agent, because this obviates the use of organic blowing agents which give rise to ecological problems. If an approximately 40% soda waterglass solution, for example, is mixed with an approximately 30% hydrogen peroxide solution, then a relatively stable mixture is obtained, i.e., only moderate decomposition of the hydrogen peroxide takes place.

It has been found that as soon as a hardener is added to the mixture, a sharp increase in the rate of decomposition of hydrogen peroxide instantly sets in and the rate of decomposition continues to increase in the course of the reaction and vigorous foaming takes place. It is also found that the gelling of waterglass solutions by means of hardeners which effect only slow gelling when used on their own was substantially accelerated by the addition of even small quantities of

H_2O_2. In this way it is possible, even with gelling agents which are only slow hardeners, to produce foam resins which have dry densities in the range from about 0.03 to 0.95. The advantages of this process are, firstly, that foaming may be carried out without the addition of organic blowing agents, although these may, of course, be used in addition to hydrogen peroxide if desired, and, secondly, that preheating of the reactants is generally unnecessary since the foaming reaction obtained by this process will set in even at temperatures near $0°C$ and liberates so much heat that the reaction mixture becomes heated.

Lime Hardener

A process described by *S. Kraemer, A. Seidl and M. Seger; U.S. Patent 3,652,310; March 28, 1972; assigned to Wasag-Chemie AG, Germany* involves the finding that excellent heat-insulating construction elements can be obtained from a method comprising the steps of mixing natural or artificial foamed silicate particles with lime, forming the elements under pressure, if such pressure is necessary for maintaining the shape, and hardening the elements in an atmosphere containing water vapor. Hardening is the result of a chemical reaction between the foamed silicate particles and the lime, during which process the silicate particles become completely or at least partially dissolved. Natural or artificial foamed silicate particles are preferably used which are absorbent or less resistant to water and are of spherical shape.

Example 1: 100 grams of foamed glass particles produced from powdered glass mixed with expanding agents, which are granulated, dried, and foamed with the addition of bonding liquids (e.g., waterglass) having a piled density of 90 g/l are mixed with 20 grams of CaO and 30 grams of diatomite in the dry state. The mixture is thereafter moistened with 200 cm^3 of water. The resulting mass is poured into a mold and left to dry for 24 hours at normal temperatures of approximately 77°F (25°C).

The dried elements are subsequently stored in a pressure tank containing a saturated water vapor atmosphere at a pressure of about 17 lb/in^2 (1.2 atm). After 36 hours both the foamed glass particles and the diatomaceous earth have reacted with the calcium oxide from which reaction a solid porous element resulted consisting mainly of calcium hydrosilicate having a specific weight of approximately 0.5 g/cm^3.

Example 2: The method is performed as described in Example 1, however, instead of applying a water vapor pressure only slightly above the atmospheric, this pressure is raised to 54.7 lb/in^2 (4 atm). The relatively long reaction time of 36 hours is thereby reduced to 18 hours.

Insolubilizer Combination

R.E. Temple and W.T. Gooding, Jr.; U.S. Patent 3,719,510; March 6, 1973; assigned to Diamond Shamrock Corporation describe a process for the preparation of an expanded insoluble aggregate from a mixture of aqueous and anhydrous alkali metal silicates, having weight ratios of 1:3.0 to 7.0 with primary and secondary insolubilizers. The anhydrous silicate is added to obtain a silicate solids content within the range of 80 to 40%. After mixing and curing, the composition is ground to a particulate form and subsequently expanded at temperatures in excess of 800°F. The primary insolubilizer, e.g., sodium silicofluoride, serves to reduce the hygroscopicity of the ground particulate material prior to expansion, while the secondary insolubilizer, e.g., calcium carbonate, reacts at expansion temperatures to provide an insoluble lightweight aggregate.

The use of two insolubilizers means that only the minimum amount of reaction necessary to prevent caking of the granulated material prior to expansion will occur, thus leaving more silicate available for expansion, while still providing for a reaction at elevated temperatures in order to achieve water insolubility. The resultant aggregate is quite lightweight, i.e., 1.5 to 2.5 lb/ft^3 and, owing to the absence of organic materials, is completely fireproof. Likewise, depending mostly upon the nature of the insolubilizer used, the melting point of the aggregate is quite high, i.e., within the range of 1000° to 2200°F. The aggregate, primarily non-load-bearing in nature, is completely insoluble in boiling water and exhibits thermal conductivities on the order of 0.3 Btu per hour per square inch of surface area per inch thickness per degree Fahrenheit at a mean temperature of 75°F.

Amide Gelling Agent

W.P. Banks, J.R. Carlson and D.E. Becker; U.S. Patent 3,933,514; January 20, 1976; assigned to Continental Oil Company describe a composition for producing a fire-resistant silicate foam having high strength, low shrinkage and high water resistance. The process involves uniformly mixing as an aqueous silicate dispersion in water on the basis of about 100 parts by weight foam a composition consisting essentially of:

(a) about 25 to 36 parts water-soluble alkali metal silicate in about 45 to 60 parts of water;
(b) about 8 to 20 parts cementing agent selected from sodium silicofluoride, zinc carbonate, magnesium carbonate, aluminum phosphate, zinc acetate and mixture thereof;
(c) about 0.2 to 6 parts gelling agent selected from low molecular weight amide and a mixture of amide and low molecular weight haloalcohol;
(d) about 0.1 to 10 parts fibrous filler selected from glass, mineral wool, asbestos, cellulose, metal fibers, ceramic fibers, carbon fibers, synthetic fiber and mixtures thereof; and
(e) about 0.1 to 20 parts particulate filler selected from vermiculite, perlite and a mixture thereof.

The basic silicate foam composition exhibits an unusual combination of properties which make use of the silicate foam practical. However, certain silicate foam compositions have exceptionally high strength and water resistance. Incorporation of certain halo- or thioalcohols into the foam produces a silicate foam which is essentially waterproof. Each of these modifications also enhances other foam properties which make the compositions of this process even more unusual.

Basically, this process comprises mixing the critical composition ingredients with water and the blowing agent, preferably air. The ingredients can be added all at once or stepwise. The cementing agent and gelling agent are preferably added last. The ingredients are mixed until the composition is relatively uniform and the viscosity begins to rise substantially. The mixing period is determined by the viscosity range required for the particular method of fabrication or application used. This simplified process produces a surprisingly uniform creamy silicate foam having unique properties in that each composition exhibits low density, high strength, good water resistance and little or no difficulty with shrinkage and viscosity stability on fabrication or spraying.

Vinyl Chloride Telomers as Strength Additives

D.D. Sparlin and C.M. Starks; U.S. Patent 3,961,972; June 8, 1976; assigned to Continental Oil Company describe a silicate foam composition having improved strength which comprises an alkali metal silicate, a cementing agent, a filler, a vinyl chloride telomer having an average molecular weight of from about 600 to 1,000, and a gelling agent. More specifically, the improved silicate foam composition contains from about 80 to 90 weight percent of the alkali metal silicate, from about 10 to 15 weight percent of the cementing agent, from about 1.5 to 2.5 weight percent of the filler, from about 0.5 to 1.5 weight percent of the vinyl chloride telomer, and from about 0.5 to 1.0 weight percent of the gelling agent.

The telomer products resulting from the telomerization of vinyl chloride with β-diketones and having a molecular weight in the range of from about 600 to 1,000 are employed in the production of the silicate foam compositions having increased strength without a substantial change in the density of composition. The production of such telomers is described elsewhere.

Example: A — A silicate foam was prepared by mixing 415 grams sodium silicate with 70 grams of sodium silicofluoride powder cementing agent, 10 grams of fiber glass filler (½" chopped glass fiber) and 4 grams of a fatty acid gelling agent (i.e., Crofatol 30, a fatty acid) in a kitchen type cake mixer. The Crofatol 30 is a distilled tall oil produced by the fractional distillation of crude tall oil and contains 25 to 30% rosin acids. The mixture was then mixed at high speed for 2 to 3 minutes and until a creamy mixture was obtained. Thereafter, the mixture was poured into a 1-quart Mason jar, sealed, and allowed to cure for 1 week.

At the end of the curing period a 1-inch diameter cylindrical core was cut and measurements as to density and compressive strength were made on the core sample. The compressive strength was determined using an Instron machine. The measurements as to density and compressive strength are listed in the table below under Example A.

B — A second silicate foam composition was prepared using the same procedure and ingredients as in A, above, with the exception that prior to the addition of the gelling agent to the mixture, 5 grams of a telomer having an average molecular weight of about 796 which was produced by the telomerization of vinyl chloride and 2,4-pentanedione was added to the reaction mixture. After the curing period, a 1-inch diameter cylindrical core was obtained and measurements as to density and compressive strength of this sample were made. The results of such measurements are set forth in the table under Example B.

C — A third silicate foam composition was prepared using the same procedure and ingredients as in B, above, with the exception that the telomer employed had an average molecular weight of 1,500. After the foam composition had cured, as previously discussed, a 1-inch diameter cylindrical core was obtained and measurements as to density and compressive strength of the sample were made. Results of such measurements are set forth in the table below under Example C.

D — Another silicate foam composition was prepared using the same procedure and ingredients as in B with the exception that the telogen constituent, e.g., 2,4-pentanedione, employed to produce the telomer employed were used in place of the telomer. After the curing period, a 1-inch diameter cylindrical core was obtained and measurements as to density and compressive strength of the sample were made. Results of such measurements are set forth in the table below under Example D.

Example	Density, lb/ft^3	Compressive Strength, psi
A	34.3	133
B	35.1	237
C	44.7	244
D	54.3	257

The above data clearly illustrate that by incorporating a vinyl chloride telomer having an average molecular weight in the range of from about 750 to 850 into a silicate foam composition, one can greatly improve the strength of the foam composition without substantially altering the density of same. For example, Example A, a control sample, had a density of 34.3 lb/ft^3 and a compressive strength of 133 psi. By incorporating about 1% by weight of a vinyl chloride telomer having an average molecular weight of 796, the density of the foam remains substantially the same (35.1 lb/ft^3) but a substantial increase in compressive strength (96 psi) was achieved.

When employing a telomer having a higher average molecular weight such as the one employed in Example C, e.g., an average molecular weight of 1,500, a substantial increase in density of the foam composition was detected. The same results were found when the telogen employed to produce the telomers was used in place of the telomer. Such is shown in the table under Example D. Therefore, the data clearly show that one can substantially increase the strength of a foam composition without substantially altering the density of same by the incorporation of from about 0.5 to 1.5 weight percent vinyl chloride telomer having an average molecular weight ranging from about 750 to 850.

Fly Ash and Powdered Aluminum

According to a process described by *P.W. Barton; U.S. Patent 3,700,470; October 24, 1972; assigned to A.C.I. Operations Pty. Ltd., Australia* lightweight foamed solid shapes, e.g., building panels, are made by mixing a ceramic filler, a powdered amphoteric metal, preferably aluminum, and aqueous sodium silicate, shaping the mix and subsequently curing it with or without facing sheets. Suitable ceramic materials include power station fly ash, blast furnace slag, pumice, red mud wastes and sand fines. Power station fly ash is a preferred material and especially preferred is fly ash derived from power stations burning black coal.

Suitable amphoteric metals include aluminum, zinc, lead, tin and chromium. It is particularly preferred to use finely powdered aluminum, for example, as used in paint pigments. In addition to its function as an active agent reacting with the metal powder, the sodium silicate also serves as a binder. Preferably the soda/silica ratio is chosen in the range 1:3.3 to 1:2.0 by weight, although slightly higher or lower ratios are also suitable. In the following Examples 1 through 5, the ceramic material was fly ash obtained from Wangi power station in New South Wales, Australia, which is a black-coal burning plant. Experimental work has shown that fly ash derived from black-coal burning plants in various locations exhibits similar properties and may be employed in the process without pretreatment. The fly ash derived from brown coal contains soluble alkaline salts and is pretreated to preprecipitate the soluble fraction before being used in the process.

Throughout the examples the modulus of rupture (MOR) quoted is a 3-point bend test taken from Australian Standards Nos. A.44 and C.20—1960, "Fibrous Plaster Products." The thickness of the composite boards produced in Examples 1 to 5 inclusive was in each case approximately one-half inch.

Example 1: A solution consisting of 680 parts of sodium silicate and water, having a solid content of 39.6% by weight, and $Na_2O:SiO_2::1:2.30$ by weight was heated to 65°C. This hot solution was then added to a mixture of 500 parts fly ash and 1 part finely divided aluminum by weight. The resulting slurry was contained between two sheets of paper and cured at 100°C for 16 hours. The resulting composite had a bulk density of 53.6 lb/ft^3 and an MOR of 1,657 psi.

Example 2: A solution of 524 parts of sodium silicate and water, having a solids content of 31.8% by weight, and $Na_2O:SiO_2::1:2$ by weight was heated to 65°C then added to a mix containing 500 parts of fly ash and 1 part finely divided aluminum by weight. The resulting slurry was contained between two sheets of paper and cured at 80°C for 16 hours. The cured composite had a bulk density of 49.7 lb/ft^3 and an MOR of 1,106 psi.

Example 3: A solution consisting of 524 parts of sodium silicate and water, having a solids content of 31.8% by weight and $Na_2O:SiO_2::1:2.3$ was heated to 50°C and then added to a mixture of 500 parts of fly ash and 2 parts of finely divided aluminum by weight. The resulting slurry was then cast between two sheets of paper and cured at 80°C for 16 hours. The cured composite had a bulk density of 65.5 lb/ft^3 and an MOR of 1,446 psi.

Example 4: A solution of 682 parts sodium silicate and water, having a solids content of 39.6% by weight, and $Na_2O:SiO_2::1:2$ was heated to 65°C. The hot solution was added to a mix consisting of 500 parts of fly ash and 1 part finely divided aluminum by weight. The resulting slurry was cast between two sheets of paper and cured at 100°C for 16 hours. The cured composite had a bulk density of 40.3 lb/ft^3 and an MOR of 1,990 psi.

Example 5: A dry mixture of 720 parts of Wangi fly ash and 2.88 parts of Alcoa atomized powdered aluminum No. 123 was added with stirring to aqueous sodium silicate at 90°C. The aqueous sodium silicate was a mixture of 508.4 parts sodium silicate solution in water having a solids content of 47.2% and $Na_2O:SiO_2::1:2.21$ by weight and 245.2 parts of extra water. This was then poured onto a bottom sheet of 0.021 inch liner board and a top sheet then applied, the top sheet being 0.040 inch thick paper which had been coated on the outer surface with 0.006 inch polyethylene film. The composite was placed in an oven at 100°C for 16 hours. Foaming commences upon mixing and continues during at least part of the curing cycle in the oven. After curing the composite had a bulk density of 40 lb/ft³.

Hollow Silicon Powder

T. Matsuda, K. Tanaka and K. Taura; U.S. Patent 3,957,501; May 18, 1976; assigned to Sekisui Kagaku Kogyo KK, Japan describe a method of producing a noncombustible lightweight shaped article which comprises shaping a mixture comprising (a) silicon powders or powders of a silicon alloy, (b) porous or hollow inorganic powders and (c) waterglass, and maintaining the shaped article at a temperature of 15° to 120°C thereby to foam and solidify the shaped article.

When the shaped article of the mixture comprising (a), (b) and (c) is maintained at 15°C to 120°C, alkali silicate is hydrolyzed by water contained in the waterglass to form alkali hydroxide which reacts with the silicon in (a) to evolve hydrogen gas. This hydrogen gas causes the shaped article to foam, and by exothermic reaction occurring at this time, the shaped article solidifies with the evaporation of water. Thus, a lightweight noncombustible shaped is obtained.

The process for continuously producing a noncombustible lightweight shaped article comprises compression molding the mixture composed of the above constituents (a), (b) and (c), and if desired, (d), a fibrous material, and/or (e), a fireproofing material, by feeding it into a compression space formed between an endless belt the inside surface of which contacts compression rolls and a feed plate moving at substantially the same speed as the endless belt and in the same direction as the movement of the endless belt, to an extent such that the bulk density of the compressed article is 1.5 to 12 times that of the shaped article before compression, and introducing the shaped article into reaction chambers maintained at 15° to 120°C thereby to foam and cure the shaped article.

These shaped article obtained by this process is easy to saw and nail, and can be used for the same construction processes as wood. Therefore, it has very high utility as a structural material requiring noncombustibility, such as roofing material, ceiling material, floor material, interior and exterior wall material, decorative material and furniture. The process is illustrated by the following examples in which all parts are by weight unless otherwise specified. The various properties appearing in the examples were measured by the following methods.

(a) Noncombustibility: The sample is burned for 3 minutes with a city gas burner with an air supply rate of 1.5 liters per minute, and then heated additionally by a nichrome wire heater of 1.5 kW/h for 17 minutes. Smoke flame and residual flame were observed, and by a synthetic evaluation of all these items, the noncombustibility of the sample is determined.

(b) Flexural Strength: JIS A1408.

(c) Compression Strength: JIS A1108.

(d) Fireproof Performance: JIS A1304 (method of fireproof testing for architectural structure), heating time 1 or 2 hours.

(e) Water Resistance: The sample is immersed for 48 hours in hot water at 80°C. The flexural strength of the sample before immersion is compared with that after immersion to evaluate the water resistance of the sample.

Example 1: 100 parts ferrosilicon (smaller than 325 microns); 300 parts pumice Microballoon (smaller than 500 microns); 60 parts asbestos (fiber length less than 30 mm) and 200 parts waterglass (No. 3) were mixed.

The mixture consisting of the above ingredients was subjected to a kneader and well kneaded. The kneaded mixture was put into a mold having a size of 150 by 150 by 50 mm and was foamed and cured at a pressure of 50 kg/cm^2 in a press-forming machine having a press plate heated at 100°C. The plate-like shaped article obtained was thoroughly dried, and its flexural strength, compression strength and bulk density were measured. The results are shown below, and indicate that the shaped article has excellent flexural and compression strengths and is light in weight and noncombustible.

Bulk density (apparent)	0.95
Flexural strength	205 kg/cm^2
Compression strength	180 kg/cm^2
Water resistance, before treatment with hot water	105 kg/cm^2
Water resistance, after treatment with hot water	102 kg/cm^2
Noncombustibility	Excellent
Thermal resistance, thermally stable up to	750°C

Example 2: Each of the following mixtures (Samples 1, 2 and 3) was treated in exactly the same manner as in Example 1. The physical properties of the resulting shaped articles are shown in Table 1 below.

Sample No. 1	
Calcium silicide (smaller than 325 microns)	100 parts
Obsidian pearlite (smaller than 500 microns)	300 parts
Asbestos (fiber length less than 30 mm)	90 parts
Waterglass (No. 3, JIS standard)	300 parts
Sample No. 2	
Ferrosilicone (smaller than 325 microns)	100 parts
Pearlite rock (smaller than 500 microns)	150 parts
Asbestos (fiber length less than 30 mm)	60 parts
Waterglass (No. 3 JIS standard)	100 parts
Sample No. 3	
Ferrosilicon (smaller than 325 microns)	100 parts
Pumice Microballoon (smaller than 500 microns)	300 parts
Waterglass (No. 3, JIS standard)	200 parts

TABLE 1

Properties	Sample 1	Sample 2	Sample 3
Apparent bulk density	1.0	0.8	0.8
Flexural strength (kg/cm^2)	89	68	38
Compression strength (kg/cm^2)	185	99	140
Water resistance (kg/cm^2)			
Before treatment with hot water	89	68	38
After treatment with hot water	87	66	35
Noncombustibility	Excellent	Excellent	Excellent
Thermal resistance, (thermally stable up to the temperatures indicated)	710°C	710°C	1150°C

Volcanic Ash

Y. Seki and M. Nakamura; U.S. Patent 3,951,632; April 20, 1976; assigned to Agency of Industrial Science and Technology, Japan describe a method for the manufacture of a foam glass which comprises mixing silicate glass, a volcanic glass material typified by shirasu (which is a whitish volcanic ash sand occurring abundantly in the Kyushu district of Japan, having as its principal component, volcanic glass and also containing crystals such as of feldspar, having a typical chemical composition of 70.0% SiO$_2$, 13.4% Al$_2$O$_3$, 2.4% NaO, 2.7% K$_2$O, 3.0% CaO and 1.3% Fe$_2$O$_3$ by weight, and possessing properties of 6.1% in ignition loss, 2.4 to 1.8 in specific gravity and 1200°C in softening point), waterglass and a sodium salt or a calcium salt of phosphoric acid, converting the resultant mixture into a slurry by addition of water, molding the slurry in a desired form, firing the molded mixture into a fused state and cooling the fused mixture.

To be more specific, two methods are available for the manufacture of foam glass from the slurry according to this process. They are a process which comprises the steps of drying the slurry and pulverizing the dried slurry into a powder, molding this powder, heating and firing the mold into a fused state and cooling the fused mold and a process which comprises the steps of casting the slurry in a mold, then drying the mold, heating and firing the dried mold into a fused state and subsequently cooling the fused mold.

The foam glass to be manufactured by this method suffers a very small increase in the diameter of bubbles in the course of manufacture. Thus the product contains bubbles not more than about 1.5 mm in diameter and, therefore, enjoys high mechanical strength. Its chemical composition is 72.2 to 74.7% of SiO$_2$, 2.6 to 9.8% Al$_2$O$_3$, 9.3 to 14.9% Na$_2$O, 1.5 to 2.2% K$_2$O, about 0.6% Fe$_2$O$_3$, 0.1 to 6.5% CaO and 0 to 1.9% MgO by weight. The product has a beautiful appearance.

The foam glass has properties excelling those of the conventionally known foam glass. It is suitable for use in interior and exterior decorative articles for buildings. The following examples illustrate the process.

Example 1: A mixture consisting of 93 grams of cullet, 9.2 cc of waterglass No. 3 of Japanese Standard (5 grams of anhydride equivalent), 2 grams of shirasu and 4.1 grams of dibasic sodium phosphate (12-hydrate) was converted into a slurry by addition of 20 cc of water. The slurry was dried, pulverized, formed, heated in an electric furnace up to 900°C at a rate of temperature increase of 150°C per hour, held at the top temperature for 5 minutes, then allowed to cool off within the furnace. Consequently, there was obtained a formed glass having a bulk density of 0.4, a foam average diameter of 1.5 mm and a bending strength of 15 kg/cm². The waterglass No. 3 used in this case was of a type having a specific gravity of more than 40°Bé at 20°C and containing 28 to 30% by weight of SiO_2 and 9 to 10% of Na_2O.

Example 2: A mixture consisting of 65 grams of cullet, 30 grams of shirasu, 9.2 cc of waterglass No. 3 and 4.1 grams of dibasic sodium phosphate was converted into a slurry by addition of water. The slurry was cast in a wooden mold (with a polyethylene fabric used as a mold release), dried at 80° to 95°C for 24 to 48 hours, heated in an electric furnace to 950°C at a rate of temperature increase of 200°C per hour, kept at 950°C for 20 minutes and then left to cool off in the furnace. Consequently, a foamed glass was obtained having a bulk density of 0.83, an average bubble diameter of 1.0 mm and bending strength of 46 kg/cm².

PROCESSING METHODS

Foaming with Microwave Energy Source

R.P. Rao; U.S. Patent 3,743,601; July 3, 1973; assigned to Fiberglas Canada Ltd., Canada describes a process for forming silicate foam comprising the steps of: (a) hydrating finely divided alkali metal silicate with water, water vapor or steam at temperatures in the range of 20° to 350°C, the resulting hydrated material containing water in an amount of 5 to 30% by weight, and (b) expanding the resulting hydrated material by rapidly vaporizing the water of hydration through the input of thermal energy.

In a preferred aspect, the process provides a water-resistant or -insoluble silicate foam or product made thereof, having the following preferred characteristics and physical properties: (a) microcellular closed cell structure; (b) density between 2 lb/ft³ and 8 lb/ft³; (c) thermal conductivity factor (K) at 75°F mean temperature of 0.200 and 0.400 Btu/hr/ft²/in/°F; and (d) compressive strength at 5% deformation from 20 to 150 psi.

The water of hydration appears to have a dual function, since it operates as a foaming agent, and also is an important variable affecting the viscoelastic behavior of the silicate hydrates before and during the process of foaming (expansion) by thermal means to form silicate foam products.

Close control and uniformity of hydrate water content, along with other process variables, affects the number, size, shape and integrity of the cells which make up the foam. With a given bulk density (lb/ft³), a foam consisting of microscopic fully closed cells will exhibit higher compressive strength, lower thermal conductivity, and less water vapor permeability than a coarse, open cell foam of the same chemical composition. Given in addition a greater freedom in choice of raw materials for the preparation of foamable hydrated silicates, the products of the process are found to offer a higher degree of insolubility than foams of prior art along with the physical characteristics reported in the examples which follow.

Example 1: Sodium silicate particles having a particle size of 80 to 100 microns were suspended in 20% water with 0.1% of Triton X-100 surfactant and blended in a mixer at a rate of 2,000 rpm for a period of 20 minutes. The resulting mixture was placed in an autoclave under a steam pressure of 15 psi. After 30 minutes, the slurry became completely hydrated and viscoplastic. The block of sodium silicate hydrate was then compressed at 100°C in a mold applying a pressure of 200 psi for 5 minutes. The monolithic block was removed from the mold and foamed by applying microwave energy at a frequency of 2,450 megacycles (mcs). The composition used in Example 1 is listed in the table below.

Composition of the Mix	Grams	Density of Form, pcf	Compressive Strength, psi
Sodium silicate SS-65 (SiO_2/Na_2O mol ratio 3.22)	653		
Water	129	7.5	25
Triton X-100	0.8		

Note: This produced a water-soluble foam.

The foam had closed pores, but when tested for moisture resistance, in an autoclave at 5 psi pressure of steam for 15 minutes, the foam collapsed and dissolved.

Example 2: A mixture containing 653 grams of powdered sodium silicate, 40 grams of boric acid, 52 grams of aluminum hydroxide, 2.5 grams of zinc dust, and 2.5 grams of titanium hydroxide was dry blended with a conventional blade type mixer. The material was then blended with 251 grams of water and continuously hydrated in an autoclave using a steam pressure of 15 psi. After 15 minutes of autoclaving, the viscoplastic hydrate was compressed at 110°C using a 200 psi pressure. The monolithic hydrate obtained was foamed in a microwave oven using 2,450 mcs frequency waves. The resulting foam was microcellular and rigid. The composition of the mix and the properties of the foam are tabulated in the table below.

Composition	Grams	Density of Foam, pcf	Moisture Absorption,* %
SS-65	653		
$Al(OH)_3$	52		
H_3BO_3	40		
Zn dust	2.5	9.9	18
$Ti(OH)_4$	2.5		
H_2O	231		
Triton-X-100	1		

*On exposure to 5 psi steam for 1 hour.

The foam on exposure to 5 psi of steam for 60 minutes maintained its original integrity and strength. It is thus considered substantially water insoluble for many practical end uses.

Finely Divided Powder as Nucleation Centers

R.P. Rao; U.S. Patent 3,756,839; September 4, 1973; assigned to Fiberglas Canada Ltd., Canada has found that the addition of a finely divided compatible powder to a hydrated alkali silicate composition promotes the uniform initiation of bubble formation throughout the mixture. During the foaming of the mixture, the particles of powder act as nucleation centers for the generation of foam bubbles by the water of hydration which constitutes the blowing agent. The powder may be chosen from nonreactive or only slightly reactive materials such as carbon black, silica, glass powder, glass fibers, ferric oxide (Fe_2O_3) metal powder and powders of organic polymers such as polyethylene and polyvinyl chloride. Alkali metal silicate foam and in particular sodium silicate foam, may be produced by the known liquid gel or solid processes, involving the rapid input of heat energy, preferably microwave energy.

The addition of finely divided metal powder to an alkali metal silicate composition may contribute heat energy by exothermic reaction during the foaming of the silicate by input of thermal energy, and further may as a result of formation of insoluble products contribute to the insolubilization of the foam article produced thereby. In particular the use of such metal powder increases the efficiency of absorption of microwave heat energy in the foaming step.

Autoclave Curing

W. Gillilan; U.S. Patent 3,951,834; April 20, 1976; assigned to Owens-Corning Fiberglas Corporation describes a process for manufacturing foamed products from sodium silicate. It has been found that curing the foam products in an autoclave after the foam has been formed, results in foam products which are free of structural damage caused by expansion or contraction. If required, after autoclaving, the foamed products can be dried by conventional means without resulting in structural damage caused by expansion or contraction. The autoclaving also shortens the time for cure.

The process thus comprises forming a mixture of a surface tension depressant and an aqueous solution of sodium silicate, subjecting the mixture to mixing with a gas at above atmospheric pressure until a wet foam is formed, blending and reacting with the wet foam an insolubilizing agent in an amount sufficient to make a foamed product rigid and resistant to being solubilized by water, prehardening the foamed product and autoclaving the resulting foamed product. Generally, the final product has a density, when dry, ranging up to 20 pounds per cubic foot, an average cell size ranging up to 300 microns in diameter, an average cell wall thickness ranging up to 16 microns and at least 50 cells per cubic millimeter.

Example 1: An aqueous sodium silicate solution of about 40% solids and having a ratio of silicon dioxide to sodium oxide of about 3.25 to 1 by weight, was placed in a pressurized, stirred container. Distilled tall oil acid was placed in the container. Air then was fed into the container at a pressure slightly in excess of 20 psi. Thereafter, the contents of the container were thoroughly mixed at this pressure. The resulting wet foam was blended thoroughly with sodium fluorosilicate in a 75% solids slurry in water.

The resulting material was placed in molds having the following sizes in inches: 6 x 6 x 2; 18 x 12 x 2; and 36 x 18 x 4. After the foam has been formed, the foam is prehardened at about the same temperature at which the foam is formed. This prehardening step generally takes 1 to 2 hours in either dry heat or steam heat. Samples of each size were oven cured at a temperature of about 200°F at atmospheric pressure. Free sodium and fluoride ions then were leached from the material by washing with hot water having a pH of about 6. The small samples (6 x 6 x 2) showed no evidence of cracking, while the remaining large samples cracked and fell apart.

Example 2: The process of Example 1 was repeated except that the samples were cured under steam at 195°F and atmospheric pressure. The small samples (6 x 6 x 2) showed no evidence of cracking, while the remaining large samples cracked and fell apart.

Example 3: The process of Example 1 was repeated except that the samples were cured in an autoclave under steam. Steam was introduced into the autoclave over a period of 30 minutes, held at a pressure of 250 psi for 3½ hours and reduced to atmospheric pressure over an additional time period of 30 minutes. All samples showed no evidence of cracking. The cured samples were dried in an oven at about 200°F and still showed no evidence of cracking or falling apart.

Frothing

According to a process described by *W.A. Mallow, R.A. Owen and E.J. Baker, Jr.; U.S. Patents 3,856,539; December 24, 1974; assigned to Fiberglas Canada Ltd., Canada and 3,741,898; June 26, 1973; assigned to Southwest Research Institute* an inorganic solidified insulating foam material having a rigid structure comprising silicic acid is made by frothing a mixture of aqueous solution of one or both of two certain alkali metal silicates, a surface tension depressant and a silicon dioxide polymer forming agent with or without the inclusion of an alkali metal silicate gelling agent. The foamed material is permitted to harden free of damaging contraction or expansion under conditions of high humidity, approaching 100% RH.

In one particular advantageous form of the process, the product will contain only up to about 10% of filler and/or reinforcement material. The entrained gas will normally be air, although other gases may be used. The hardening should preferably be effected at a temperature not appreciably different from the temperature at which the foam reaches the polymerizing zone.

Such hardening should preferably be effected at a temperature not more than about 10°F cooler than the temperature at which the foam reaches the polymerizing zone, to minimize damaging contraction. The hardening temperature should preferably be not more than about 5°F above the temperature at which the foam reaches the polymerizing zone, to avoid damaging expansion unless the foam is physically restrained from expanding.

The polymer forming agent may be added to the composition after the composition is frothed into a foam. Where the frothed product is at an elevated temperature, the polymer forming agent will preferably be added to the foamed product just prior to its arrival in the polymerization zone.

The frothing is preferably effected under superatmospheric pressure with the frothed product being permitted to expand to ambient pressure before hardening by polymerization. The temperature of a frothed product will preferably be below about 200°F when it reaches the polymerization zone, and most preferably in the range of about 135° to about 200°F. The process provides a low density, high strength, shaped foam product prepared from alkali metal silicate, generally having a dry density not greater than about 25 lb/ft^3.

Example 1: A foam having a dry density of 3 lb/ft^3 is obtained by frothing at 165°F for 2 minutes at 80 to 100 psi air pressure in a Binks pot with a stirring speed of 700 rpm, a mixture which consists of 60 pounds of 3.22 to 1 ratio sodium silicate of approximately 100 cp viscosity at 165°F (600 cp at 74°F), 0.5 part tall oil acid of low rosin content (10% or less) per 100 parts of alkali silicate and 0.5 part of oleic acid of 93% purity per 100 parts of alkali silicate. The foam is then blended with 8.4 pounds of slurried sodium fluorosilicate of fine particle size (90% through 200 mesh sieve) by rapid injection and mixing with a short contact to prevent premature rigidization.

The resulting foamed mixture is then confined in a mold with the mold being completely filled with the mixture so that no moisture can escape. The mixture is maintained in this mold at approximately 180°F until the foam is completely reacted and rigidized by the polymer forming agent. The foam is cured in 20 to 25 minutes at this temperature and is rigidized within 10 minutes. Gelation occurred within 5 minutes. After curing, the wet foam is dried to any desired degree of residual water content. The rate of drying should be controlled to prevent undue internal stress during water removal. A drying rate of 40 pounds water extraction per 100 pounds of original foam each hour is preferred.

Example 2: This is the same as Example 1 except that the alkali metal silicate is a mixed salt composed of sodium and potassium silicates at approximately 3/1 to 20/1 molar ratio of sodium oxide/potassium oxide and a silica/metal oxide ratio between 3/1 and 4/1 in a 39 to 42% solid solution in water. The temperature during frothing and in the mold was 180°F. When the sodium and potassium silicates are present at approximately 4/1 molar ratio of sodium oxide/potassium oxide, the foam will cure in about 5 minutes, become rigidized in about 2 minutes and gelation will occur in about 1 minute.

Example 3: This is a comparison to show the effect of the presence of filler in the solidified foam product. In each of the tests set forth in the table below, the fluid foam was made in a Binks pot with 100 parts sodium silicate, 1 to 2 parts fatty acid, 14 parts sodium fluorosilicate at room temperature with or without the addition of glass fibers in the form of one-fourth inch chopped fibers. The various densities of the final foam product were controlled by the conditions of frothing, that is, the greater the air pressure and the greater the time involved in frothing, the smaller the density of the final product. The fluid foamed mixture was confined in a mold with the mold being completely filled with the mixture so that no moisture could escape. The mixture was maintained in this mold at room temperature until the foam completely hardened.

In the table, the listed percentage of glass fiber is percentage of the weight of the dry solidified foam. The densities of 8, 10 and 15 lb/ft^3 are on a dry basis and the comparative strengths listed for those densities are in the pounds per square inch which cause 10% deformation under compression.

Percent Glass Fibers	Compressive Strength 8 lb/ft³ Density	Compressive Strength 10 lb/ft³ Density	Compressive Strength 15 lb/ft³ Density
0	45	70	120
2.5	30	55	80
5.0	15	40	80
5.0	15	40	80
7.5	10	30	60

Tests have also been run making products the same as the 8 and 10 lb/ft³ density materials of the above table, except that these products were allowed to harden while exposed to ambient humidity rather than being confined to 100% relative humidity. The resulting foams have markedly inferior properties. The 8 lb/ft³ product containing no glass fiber and which hardened while exposed to ambient humidity has a compressive strength of 10 psi compared to the 45 psi of the table. Similarly, a 15 lb/ft³ density product containing no glass fiber and which hardened at ambient humidity has a compressive strength of 45 psi compared to the 120 lb/in² of the above table.

Extrusion

F.W. Maine; U.S. Patent 3,839,517; October 1, 1974 describes an apparatus and process for producing a foamed product from a hydrated alkali metal silicate. The alkali metal silicate starting material is hydrated preferably in an extruder apparatus, and is physically converted to a fluid-like viscoelastic mass containing water of hydration under pressure. Physical conversion to a viscoelastic condition takes place preferably in a thermoplastics extrusion device in which the alkali metal silicate is subjected to temperatures in the range from about 100° to about 400°C and a pressure of less than about 10,000 psi.

The hydrated alkali metal silicate is preferably insolubilized by chemically reacting at least one selected insolubilizing additive therewith to render the silicate sparingly water soluble. The insolubilizing additive can be one or more commonly known agents forming insoluble silicates, i.e., typically compounds of alumina, alkali earth metals and magnesium metal powder. Hydration and insolubilization of the alkali metal silicate is more preferably carried out in a screw-type extruder device which subjects the mixture of alkali metal silicate and insolubilizers to a temperature in the range from about 100° to 400°C and a pressure in the range from about 20 to 10,000 psi in the presence of water thereby mixing, chemically reacting, and hydrating the insolubilizing material and alkali metal silicate.

The silicate hydrate in a fluid-like viscoelastic condition is extruded and depressurized rapidly such that the water of hydration, previously under pressure, is caused to expand and cause foaming of the extrudate. The extrudate is preferably confined during foaming to cause shaped foaming.

OTHER PROCESSES

Fluoride-Free Composition

R.P. Rao; U.S. Patent 3,855,393; December 17, 1974; assigned to Fiberglas

Canada Ltd., Canada describes a process for the production of fluoride-free silica foam having thermal stability in excess of 2000° and up to 2500°F and thermal conductivity as low as 0.25 Btu in/hr/ft^2/°F at 75°F mean temperature. An added object is to remove the highly conductive alkali metal fluoride such as sodium fluoride from silica foam. The process involves leaching the rigid silica foam sodium fluoride composite with an aqueous solution of an extracting agent. The leaching can be carried out at ambient or elevated temperature. The reagent is chosen so as to form complexes or stable compounds with the fluoride present in the foam. Commercially valuable by-products can be recovered by conventional means, such as precipitation, crystallization, evaporation, etc., followed by gravitational or mechanical separation. If desired, the resulting foam can then be washed with water and dried to obtain the product.

Preferred leaching agents include ammonium sulfate, aluminum sulfate, calcium chloride, aluminum chloride, boric acid and hydrochloric acid, in aqueous solution. The concentration of the leaching solution or suspension may be varied, so as to optimize the rate of leaching and other process considerations. As a matter of convenience, it will normally lie in the range of 5 to 45%. It is found that water-soluble or water-dispersible complexes are formed, which drain out of the foam. This is true even in the case of such materials as calcium chloride and aluminum chloride, which could have been expected to react with the sodium fluoride to form insoluble calcium fluoride or sodium aluminum fluoride in the foam.

Figure 8.1a illustrates a lab-scale apparatus for the extraction of square foam slabs measuring approximately 8" x 8" x 2" thick. A rubber gasket **10** having an inner opening somewhat smaller than the dimension of the foam slab **11** assists in maintaining a negative pressure at the bottom face of the foam slab.

FIGURE 8.1: HIGH STRENGTH SILICA FOAM

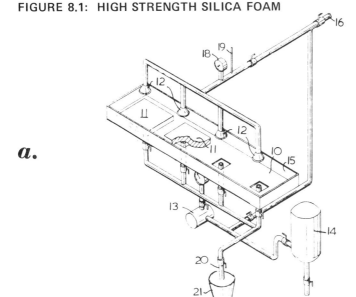

a.

(continued)

FIGURE 8.1: (continued)

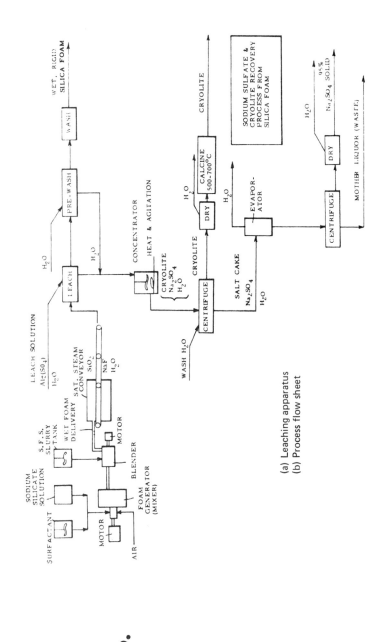

(a) Leaching apparatus
(b) Process flow sheet

Source: U.S. Patent 3,855,393

A typical leaching process involves the application of the leaching solution through the spray nozzles **12** to the surfaces of foam slabs while a pump **13** is employed to pull a suction on the opposite surface as well as to recirculate the leachate back through the spray nozzles. The leachate can be recirculated for the desired period of time, and then returned to a storage reservoir and recovery tank **14**. The leaching tank **15** is also connected to a water supply line **16** so that the leaching cycle can be followed by a spray-suction washing cycle. A 0 to 30 inch of mercury vacuum gauge is in line with the pump **13** used for circulating the leachates, as is pressure gauge **18**.

A thermometer **19** measures leachate or rinse water temperature. A waste line **20** and waste receiver **21** is also provided. The temperature maintained during leaching is generally in the range from 10°C up to the boiling point of the solution, preferably above about 40°C. The leaching or extraction is normally carried out for a period from about 3 minutes to about 10 minutes, and normally will be terminated only when substantially all of the fluoride has been removed. This can be determined by the use of a specific fluoride ion electrode, by standard analytical methods. Longer periods are used where lower leaching temperatures are employed and in those instances where dilute solutions of the leaching agent are circulated.

Figure 8.1b shows an overall process for the mixing, foaming, curing, leaching and washing of a silica foam material. This flow sheet in Figure 8.1b also shows the disposition of the leachate and its treatment for the removal of the cryolite and sodium sulfate as useful by-products.

Depending on the leaching agent employed, the fluoride is recovered directly as a precipitate or as a soluble fluoride complex of the leaching agent, which is further treated to obtain, for example, a precipitate of sodium aluminum fluoride (cryolite) or aluminum fluoride. The slurry containing the precipitate is passed from the reservoir to a settling tank and retained for a period of from 5 to 60 minutes, where the fluoride precipitate is caused to settle in the tank while the clarified liquor consisting of substantially defluorinated leachate is drawn off for subsequent use. The volume and cross sectional area of the settling tank are selected to be sufficient to achieve a separation of the fluoride precipitate at the settling velocity over the desired period of time.

Example 1: An 8" x 8" x 2" rigid foam sample weighing 660 grams was cut from a slab of foam which had been prepared by mechanical frothing methods, indicated in Figure 8.1b. It comprised by weight 56% H_2O, 16% NaF and 28% silica, and had a wet density of 18.2 pcf, equivalent to a dry density of 8 pcf. It was spray-leached with a 10% solution of $Al_2(SO_4)_3$ using the assembly described in Figure 8.1a and the leachates were recycled through the foam five times at ambient temperature. After 5 minutes the foam was washed with water and dried under vacuum at 280°F; and heat-aged for 1 hour at 700°C.

The resulting silica foam had the following characteristics: SiO_2, 99.99%; fluoride, <5 ppm; density, 5.1 pcf; porosity (by gas displacement), 98%; compressive strength, 52 psi at 5% deformation; thermal conductivity at 75°F, 0.28 Btu per inch per hour per square foot per degree Fahrenheit; flexural strength, 40 psi; and thermal stability, 2140°F with less than 0.5% shrinkage. The leachate from the foam was concentrated and the solute was precipitated. The precipitate was filtered to obtain 80 grams of commercial purity cryolite.

Example 2: The recovery of sodium fluoride as the commercially pure cryolite shown in Example 1 above is dependent on the proportion of the aluminum sulfate salt added to the wet silica-sodium fluoride foam matrix. This is indicated by the various tests utilizing the process described in Example 1, but varying the quantity of aluminum sulfate salt added. Results are summarized in the following table.

$Al_2(SO_4)_3$, %*	Percent Cryolite Recovery	Total Fluoride Recovery, %**
100	99	100
150	70	100
200	30	100
300	50	100
400	10	100

*Percent of stoichiometric requirement.
**Including salts as $Al(OH)F_2$, $Al(OHF)_3$, $Al(OH)_2F$ and cryolite.

When more than stoichiometrically required aluminum sulfate is used, the efficiency of recovery for cryolite decreases. A postulated stoichiometric for cryolite formation is indicated in the following equation:

$$Al_2(SO_4)_3 + 12NaF \longrightarrow 2Na_3AlF_6 + 3Na_2SO_4$$

The remaining fluorine while extracted effectively from the silica foam, reacts with the excess aluminum sulfate to form an unrecoverable aluminum difluoride complex AlF_2^+. This complex remains in solution, and can be converted to pure cryolite by the addition of NaCl or NaF. The cryolite obtained according to Examples 1 and 2 is dried and calcined at 500°C to obtain a commercially pure product.

Silica Gel

K. Fukumoto, T. Nakamura and K. Kadota; U.S. Patent 3,717,486; February 20, 1973; assigned to Shikoku Kaken Kogyo KK, Japan describe a method for manufacturing a foamed product of silica from silica gel which comprises prefiring silica gel at a temperature ranging from 500° to 900°C so as to obtain a prefired product which can be ignited with a loss of not more than 5% by weight. The silica gel must have a specific surface area of at least 500 square meters per gram and a size not passing through an 80-mesh sieve and substantially free of adsorbed water. The prefired product is then fired at a temperature ranging from 1000° to 1450°C to effect foaming.

For a better understanding of the process examples are given below, in which foaming degree, bulk density, hygroscopicity and surface strength are determined in accordance with the following methods.

Foaming degree: Foaming degree is determined by the following equation:

$$\text{Foaming degree} = \frac{\text{Volume (ml) of the foamed silica product}}{\text{Volume (ml) of the starting silica gel}}$$

Bulk density: The foamed product obtained in each example is weighed and the volume thereof is measured by soaking the sample in water to determine the bulk density shown by the equation on the next page.

$$\text{Bulk density} = [\text{weight (g)}] / [\text{volume (cm}^3)]$$

Hygroscopicity: The sample is left to stand under a saturated steam pressure at 30°C for 24 hours to measure the amount of sorbed water, and hygroscopicity is determined. The results are shown in following designations. (–) = 1% or less of sorbed water; and (+) = more than 1%.

Surface strength: In a porcelain ball mill, 120 mm in inner diameter, are placed 200 ml of the foamed product obtained in each example and having a particle size of 5 to 10 mesh and 50 steel balls having a diameter of 4 mm. The ball mill is rotated at a rate of 30 rpm for 10 minutes. Thereafter the amount of the sample not passing a 10-mesh sieve is measured for the determination of surface strength evaluated from the following equation.

$$\text{Surface strength (\%)} = \frac{\text{Amt (g) of sample not passing 10-mesh sieve}}{\text{Amt (g) of foamed product to be tested}} \times 100$$

Example 1: In this example were used eight kinds of silica gel, previously dried at 150°C for 2 hours and having specific surface areas shown in the table below and a particle size of 6 to 10 mesh.

Number	Specific Surface Area, m^2/g
1	380
2	508
3	612
4	690
5	746
6	796
7	837
8	897

Each silica gel was placed in an electric furnace and prefired at 650°C for 30 minutes to produce a prefired product, the ignition loss thereof being shown in the table below. The temperature was raised to 1150° to 1200°C in 5 minutes and the prefired product was heated at that temperature for 4 minutes for firing, whereby a foamed product having the properties shown below was obtained.

	. Prefired Product .		. . . . Foamed Product				
Sample No.	Specific Surface Area (m^2/g)	Ignition Loss (%)	Foaming Degree	Bulk Density (g/cm^3)	Specific Surface Area (m^2/g)	Hygroscopicity	Surface Strength (%)
1	335	1.21	0.5	1.9	60.5	+	–
2	453	1.09	2.0	0.57	0.31	–	96
3	544	1.36	2.5	0.50	0.28	–	98
4	531	1.48	3.0	0.42	0.18	–	97
5	586	1.34	3.6	0.39	0.21	–	97
6	650	1.50	4.1	0.32	0.28	–	96
7	748	1.51	3.5	0.42	0.21	–	97
8	726	1.65	3.5	0.40	0.26	–	96

The foamed products No. 2 to 8 were free of cracks and high in foaming degree, while no foaming was observed in the product No. 1 prepared from silica gel having a specific surface area of less than 500 m^2/g with many cracks produced and marked volume contraction. The foamed products No. 2 to 8 were granules covered with a lustrous vitrified surface free of cracks and by microscopic observation of the broken surface thereof were observed many open or semiclosed pores defined by vitrified partitions.

Example 2: In this example were used 10 kinds of silica gel, previously dried in the same manner as in Example 1 and having specific surface area of 796 m^2/g and granular size shown in the table below.

Number	Average Granular Size (mm)
9	0.11*
10	0.20
11	0.28
12	0.53
13	0.85
14	1.32
15	2.49
16	3.65
17	4.33
18	5.69

*Passing 80-mesh sieve.

The above silica gels were prefired and then fired in the same manner as in Example 1 to produce foamed products having the properties shown in the table below.

| | . Prefired Product . | | . . . . Foamed Product. . . . . | | | |
Sample No.	Specific Surface Area (m^2/g)	Ignition Loss (%)	Foaming Degree	Bulk Density (g/cm^3)	Specific Surface Area (m^2/g)	Hygroscopicity
9	664	1.11	0.6	1.6	61.9	+
10	656	1.21	2.0	0.57	0.29	−
11	648	1.30	2.0	0.57	0.28	−
12	632	1.42	3.0	0.43	0.23	−
13	669	1.48	3.5	0.38	0.31	−
14	672	1.69	4.0	0.35	0.20	−
15	650	1.50	4.1	0.32	0.28	−
16	675	1.45	3.7	0.39	0.24	−
17	702	1.82	3.2	0.43	0.23	−
18	699	1.60	3.4	0.41	0.21	−

The foamed products No. 10 to 18 were granules covered with a lustrous vitrified surface free of cracks and by microscopic observation of the broken surface thereof were found many open or semiclosed pores defined by vitrified partitions, while no foaming was observed in the product No. 9 prepared from silica gel having a particle size passing an 80-mesh sieve with many cracks and volume contraction.

Foamable Fibers

According to a process described by *K.P. Gladney and R.P. Rao; U.S. Patent 3,692,507; September 19, 1972; assigned to Fiberglas Canada Limited, Canada* foamable fibers of alkali metal silicate glass are prepared by dispersing flowing fine streams of molten glass to centrifugation or to a high velocity blast of gas thereby forming fibers of random length and average diameters in the range 0.1 to 1.0 mil, and hydrating the resulting staple fibers to a moisture content of 5 to 40% by treatment with steam or a water spray. If desired, fiberizing may be effected by centrifugally dispersing the molten streams in a first step and then subjecting the dispersed streams to the high velocity gas blast. The gas is preferably superheated steam, hot compressed air or a fuel gas flame. The thus-hydrated product may be foamed by conventional foaming techniques.

Alkali Metal Aluminoborosilicates

D. Rostoker; U.S. Patent 3,793,039; February 19, 1974; assigned to Corning Glass Works describes a process for the production of inexpensive, lightweight articles of foamed glass which will be essentially free from devitrification and demonstrate good resistance to thermal shock, i.e., will exhibit a coefficient of thermal expansion less than $50 \times 10^{-7}/°C$.

The base material consists of rock selected from the group of finely divided clay, volcanic ash, weathered volcanic ash, and mixtures thereof and comprising at least 50% by weight of the total batch. Essentially all of the batch particles will pass a standard U.S. 200 mesh screen with at least 10% thereof being finer than 5 microns in diameter. The cellulating agent comprises about 0.1 to 2% carbon and is selected from the group consisting of carbon, an inorganic compound which thermally decomposes below the cellulating temperature to yield carbon, and mixtures thereof.

The flux comprises about 3 to 10% alkali metal oxide (R_2O) where K_2O constitutes at least 60% by weight of the total. K_2O rather than Na_2O is used as the flux to assist in maintaining the coefficient of thermal expansion at a low level and to stabilize the glass against devitrification. A B_2O_3-containing material is included in amount to yield B_2O_3 contents ranging between about 7 to 18% by weight. The combination of base material, cellulating agent, flux and B_2O_3 constitutes at least 75% by weight of the anhydrous batch with up to 25% of the batch being added silica, alumina, and/or grog.

Grog is waste fired material that has been finely ground. The principal batch constituents are silica, alumina, alkali metal oxide, and boric oxide. These are present in weight percent as calculated from the batch in approximately the following amounts: 50 to 80% SiO_2, 10 to 20% Al_2O_3, 3 to 10% R_2O and 7 to 18% B_2O_3 with the $SiO_2:Al_2O_3$ ratio ranging between about 3:1 to 7:1. The total of all impurities should not exceed about 10% by weight.

Example: A batch was prepared from the following ingredients: (a) 84.4 lb of air floated Gonzales bentonite clay, having as its principal clay mineral montmorillonite with an approximate oxide composition, in weight percent, of 77% SiO_2, 16% Al_2O_3, 0.8% Na_2O, 0.5% K_2O, 1.3% CaO, 3% MgO and 1.4% Fe_2O_3 (The clay also contained about 10% adsorbed water and 6% water of hydration.);

(b) 10.8 lb of anhydrous B_2O_3; (c) 3.8 lb of caustic KOH; and (d) 1 lb of anhydrous sodium acetate. The composition of the overall batch, as calculated on the oxide basis in weight percent, was 63.5% SiO_2, 13.8% Al_2O_3, 12.4% B_2O_3, 1.6% Na_2O, 4.2% K_2O, 2.5% MgO, 1.3% CaO and 0.7% Fe_2O_3.

The ingredients were blended together and then ball milled for 2 hours to secure uniform intimate mixing and grain sizing. A screen analysis demonstrated that all the material passed a standard U.S. 200 mesh screen and more than 25% thereof was finer than 5 microns. About 6 lb of water was added thereto, through spraying, to yield some granulation.

A tunnel kiln was employed for firing which had a moving 6 foot wide stainless steel belt running therethrough. A thin (0.025 inch) sheet of asbestos paper perforated with $\frac{1}{32}$ inch diameter holes on $\frac{1}{4}$ inch centers was placed on the belt as protection from the foam. The batch was spread over the asbestos paper in an even layer 27 inches wide and $\frac{3}{8}$ inch deep and then passed through the kiln which was operating according to the following schedule: kiln entrance at 400°C; temperature raised to 950°C in 20 minutes; held at 950°C for 10 minutes; temperature reduced to 650°C in 35 minutes; temperature reduced to 520°C in 55 minutes (annealing step); temperature reduced to exit temperature of 100°C in 30 minutes and total time in kiln, 150 minutes.

The resultant fired foam glass was annealed and shaped to an even 1½" thick. The bulk density thereof was about 0.42 g/cc, and the coefficient of thermal expansion (25° to 300°C) was 46 x 10^{-7}/°C. No crystallization within the structure was observed through x-ray diffraction analysis and no deviations from the linear were observed in the plot of thermal expansion. A chemical analysis of the foam indicated 11.85% B_2O_3, thereby demonstrating the essentially complete retention of B_2O_3 from the batch. The average modulus of rupture was about 240 psi and the thermal conductivity was 0.59 Btu/in/hr/ft²/°F. This latter very low value is believed to be due, at least in part, to the very fine uniform pore size of the body.

Calcium Sodium Aluminodisilicate

W.H. Flank, J.E. McEvoy and J.R. Stuart; U.S. Patent 3,775,136; November 27, 1973; assigned to Air Products and Chemicals Inc. describe a nonzeolitic crystalline calcium sodium aluminodisilicate having low density, low surface area, minimized moisture sorption propensities, and open pore structure which is prepared by heating at 900° to 1150°C for 10 to 120 minutes a zeolite corresponding to $(CaO)_{0.5-0.9} \cdot (Na_2O)_{0.1-0.5} \cdot Al_2O_3 \cdot 2SiO_2 \cdot yH_2O$ in which y has a value such that the water content is about 0.2 to 3% by weight of the zeolite.

Foams produced in accordance with this process have a volume shrinkage of less than about 0.7% when maintained at 1250°C for 24 hours; a moisture absorption of less than 0.1% when maintained at 95% relative humidity for 24 hours at ambient temperature; and a compressive strength of at least 1.5 kg/cm².

Foam-Metal Composite

A process described by *K. Oshima, T. Yasumoto, M. Higashikuze and M. Kawamoto; U.S. Patent 3,922,414; Nov. 25, 1975; assigned to Toray Industries, Inc., Japan* provides a composite structure of alkaline silicate foam possessing the properties of higher mechanical strength, lower thermal conductivity, higher sound transmission loss, light weight, fire resistance, thermal stability and nonflammability at a lower cost.

In the process of foaming alkaline silicate by heating, alkaline silicate, whose molar ratio of SiO_2/R_2O (R is Na and/or K) is 4 to 8 and whose water content is 20 to 170% on a dry basis, and metal sheet, in a thickness of 0.05 to 5 mm and linear thermal expansion coefficient of 5×10^{-6} to $20 \times 10^{-6}°C^{-1}$, are stacked into the formation of metal sheet/alkaline silicate layer, or metal sheet-alkaline silicate layer/metal sheet.

By utilizing the above process in the reinforcement of brittle board: (1) less equipment and adhesion time is needed; (2) the desirable characteristics of alkaline silicate such as fire resistance, thermal stability, nonflammability, etc. are retained as the result of not using organic adhesive; (3) ease and efficiency in handling with few destructions occurring during production and construction are achieved as a result of the foam being reinforced during production; and (4) after production of the alkaline silicate foam is completed, release of the foam from molds is very easy and no problems are presented.

FOAMED CLAY, CEMENT
AND OTHER PROCESSES

FOAMED CLAY

Hydrogen Peroxide Gas Generator

J. Magder; U.S. Patent 3,944,425; March 16, 1976; assigned to Princeton Organics, Inc. describes the production of foamed clay compositions by blending of a formulation comprising:

Constituent	Wt % of Dry Ingredients
Clay	~20-95
Hydraulic cement	~ 4-35
Inert particulate lamellar foam stabilizer	~0.2-30
Water	~21-70
Gas generating agent	*

*Sufficient to provide a final density of ~10–105 pcf.

The order of addition of the ingredients may be varied, and the preferred order depends mainly on the specific equipment used for mixing and the particular gas generating agent used. After the gas generating agent and the hydraulic cement have been incorporated, the blending is performed as rapidly as possible and the resulting dispersion is removed from the blending zone before an appreciable proportion of the gas has been generated. The dispersion is then discharged into a mold, or onto a conveyor or other next point of application. The mixture is then allowed to foam and set. With suitably chosen formulations foaming is essentially completed within a few minutes.

The foamed clay compositions so produced exhibit good early green strength, and can be sufficiently hard in about 15 minutes or less to permit demolding and other handling operations. The green foamed clay articles can then be fired either with or without a prior drying step, so as to develop full ceramic maturity and achieve high strengths. The following examples illustrate the process.

Examples 1 through 6: A dry batch was prepared by thoroughly blending the dry ingredients in the proportions by weight given in Example 1 in the table below using an 8 quart twin shell Patterson-Kelly blender, and 3 kg of blender charge. The water mix was prepared by stirring the sodium silicate solution and then the hydrogen peroxide solution into the water in the proportions shown in the table.

One hundred grams of the dry batch was thoroughly and uniformly blended into 38.3 grams of water mix, using a high shear laboratory stirrer. Total mixing time was 20 to 30 seconds. The dispersion, which had the flow characteristics of a deflocculated kaolin casting slip, was poured into an open mold 3 inches square in area and 3 inches deep. The formulation foamed for about 8 minutes, and then hardened. About 1½ to 2 hours after casting, the foamed clay was hard enough to handle without breakage. The foam was demolded, and allowed to dry in air overnight. Air dried density was 21 pcf and linear drying shrinkage less than 1%.

The dried foam was then fired by heating in an electric muffle furnace to 2350°F over 3.5 hours, soaking at 2350±25°F for 1 hour, and then allowed to cool down over about 6 hours.

Fired density was 27 pcf and linear firing shrinkage 14%. The foam decreased in weight a total of 19% between drying and firing. In another experiment, a sample of green foam prepared as in Example 1 was fired on the same schedule; the firing was begun 2.5 hours after casting without any prior drying of the foam. There was no apparent difference in fired properties from those of the predried fired foam.

The fired clay foam satisfied the requirements for Group 23 insulating firebrick, ASTM Classification C155-70. The pores were substantially spherical in shape, only partially interconnected, and extremely narrow in size distribution. Average pore diameter was about 1.3 mm (1,300 microns). X-ray diffraction analysis indicated mullite and cristobalite were the crystalline phases present in the largest proportions. Cold crushing strength was 380 pounds per square inch by ASTM test C93-67, as compared to 145 pounds per square inch for a commercial Group 23 insulating firebrick having a density of 32 pcf.

The formulations of Examples 2 through 6 were blended, foamed, dried and fired similar to the procedure used for Example 1. In each case the fired clay foams were white or off-white in color, extremely uniform in appearance and narrow in pore size distribution. Compressive strengths were high for the densities obtained.

Examples 4, 5 and 6 illustrate the replacement of raw clay by calcined clay to reduce firing shrinkage. Low firing shrinkage allows the production of relatively large unit dimensions on fast firing cycles, without excessive warpage or cracking.

The gas generating agent in each of the Examples 1 through 6 is hydrogen peroxide catalyzed by manganese dioxide ore. The lamellar foam stabilizer is platey talc. If the lamellar foam stabilizer is omitted from any of the formulations, the gas evolved by the gas generating agent is not retained by the dispersion, but escapes and the foam collapses.

	Parts by Weight					
EXAMPLE	1	2	3	4	5	6
Kaolin, CW-L	76.6	86.1	90.9	56.3	37.5	—
Calcined Kaolin −35 M	—	—	—	18.8	37.5	73.9
Calcium aluminate, CA-25	19.1	9.6	4.8	18.8	18.9	18.5
Platey talc	3.8	3.8	3.8	5.6	5.6	7.4
Manganese dioxide	0.5	0.5	0.5	0.5	0.5	0.23
Total dry batch:	100.0	100.0	100.0	100.0	100.0	100.0
Sodium silicate, Type N	—	—	0.5	0.14	0.28	—
35% Aqueous hydrogen peroxide	2.2	2.2	2.2	2.2	2.2	2.1
Water	36.1	36.1	35.6	30.6	30.4	25.6
Total water mix:	38.3	38.3	38.3	32.9	32.8	27.7
Green Foam Properties						
Air dry density, pcf	21	23	23	23	25	24
Drying shrinkage, linear %	<1%	<1%	<1%	<1%	<1%	<1%
Handling time, hr.	2.	3.	5.5	3.5	8.	16.
Fired Foam Properties						
Average pore diameter microns	1300	1300	1000	800	700	1500
Density, pcf	27	30	25	23	29	24
Color	white	white	white	light grey	white	white
Firing shrinkage, linear %	14	15	9.0	2.3	5.5	2.8

Flue Dust, Clay and Oxidizing Agent

C.B.A. Engström, H.G. Klang and G. Persson; U.S. Patent 3,942,990; March 9, 1976; assigned to EUROC Administration AB, Sweden describe a method for producing foamed ceramics from a starting material which contains one or more components which when heated form a viscous, sintered and porous body, and a relatively difficultly oxidized pore-forming agent for creating the porous structure.

The process is mainly characterized by using as a starting material for producing foamed ceramics a composition containing (1) at least one waste product rich in silica and containing readily oxidizable substances which, when heated, are themselves capable of producing uncontrollable pore formation and/or an undesirable melt, and (2) a strong oxidizing agent, the quantity of oxidizing agent being such that the oxidizable substances are oxidized to eliminate or to reduce the uncontrollable pore formation and/or the melting. The desired pore structure is obtained by oxidation of the difficultly oxidizable carbonaceous pore-forming agent.

One common feature of the starting materials used in the process, and having the character of waste material, is that they shall have a relatively high silica content and be very finely divided. The waste materials, i.e., dust removed from waste gases emanating from electrometallurgical processes and dust trapped in stone crushing plants, for example in conjunction with the manufacture of macadam, possess the desired combination of properties.

In order for an acceptable foamed ceramic to be obtained when sintering in accordance with the process, the starting material shall have the following chemical composition. With regard to each specific raw material used, however, a test must be made for the purpose of determining in each particular case the suitable analyses.

60–75% SiO_2	(glass former)
5–13% Al_2O_3 + Fe_2O_3	(glass stabilizer)
~2% CaO	(glass stabilizer)
0–6% MgO	(glass stabilizer)
10–15% Na_2O + K_2O	(fluxing agent)

It has been found that the reaction rate can be increased in the melting process by using alkali hydroxide as the fluxing agent. More specifically, it has been found that at least 80% of the fluxing agent should be alkali hydroxide. This hydroxide is very active, since it dissolves to a concentrated aqueous solution and is able to retain its activity when heated. The alkali hydroxide is very reactive at those temperatures envisaged for forming the foamed ceramic according to the process.

Further, alkali hydroxide forms eutectic melts with other components, primarily with silica and Al_2O_3. The quantity of "active alkali" should be carefully adjusted in relation to the other components of the composition, since excessive quantity of active alkali can result in an impaired resistance of the product to water, while an insufficient quantity prevents a sufficient quantity of molten phase from being formed at the desired low temperatue, at which products having a relatively high volumetric weight are obtained.

The following procedure can be applied when producing foamed ceramics: a finely divided waste dust rich in silica and optionally admixed with natural mineral (together approximately 80% of the dry weight of the composition) may be mixed, e.g., by wet grinding in a ball mill, with an approximately 10 to 15% aqueous solution of alkali hydroxide, 3 to 4% hydraulic or latent hydraulic binding agent, e.g., Portland cement binders or blast furnace slag, 1 to 3% of a strong oxidizing agent, e.g., silicon carbide, whereafter the thus obtained slurry may be dewatered to produce thin walled, porous nodules or a fine-grained product. The product is then heated very rapidly to approximately 800° to 900°C in special molds, at which temperature a melt is formed.

The reaction between the pore-forming agent and the remaining surplus of oxidizing agent causes a gaseous product to be formed, which forms pores in the melt. The porous product obtains a very uniform pore structure. The product is then rapidly chilled to approximately 600°C, whereafter it is allowed to cool slowly in a cooling furnace.

Example 1: An aqueous raw material mass having the following composition was produced:

Flue dust	55 kg, SiO_2 content approximately 90%	Specific surface, approximately 30,000 cm²/g
Dust from crushed stones	10 kg, SiO_2 content approximately 75%	Specific surface, approximately 4,000 cm²/g
Glacial clay	17 kg, SiO_2 content approximately 58%	—
Granulated blast furnace slag	3 kg, SiO_2 content approximately 40%	Specific surface, approximately 3,000 cm²/g
Sodium hydroxide	13 kg, SiO_2 content approximately commercial quality	—
Pyrolusite	1.8 kg MnO_2, commercial quality	—
Silicon carbide	0.2 kg, particle size <5 μ	—
Water	55 liters	

The mass of raw material was permitted to set at 80° to 90°C, and the hard mass was disintegrated in a suitable manner to a particle size <4 mm, whereafter the material was transferred to a furnace and heated to 600°C, since practically all water must be removed from the mass in order for the final pore formation not to be destroyed. During this process a primary pore forming process takes place in the material, as the result of a reaction between alkali hydroxide and finely divided metal particles originating from the flue dust.

The primarily expanded material is transferred to molds without being crushed to smaller particle size, either immediately or subsequent to being cooled, for continued heating to 800° to 850°C for 1 to 2 hours, during which time the final formation of pores takes place.

Subsequent to being chilled quickly to a temperature of approximately 600°C, the product was allowed to cool.

The properties of the foamed glass product thus produced are found to be very satisfactory: uniform pore distribution with closed pores, low volumetric weight of 250 kg/m³, good compressive strength and very good resistance to water.

Example 2: A slurry was prepared by wet grinding the following components:

Flue dust	29.8 kg	—
Sand	34.5 kg, SiO₂ content 74%	Specific surface 4,000 cm²/g
Blast furnace slag	3.0 kg	—
Sodium hydroxide	11.6 kg	—
Pyrolusite	0.9 kg	—
Silicon carbide	0.2 kg	—
Water	25 liters	—

The slurry was spray dried to a moisture content of 0.2%. The dried material was transferred to molds and heated directly to 830°C, and afterward cooled as above.

The foamed glass product obtained presented a uniform and well-formed pore structure, while the volumetric weight was measured to 270 kg/m³ and the water resistance was found to be good.

Cooling Method for Heat Bloated Particles

A. Jebens and R.H. Westergaard; U.S. Patent 3,850,715; November 26, 1974; assigned to Sentralinstitutt for industriell forskning, Norway describe a method for producing an element of inorganic foam material from pellets, granules or powder, e.g., a clay-containing material, glass powder or aggregates of slag, flotation waste and similar expandable material. The element is built up by depositing a number of layers successively. Each layer is bloated before the next layer is deposited. After two or more layers have been bloated, a forced cooling is performed from the bottom part of the bloated material for the remaining bloating process. The temperature difference between the top and bottom parts rises to between 80° and 600°C and preferably 450°C. The forced cooling is carried out by drawing cooling air past the bottom part. The following example illustrates the process.

Example: In order to illustrate the process there is given below an example of how the process may be performed as a batch operation. As raw material is used a natural occurring clay containing iron oxides (for example between 4 to 8% by weight) and some organic matter (for example corresponding to a carbon content about 0.5% by weight). The clay used expanded during ten minutes when a dry pellet (7 to 9 mm diameter) was brought into an oven kept at 1130° to 1150°C, and it has been found that many Norwegian clays are suitable.

The kiln used was heated by electric heating elements in the ceiling, but blocks have also been made in the same kiln using an oil burner as heat source. The top of the kiln could be lifted up in order to unmold and remove the block. The bottom of the kiln was made of thin plates of refractory material. Under these plates were cooling channels for circulating cold air, and under these channels was thermal insulation. Upon the bottom plates was a mold consisting of loose heat resisting bricks covered with slip agent in order to prevent the block from sticking to the sides. Upon the plates inside the mold was a layer of sand preventing the block's sticking to the bottom. Before the process started, the kiln was heated to 1200°C and kept at this temperature for several hours in order to establish stable conditions. Then a first layer of 7 to 9 mm diameter dried pellets of clay was introduced into the kiln. This was done by means of a device making it possible to deposit one single layer of pellets.

When the area inside the mold was covered by a pellet layer the temperature in the oven decreased considerably but increased again to about 1200°C in approximately 10 minutes. The pellets had then expanded and melted together to a foamed material of 2.5 to 3.0 cm thickness. At this moment a new layer of pellets was introduced into the kiln, and at 10 minute intervals the procedure was repeated.

In order to stabilize the pore structure forced cooling from below was started about the time when the 4th layer of pellets was fed into the kiln. When 10 layers of pellets had been fed into the kiln, that is after 100 minutes approximately, the dimensions of the block were 60 x 80 cm and 25 to 30 cm thick. The density of the block could vary between 0.30 and 0.40 g/cm³. At this moment the temperature of the top of the block could be 1150° to 1180°C and the bottom 700° to 800°C depending on the cooling from below. After removal of the mold the block was transferred to an electrically heated cooling chamber, the cooling rate of which could be programmed.

During the first 6 hours the temperature in the cooling chamber is kept at about 680°C, the solidifying temperature of the material, and when the temperature in the middle of the block is for example 50°C higher than the temperature on the surface of the block, the temperature in the cooling chamber is decreased at a rate of 40°C per hour, corresponding to approximately 80°C higher temperature in the middle of the block than on the surface.

After 2.5 hours the temperature in the middle decreased from approximately 730° to 630°C and it is certain that the interior of the block has solidified and the block is tension free. From now on the temperature difference between the interior and the surface may be increased, for example to 100°C or as much as can be done without forming preliminary tensions that result in cracks, corresponding to using an increased cooling rate. To put it more precisely, it is possible to apply a higher cooling rate than the cooling rate which will maintain the block absolutely tension free during the continued cooling. With a cooling rate of 35°C per hour at this moment (630°C in the middle of the block) the cooling rate during the continued cooling is maintained proportional to the thermal conductivity of the material. This means that the cooling rate will be approximately as follows: 25°C per hour at 500°C; 22°C per hour at 400°C; 18°C per hour at 300°C; 15°C per hour at 200°C and 13°C per hour at 100°C, the temperatures referring to the middle of the block. By using the above procedure it is possible to obtain crack-free blocks with a cooling time of less than 36 hours.

In one case when this procedure was not used one could hear the block crack due to permanent tensions being built up after the block had been taken out of the cooling chamber and while the temperature equalization was taking place. The temperature in the middle of the block was then measured immediately (through a hole made when the block temperature was 700°C) and found to be 25°C higher than the surface temperature. Thus a test to check whether the cooling rate is too high, is to listen to the block as it cools.

The blocks were taken out of the cooling oven when the surface temperature was 25° to 40°C and the temperature in the middle 70° to 100°C, and the cutting was done before the temperature difference between the interior and the surface had dropped below 50°C. The cutting to precise dimensions was performed by cutting off approximately 5 cm of the four (imperfect) edges, dividing the block in two lengthwise and cutting off one or two centimeters of the top and the bottom. A blank of 80 x 60 x 28 cm would thus give two blocks of 70 x 25 x 25 cm.

Bloating Process

B.D. Brubaker and N. Waldman; U.S. Patent 3,673,290; June 27, 1972; assigned to The Dow Chemical Company describe a foamed ceramic material having a relatively small and uniform cell structure. The material is suitable for structural or insulating purposes. The continuous process for producing the foamed ceramic material comprises dropping a bloatable ceramic composition, in particulate form through a heated zone thereby fusing and bloating the particles, and subsequently collecting the fused particles. To produce the desired cellular structure, the process does not fully bloat the particles during dropping, thereby delaying part of the bloating of the particles until after they have been collected as an agglomerated slab.

It is important in carrying out the process that during the passage downward through the heated zone, the particles are not bloated to full capacity, i.e., the particles should be bloated only to from about 50 to 90% of capacity. The maximum bloating capacity (minimum density) of the particles is determined by a series of runs in which aliquots of the foamable ceramic are dropped through the heated zone at incrementally increased temperatures.

A plot of the bulk densities (as ordinate) of the resulting bloated ceramic particles against temperatures (as abscissa) yields a cup-shaped curve, the minimum of which defines the particle density at maximum bloating capacity. Particles bloated within the above range will retain a vestige of their original shape. For example, substantially all of the particles will be nonspherical in shape, assuming that the particles were produced by extrusion or "slinging" and were therefore nonspherical before they were dropped. A slinging process and apparatus is described in U.S. Patent 3,259,171, which is one means suitable for the preparation of particles.

It has been found that if the particles have fused and bloated to hollow essentially unicellular spheres, the point of minimum density will have been passed and the particle density will have started to increase above the minimum point. Thus, further expansion after agglomeration is unlikely. By not utilizing the full bloating capacity of the particles during dropping, the particles can be bloated further after they are collected. The gases which normally escape as the particles are bloated to maximum potential are trapped within the mass of collected particles to produce the desired low density foams.

Fluxing Method

J.W. North; U.S. Patent 3,967,970; July 6, 1976 describes a method of preparing cellulated or bloated products from natural material sources using clay or a clayey material as a starter and using treating agents capable of being placed in aqueous solutions for absorption into ultimate particles of the material.

It has been found that a clayey material such as clay, shale, or bentonite provides a natural and widely available material in which there is a uniform and wide dispersion of a natural amount of iron in the form usually of iron oxide (Fe_2O_3). The natural size of the particles or pellets that are present or can be obtained by a mulling operation are of 1 to 2 mm in diameter down to fine dust and provide a satisfactory starting material.

This starting material is prepared, for example, by dissolving clay in water and then screening it to remove certain impurities. Then the clay solution is subjected to a dewatering filter after which the material is dried. A water solution containing the flux and the carbon material is then added to the clay starting material including the widely dispersed iron component and mixed as described below. For example, a water solution of hydroxide acts as flux and the sugar acts as a source of carbon. The mixture is put into a muller and mulled about 15 minutes to produce moist macroscopic pellets or particles about one to two millimeters in diameter, as a starting material.

The flux used should be one which is soluble in water and which is meltable at a relatively low temperature (at about 700°F in the case of sodium and potassium hydroxide) which is capable of being reabsorbed in its molten state at a temperature below its reaction temperature so that it is capable of coating the ultimate particles or plates of the clay. Both hydroxides and phosphates of sodium, potassium and lithium are suitable but sodium and potassium hydroxide are preferred because of cost. Sugar, molasses or urea also can be added to the solution as a source of carbon and when in solution will penetrate and coat the ultimate particles or pellets.

In subsequent heating when the water of the solution evaporates, sugar is left on the surface of the pellets or particles and the sugar melts during continued heating and reabsorbs into the pores, when the sugar is further decomposed by heating it leaves a carbon residue on substantially each ultimate particle. This provides for a more uniform fluxing action and enables the fine ultimate particles to sinter more evenly and trap gases in smaller more regular uniform bubbles as a result of the steady access of the carbon and iron to each other.

The prepared material is put directly into a kiln at full temperature of about 2100° to 2300°F, and carried directly up to a melting temperature as fast as it can absorb heat. Directly upon melting, the flux can reach the iron in the clay and react to form gas so that cellulation or bloating occurs so that a very uniform cellulation or bloating action is obtained. The bloating occurs progressively from the outer surface of the material inwardly as melting occurs.

The time of processing in a kiln may be for from ten minutes to two hours with the ten minutes being used for a one inch thick tile, for example, and the two hours being used for a thicker 8 inch thick slab. The material may be introduced into a conventional roller kiln as a layer and is carried on thin plates either with or without a layer of sand bonded thereon. Side angles loosely fixed in place may be employed to confine loosely the layer of material.

The following are examples of bloated material prepared in accordance with the above described process. The firing is done in a nonoxidizing atmosphere.

Example 1:

	Parts by Weight
Black Hills bentonite	100
Granulated cane sugar	1
NaOH	5
Tap water	30

Dissolve the sugar and then the NaOH in water and then mix with the aggregate in a muller-mixer. Rub through a 10 mesh screen to pelletize. Put in pan as described above and place in hot kiln at about 2160°F for 30 minutes. Take the pan from the kiln and place in an annealing oven and cool slowly to below 300°F before removing to cool completely.

The resulting block has well formed cells, mostly about 1 mm in diameter, and a density of about 12 pounds per cubic foot, and glassy appearance. Approximately the same results are obtained with the equivalent mol percent of KOH replacing the NaOH, or with an equivalent mixture of NaOH and KOH.

Example 2:

	Parts by Weight
Surface clay*	100
Granulated sugar	1
NaOH	9
Tap water	17

*From near Corona, California, reddish colored

Mix and fire as in the first example, except firing temperature is about 2300°F. This gives a block of about 30 pounds per cubic foot, and fine cells.

Cellulated Glassy Bodies

A process described by *J.H. Cowan, Jr and D. Rostoker; U.S. Patent 3,666,506; May 30, 1972; assigned to Corning Glass Works* relates to cellulated glassy bodies, having a uniform closed cellular structure and a density on the order of 0.4 g/cc. These are made by preparing a mixture of clay, sodium hydroxide, other fluxes, and an organic cellulating agent, and then firing the mixture, preferably at temperatures less than about 950°C (1740°F).

Grog is fired batch material which has been reground to a granular form. The addition of this grog aids in producing a crack free body. The amount of grog which may be introduced is related to the density which is desired. For a given set of conditions, large additions of grog will generally increase the resultant density. In this process, wherein it is desired to produce products of density on the order of 0.4 g/cc, additions of more than 35% by weight grog are undesirable.

The special features of the fired product are that it is strong, lightweight, inexpensive to produce, has an extremely uniform cell size and distribution thereof, has a closed cell structure and maintains the configuration of the pressed body.

Normally, the cells are of a generally spherical shape with few, if any, cells being larger than about ¼" diameter. This is highly advantageous since if the sphere size is increased or the sphericity decreases, thin spots in the walls can be created which weaken the body as well as reducing the desired impermeability. Due to an almost total loss of foaming agent at the surface of the body, the surface is somewhat denser than the interior portion. The density of the resultant product is normally less than about 0.5 g/cc, when the body is fired at temperatures less than 950°C. However, with changes in the firing schedule and changes in the batch composition, densities as low as 0.15 g/cc or as high as 0.65 g/cc can also be produced. The bulk thermal coefficient of expansion of the resultant bodies is on the order of $65\text{-}95 \times 10^{-7}/°C$ (25° to 300°C).

Since the body is a foamed glassy body with residual undissolved compounds, a reproducible exact expansion is difficult to obtain and thus the coefficients of expansion are averages. The glassy portion of the cellulated body is basically an alkali aluminosilicate having minor amounts of the alkaline earth metal oxides therein.

The modulus of rupture (MOR) for bodies having a density of about 0.65 g/cc is on the order of 550 psi, while when the density is about 0.15 g/cc the MOR is about 150 psi. The gas content of the cells was analyzed and determined to be primarily CO_2, CO, N_2, and H_2O. The glassy portion of the cellulated body is a partially melted glass containing minor amounts of nonglassy phases including alpha quartz, cristobalite, and plagioclase feldspar.

Examples 1 through 7: A batch to be foamed was prepared from the following batch ingredients: (1) 83 lb of an air floated Gonzales bentonite clay, having as its principal clay mineral montmorillonite, and an approximate oxide composition in weight percent as follows: 77% SiO_2, 16.0% Al_2O_3, 0.8% Na_2O, 0.5% K_2O, 1.3% CaO, 3.0% MgO, and 1.4% Fe_2O_3. This clay also contained about 10% absorbed water and 6% water of hydration; (2) 8 pounds anhydrous granular sodium hydroxide; (3) 8 pounds anhydrous granular sodium carbonate; (4) 1 lb anhydrous powdered sodium acetate.

The above ingredients were mixed and then ball milled for two hours to insure adequate mixing. The mixed batch was then blended with 25 pounds of –14 mesh grog and 6 pounds of water was added by spraying to cause some granulation.

The granulated batch was fed into a mold and pressed at 5,000 psi to form a green body ½" x ½" x 4". The bar was then placed in a furnace at 900°C for 15 minutes to foam. The cellulated bar was removed, placed in another furnace and slow cooled to room temperature. The foamed body was of the same configuration as the green body and had dimensions of $^{11}/_{16}$" x ¾" x $5^5/_8$". The body had a uniform cellular structure and a density of 0.30 g/cc. Several bodies were prepared, but without grog, and fired at the temperatures shown. The bodies all foamed and yielded the densities shown.

Example	Firing Temperature, °C	Density, g/cc
2	800	0.60
3	850	0.40
4	875	0.30
5	900	0.26
6	925	0.25
7	950	0.22

Aluminum Oxide, Glass Frit, Bentonite and Phosphoric Acid

S.L. Leach; U.S. Patent 3,762,935; October 2, 1973 describes a composition
for making a foamed-in-place shaped article having high dimensional stability at
extremes of heat and cold, good insulating properties, and high tensile strength.

The composition includes aluminum oxide, a glass frit, bentonite, and phosphoric
acid. Under certain conditions, the composition could contain aluminum hydroxide and a powdered metal above hydrogen in the electromotive force series, the
powdered metal preferably being aluminum. The dry ingredients are thoroughly
mixed and the phosphoric acid is then added to make a slurry. The slurry is
charged to the mold in the desired amount to produce a product having the
desired characteristics and allowed to foam at ambient temperature. When the
foaming is complete, or has reached the desired degree, the resulting foamed-in-
place shaped article is cured at a temperature below about 200°F.

In the preferred case, after curing the object is heated to between 1000° and
2000°F for a period of time sufficient for the heat to penetrate throughout
the cellular mass. This causes the glass frit to "flux" or melt and flow thereby
forming a thin film over the entire surface area of the cellular mass, that is,
inside the cells as well as outside, to provide a glass coating. The material is
then air quenched.

The uses to which the material of the process can be put include ceiling panels,
structural wall panels, shingles, pipe covers, air duct covers, electrical components,
fireproof vaults, heat shields, containers for keeping foods and other products
either hot or cold, handles for cooking utensils and the like, flower pots, and
shipping containers. The fact that the material can be used at temperatures of
greater than 2000°F on a continuous basis makes it particularly useful.

Example 1: A slurry was made of the following ingredients:

	Grams	Percent
Al_2O_3	150.0	41.06
Glass frit	40.0	10.95
Bentonite	5.0	1.37
Al powder	0.3	0.008
H_3PO_4	70.0	46.54

After thoroughly mixing the ingredients the resulting slurry was charged to a
mold and allowed to foam in place for 30 minutes at room temperature. The
mold was then placed in an oven at about 200°F for 2 hours. After cooling
the shaped article was removed from the mold and exhibited desirable mechanical
and thermal properties.

Example 2: A shaped article was made according to the procedure of Example
1, but after curing at about 200°F for 30 minutes, it was heated to approximately
100°F and maintained at that temperature for 7 minutes.

The product was then air quenched and cooled. The product had a nominal
specific gravity of 2.567 and a nominal molecular weight of 121.95. It was
unaffected by 2500°F for 2 hours. The product was also unaffected by water
and organic solvents. It exhibited a hardness of between 7.0 and 9.0 on the
Mohs scale. The K-factor was 0.25.

FOAMED CEMENT

Magnesium Oxychloride Cement

H.C. Thompson; U.S. Patents 3,969,453; July 13, 1976; and 3,963,849; June 15, 1976; both assigned to Thompson Chemicals, Inc. describes a method of making a foamed fireproof product of magnesium oxychloride cement. A porous substrate is impregnated with a foaming mixture of magnesium chloride, magnesium oxide, and frothing agent in water. The mixture hardens with small voids throughout the porous substrate, thus providing a fireproof product of low density. The foamed product may be formed in a mold with fibrous elements introduced into the foaming mixture. The fireproof products of relatively low density are particularly valuable for building and construction purposes. Magnesium powder is the preferred frothing agent which, in combination with a surfactant, induces a large volume of small bubbles that remain in the composition as it is set.

Insoluble fillers may be added where desired. For example, insoluble calcium carbonate, such as powdered limestone or marble dust, silica flour, magnesium carbonate, and other insoluble materials may be introduced. Ordinarily, soluble salts should be avoided because they can be leached out and reduce the strength of the foamed cement.

In certain instances, the material used as a filler may also have ingredients contained therein which also functions as a frothing agent. For example, black slate from the Chili Bar Quarry in Placerville, California, has the following analysis upon calcining:

SiO_2	59.70%	K_2O	3.77%
TiO_2	0.79%	Na_2O	1.35%
Al_2O_3	16.98%	CO_2	1.40%
Fe_2O_3	0.52%	FeS_2	3.82%
FeO	4.88%	Carbon	0.46%
CaO	1.27%	Sundry others and water	
MgO	3.23%	present below 110°C	0.70%

While it is not precisely understood what component of slate having the foregoing analysis serves as a foaming agent, it has been determined in practice that the addition of a foaming agent is unnecessary where substantial amounts of finely divided slate of the foregoing analysis are used in the manufacture of foamed products according to the process.

The slate may be added in a variety of amounts depending on the product desired. Preferably, in manufacturing simulated roofing shakes, between 10 and 25 pounds of slate is added per gallon of $MgCl_2$ solution. The following examples illustrate the process.

Example 1: In this example simulated roofing shakes were prepared using molds having the configuration of wooden shakes in two courses of several shakes per course. 72.2 pounds of 20 mesh slate having the analysis indicated in the table was blended with 24.8 pounds of magnesium oxide and 3 pounds of iron oxide pigment. After these ingredients are thoroughly mixed, 21.4 pounds of the blended mixture is added to one gallon of 22°Bé magnesium chloride liquid. The liquid and powder are thoroughly mixed and the mixture is sprayed onto plastic molds simultaneously with the application of chopped fiber glass strands at the rate of 0.8 ounce per square foot of mold area. The slate reacts with the cement ingredients to foam the cement in the same manner as magnesium powder does. The fiber glass strands are cut into lengths of between 1 and 2 inches and applied to the molds at the same time as the mixture is sprayed to give simultaneous application.

The sprayed molds are stacked and held for two hours until they are hard enough to be removed. They are then trimmed and packaged for shipment. The molded product has the appearance of wooden shakes and can be easily applied as a roofing material with hammer and nails. The product is completely fireproof, as contrasted to wooden roofing shakes.

Example 2: In the preparation of the froth, 19¼ pounds of magnesium chloride were added to 4¼ gallons of water. The specific gravity was measured and additional magnesium chloride was added to adjust the specific gravity to 22°Bé. The combined total makes about 5 gallons. 200 grams of lactic acid were added to the aqueous mixture. Magnesium oxide was then measured in an amount equal to 5.5 pounds magnesium oxide to each gallon of magnesium chloride in water. Then, 32 grams of powdered magnesium, which is about 0.1% of the dry weight of the magnesium oxychloride cement mixture and 100 grams of surfactant (Rohm & Haas X-100) were added to the magnesium oxide component. The surfactant constituted about 0.5% of the dry weight of the total mixture. Chopped glass fibers, in an amount equal to 3% of the total were added to the dry component and mixed thoroughly.

The dry magnesium oxide component was added to the aqueous magnesium chloride component and the mixture immediately began to froth. It was poured into the female part of a mold having a decorative configuration for a building ornament. The male part of the mold was closed over the female part and the cement permitted to set. After 2 hours the molded fireproof product was removed from the mold.

Blowing Agent and Foam Stabilizer Combination

S. Uogaeshi; U.S. Patent 3,834,918; September 10, 1974; assigned to Teijin Limited, Japan describes a raw batch containing fine bubbles which may be used for the formation of porous architectural structures. The batch comprises a homogeneous mixture of 100 parts by weight of a hydraulic substance, 0.0001 to 0.1 part by weight of a blowing agent, 0.001 to 1.0 part by weight of a foam stabilizer and 70 to 200 parts of water.

Typical examples of the hydraulic substances are the cements such as Portland cement and aluminous cement. Blast furnace slag can also be used as hydraulic substance in this process, though its hydration reaction is mild and much time is required for its setting. Further, though a mixture of quicklime and siliceous sand only sets up a hydration reaction at elevated temperatures and high pressures such as 180°C and 10 atmospheres, this also is included in the hydraulic substances.

By the term "blowing agent" is meant a substance that can introduce a multiplicity of fine air bubbles into a slurry composed of a hydraulic substance and water when the blowing agent is added to the slurry, and the slurry is stirred in air. Typical examples of this blowing agent are saponin, resined soap, colloid gelatin, etc. Of these, saponin is the most preferred blowing agent.

Typical examples of the foam stabilizer are the water-soluble cellulose derivatives such as methyl cellulose, ethyl cellulose, hydroxypropylmethyl cellulose, etc., as well as the water-soluble synthetic polymers such as polyvinyl alcohol, the salts of polyacrylic acid, the salts of polymethacrylic acid, etc. Of these, the preferred foam stabilizers include hydroxypropylmethyl cellulose, polyvinyl alcohol and the salts of polyacrylic acid. In the examples, the parts are in all cases on a weight basis.

Example 1: A mixing tank equipped with a stirrer was charged with 100 parts of Portland cement, 30 parts of sand, 0.01 parts of saponin, 0.5 part of hydroxypropylmethyl cellulose and 100 parts of water, following which the mixture was stirred and mixed at a stirring speed of 250 rpm to prepare a raw batch. This raw batch had stably incorporated therein fine bubbles which were homogeneously dispersed throughout the batch. In addition, this raw batch possessed good fluidity.

When this raw batch was poured into a form and allowed to set, a porous structure was obtained without sinkage of its placed height. On microscopic examination of this structure, fine voids were found to be homogeneously distributed throughout the structure. Hence, there was also uniformity in the mechanical strength of this structure.

Control 1: When, by way of comparison, the experiment was carried out exactly as in Example 1 but without using the hydroxypropylmethyl cellulose, a raw batch having fine bubbles homogeneously dispersed throughout the batch could not be obtained. The reason for this was that the fine bubbles only formed in the upper portion of the mass that was being stirred and mixed, and moreover these bubbles tended to become disintegrated and disappear, with the consequence that the lower portion of the mass was maintained in a pasty or slurry state not containing any bubbles. When this raw batch was placed to a height of 300 mm, a separated layer of a height of about 65 mm was formed.

Control 2: Again, by way of comparison, the experiment was carried out as in Example 1, except that saponin was not used. In this case, there was a drop in the air entrainment capacity of hydroxypropylmethyl cellulose, and fine bubbles could not be formed adequately. Hence, a raw batch such as obtained in Example 1 could not be obtained.

Example 2: A slurry was prepared by mixing 100 parts of Portland cement, 50 parts of sand and 75 parts of water with stirring. Separately, a homogeneously foamed mixture was prepared by adding 0.012 part of saponin and 0.6 part of hydroxypropylmethyl cellulose to 75 parts of water followed by vigorous stirring. Next, the foregoing slurry was introduced all at once to this foamed mixture, and the resulting mixture was kneaded together with stirring to prepare a raw batch containing fine bubbles homogeneously dispersed therein.

When this raw batch was placed to a height of 300 mm and allowed to set, only a sinkage of 1 mm was noted. The resulting structure had a compressive strength of 40 kg/cm^2. On microscopic examination of this structure, fine voids were found to be homogeneously distributed throughout the structure.

Examples 3 and 4 and Controls 3 and 4: Various types of raw batches and porous structures were prepared by operating as in Example 2. The details are shown in the following table.

	Raw Batch: Slurry: Portland Cement (parts)	Sand (parts)	Water (parts)	Saponin (part)	Foamed Mixture: Hydroxypropyl Methylcellulose (part)	Water (parts)	Water-Cement Ratio	Porous Structure: Sinkage from 300 mm Placed Height (mm)	Time Required for Setting (hr)	Compressive Strength (kg/cm²)
Example 3	100	50	35	0.012	0.6	35	0.7	2	5.5	38
Example 4	100	40	50	0.012	0.6	150	2.0	1	5.5	40
Control 3	100	50	30	0.012	0.6	20	0.5	10	6.0	30
Control 4	100	50	80	0.012	0.6	220	3.0	7	7.0	27

Lightweight Mineral Aggregate

According to a process described by *M. Plunguian and C.E. Cornwell; U.S. Patent 3,989,534; November 2, 1976* lightweight foamed compositions are produced from:

(a) A mineral cement such as gypsum cement, Portland cement, calcium aluminate cement or magnesia cement;
(b) A lightweight mineral aggregate, such as perlite, vermiculite, or hollow silicate spheres, in an amount of 10 to 50 parts to 100 parts of the cement;
(c) A film former and viscosifier foam stabilizer which may be either an organic film former such as guar gum and/or a collagen protein colloid, or an inorganic film former such as bentonite or montmorillonite, in an amount of about 1 to 20 parts per 100 parts cement;
(d) A synthetic surfactant which is preferably a combination of anionic and nonionic surfactants in an amount of about 0.1 to 3.0 parts to 100 parts cement;
(e) Water which is used in processing in an amount of 30 to 150 parts per 100 parts cement; and
(f) Air incorporated into the cementitious composition by aeration (i.e., exclusive of the amount of air contained in the lightweight aggregate) in an amount to increase the volume of the cement 5 to 400%, and preferably 20 to 200% for most applications.

The foam compositions may be formed into final products by known techniques such as pumping, spraying, casting, or trowelling to provide settable lightweight foamed compositions whose densities are appreciably less than the density of the employed mineral cement and aggregate composition per se.

The stability of the foams can be improved by the in situ production of a colloidal or structural precipitate from salts in the cementitious compositions, particularly the calcium salts, which will cause the precipitation or salting out by reaction with such organic compounds as sodium abietate, sodium alginate, or sodium carboxymethyl cellulose.

Polyethylene Glycol Ether Surfactant

C.E. Cornwell; U.S. Patent 3,819,388; June 25, 1974 describes a method for the preparation of a cementitious composition having a cellular structure for protecting steel and other structural elements from fire. The process involves spraying directly upon the desired structural substrate a composition containing gypsum cement, filler or lightweight aggregates, water, and ½ to 2% by weight of a nonionic detergent such as a polyethylene glycol ether combined with 9 mols of ethylene oxide. This same composition is suitable and useful for thermal insulation and acoustical soundproofing.

A critical feature of the process involves adding within the mixture the precise amount of a bubble forming additive known as a nonionic detergent, Tergitol 155S9, which consists of a polyethylene glycol ether combined with 9 mols of ethylene oxide. This bubble forming agent provides the means for mixing a cementitious composition that forms a cellular structure with equal distribution of the cells throughout the composition.

A uniform consistency of the density is maintained at all times with the use of this nonionic detergent when mixed with the slurry. This composition is shown in the examples listed below. The first example uses gypsum cement, the second uses Portland cement and the third uses magnesia cement. The lightweight aggregates can be substituted as availability warrants. The use of vermiculite, perlite, a product called Snowdon, or other aggregates may be used for desired strengths and densities.

Example 1:

Ingredient	Parts by Weight
Gypsum cement	50-100
Vermiculite	10-45
Water	100-200
Nonionic detergent	0.25-2.0
Optional (polyvinyl acetate)	1.0-5.0

Example 2:

Ingredient	Parts by Weight
Portland cement	50-100
Vermiculite	10-45
Water	100-200
Nonionic detergent	0.25-2.0
Optional (polyvinyl acetate)	1.0-5.0

Example 3:

Ingredient	Parts by Weight
Magnesia	5-35
Magnesium salt	5-35
Vermiculite	10-45
Water	100-200
Nonionic detergent	0.25-2.0
Optional (polyvinyl acetate)	1.0-5.0

By adding appropriate proportions of the nonionic detergent to such slurry compositions the material will air dry having densities as little as 10 pounds per cubic foot to approximately 35 pounds per cubic foot. The cement to water ratios determine the actual strength of the composition. The amount of the nonionic detergent determines the cellular content.

The liquid polyvinyl acetate is added to the composition when required for the purpose of adding greater adhesion to the surface being sprayed; i.e., overhead structures. The addition of polyvinyl acetate in appropriate proportions does not deter the inherent fireproofing qualities or cellular structure.

The cellular structure of this material has inherent insulating and acoustical values not generally available in other products that are currently used. The spraying, casting or troweled application of this composition lends itself to many applications. Spraying an appropriate thickness of this material over steel will meet ASTM fireproof ratings. When cast or pumped into wall structures, as the inner core, the material provides a "fire wall" barrier. It further provides acoustical soundproofness that is valuable in homes and commercial buildings, as well as in helping the control of ecological sound transmission or suppression.

The equipment required to form this slurry composition is all standard. The mixer can be either the standard paddle mortar mixer or the ribbon type. The rotating drum type is not recommended. Mixing times need not exceed 2 to 3 minutes per batch.

Continuous High Shear Mixing Technique

J. Magder; U.S. Patent 3,729,328; April 24, 1973; assigned to Princeton Organics, Inc. describes a method for continuously producing foamed cement-based cellular material by subjecting specific liquid and solid components to high-shear mixing for very short periods of time and then discharging the mixture so that it can foam and set promptly. The liquid material which is mixed consists principally of water, and preferably has a viscosity of less than about 2.5 centipoises. The solid component which is mixed contains a particulate water-settable (hydraulic) cement and an inert particulate lamellar foam stabilizing agent. A gas-forming agent will also be included as part of either or both of the liquid and solid components.

The time of mixing is not greater than about four seconds and preferably less than about one second and the high-shear mixing preferably provides a maximum nominal velocity gradient greater than about 500 seconds^{-1} for substantially all of the material. The maximum nominal velocity gradient is defined for this purpose as the maximum value of the ratio of relative speed of two surfaces of the mixer between which substantially all of the mixture is passed, to the distance between the two surfaces.

Preferably the mixer comprises a rotor disposed along the axis of the tubular chamber into which the ingredients to be mixed are continuously fed, the rotor comprising helical screw means disposed along the length of the rotor for moving the mixture through the chamber to its outlet and for providing at least part of the mixing action, together with mixing pins extending radially from the axis of the rotor to within a small distance of the chamber wall to effect further mixing action.

The resultant product is of increased compressive strength and characterized by the fact that substantially all of the pore volume is provided by pores having effective diameters outside a range extending from about 2 microns to about 300 microns, and preferably at least about 97% of the pore volume is provided by pores outside the range.

CARBON FOAM

Lignin Solution

Moldings of carbon of high pore volume, so-called foamed carbon, have for some time been commercially available. They are used above all in the fields of heat insulation up to very high temperatures, filtration of chemically reactive and corrosive gases and liquids and, finally, catalysis, where they act as a carrier material of high available surface area. They are manufactured by carbonization of rigid synthetic resin foams according to two methods: a foam which on carbonization only gives a low yield of carbon, such as, for example, a polyurethane foam, is impregnated with a synthetic resin which gives a high carbon residue, for example with a phenolic resin, or the foam to be carbonized consists directly of a plastic which gives a sufficient yield of carbon, such as, for example, a phenolic resin foam. In each case, a relatively expensive organic synthetic material has to be used, which furthermore must not soften or melt on heating.

M. Mansmann and G. Winter; U.S. Patent 3,894,878; July 15, 1975; assigned to Bayer AG, Germany describe a process for making carbon foams starting with lignin. In the process, an aqueous solution of lignin is formed into a shaped structure such as a molding or a film having a thickness of about 500 μ or more, preferably in excess of about 1 mm, and subjected to heat or other conditions to form a foam. The foam is thereafter consolidated, i.e., fixed in physical structure, and is subsequently carbonized.

Particularly suitable starting materials for the manufacture of foamed carbon which is as pure as possible are free ligninsulfonic acid and ammonium ligninsulfonate. Ammonium ligninsulfonate gives off NH_3 at elevated temperature and on carbonization and/or graphitization yields pure foamed carbon free of metal ions.

For the manufacture of foamed carbon articles which are as free as possible of cations it is possible conveniently to begin by converting any ligninsulfonate solution into the free ligninsulfonic acid by reaction with cation exchangers in the H^+ form. Contaminations due to contained sugar do not interfere.

Example 1: A 60% strength aqueous solution of ammonium ligninsulfonate (of 43% C; 4.8% NH_3; 7.6% total S; 0.6% Ca) was poured into a glass mold to give a 5 mm thick layer and placed in a drying cabinet heated to 120°C. Thereupon, the lignin foamed up to five times its volume. The foam was heated to 210°C over the course of 2 hours and left for 5 hours at this temperature. The foamed lignin article had an apparent density of 0.08 g/cm³ and was resistant to aqueous acids and ammonia.

A part of the foamed lignin article was heated, under nitrogen, to 400°C at a heating rate of 60°C/hr and to 1000°C at a heating rate of 600°C/hr. The residue amounted to 37%, relative to lignin employed. The apparent density of the foamed carbon was 0.15 g/cm³. The article showed good strength. Heating to 2600°C under argon did not produce any externally visible change in the article; x-ray examination showed a diagram of turbostratic carbon. The residue after graphitization amounted to 35%, relative to lignin employed.

Example 2: 300 grams of aqueous 50% strength ammonium ligninsulfonate solution, as in Example 1, were mixed with 100 grams of a 2% strength aqueous polyethylene oxide solution of degree of polymerization 100,000, and 45 grams of water. The solution was homogenized while passing in ammonia gas until a pH value of 10 was reached. Its concentration was 33.7% of ammonium lignin-sulfonate and 0.45% of polyethylene oxide.

Thirty grams of 5 cm long lignin-carbon fibers were uniformly distributed in 200 grams of this solution. The dispersion was introduced into a steel vessel which was closed on all sides apart from venting slits and placed in a drying cabinet, heated to 120°C, where the composition consolidated to give a porous article of apparent density 0.28. The molding was subsequently heated to 100°C under nitrogen (heating rate up to 400°C, 100°/hour; up to 600°C, 200°/hour; up to 1000°C, 500°/hour). A highly pressure-resistant, porous carbon article of apparent density 0.16 g/cm^3 resulted.

Hollow Carbon Microspheres

A process described by *Y. Amagi, K. Noguchi and S. Inada; U.S. Patent 3,830,740; assigned to Kureha Kagaku Kogyo KK, Japan* provides a high quality carbon or graphite foam by adding hollow carbon microspheres to a phenolic or urethane resin, foaming the mixture, and baking the foam to carbonize or graphitize same.

The hollow carbon microspheres used in this process have an average diameter of from 10 to 500 μ and a bulk density of from 0.05 to 0.50 g/cc. The hollow carbon microspheres used in the process should preferably have an average diameter of from 50 to 500 μ and a bulk density of from 0.1 to 0.3 g/cc. The hollow carbon microspheres may be added to the starting resin in a ratio of 5 to 90 parts of microspheres per 100 parts resin, and preferably 10 to 80 parts of microspheres per 100 parts resin (by weight).

It has been found that the shrinkage of the resin foam during baking can be reduced to about one-half that of the unfilled urethane foam by adding hollow carbon microspheres to the starting urethane resin in an amount of 20% by weight of the starting resin. Furthermore, the volumetric contraction of the material can be reduced to 10% of the original volume by adding hollow carbon microspheres to the starting resin of a phenolic foam. Even if the resin foam is subjected to an oxidation treatment carried out at 200°C for 16 hours, the addition of 10% by weight of hollow carbon microspheres to the starting resin of a urethane foam eliminates the fusing phenomenon and deformation.

In the prior art processes, the cells of the synthetic resin foam are likely to be ruptured during the baking operation, due to the low carbonization degree of the synthetic resin, with the result that the carbon foams so produced are 70 to 90% by volume open-celled. The rupturing of the cell walls during carbonization results in lowered heat insulating values at the elevated temperature applications in which carbon foams are commonly used. In contrast, the carbon foam produced according to the process, has a high content of independent minute balloons, which feature provides excellent heat insulating values at high temperatures.

Another advantage of a carbon foam having hollow carbon microspheres is its low hygroscopicity because its ash content is as low as 0.3% and because the hygroscopicity of the microspheres may be minimized by regulation of the baking temperature. The following examples illustrate the process.

Example 1: Hollow carbon microspheres (average diameter of 100 μ, a bulk density of 0.15 g/cc) were added to a biliquid type hard urethane liquid reactant (Air Light Foam) and a foam was produced. More specifically, the hollow carbon microspheres were added in a given amount to each of the two liquids, and thereafter the two liquids were mixed together and poured into a mold 150 x 150 x 50 millimeters. After the mixture had been foamed in the mold, the foamed body was cut to the dimensions of 90 x 30 x 30 mm and subjected to oxidation in air at 200°C for 16 hours and then baked in a nitrogen atmosphere at 1000°C for 16 hr. The bulk density of the foamed body, after the baking, was 0.12 g/cc in average, ranging from 0.10 to 0.14 g/cc.

Table 1 indicates that the volumetric contraction and weight loss are decreased in proportion to the amount of hollow carbon microspheres added. In the runs where urethane was used alone or with the addition of 5% by weight of carbon microspheres, the center portion of the carbon foam product was found fused. Oxidation at 200°C for 48 hours resulted in the fused center portion of the urethane foam. The time required for the "infusibilization" treatment was greatly shortened.

TABLE 1: THE EFFECT OF THE ADDITION OF HOLLOW CARBON MICROSPHERES TO THE URETHANE FOAMS

Composition of Starting . . Material (by weight). .		Volumetric Contraction Rate on Baking (%)	Weight Loss on Baking (%)	Compressive Strength of Carbon Foam (kg/cm²)	Condition of Carbon Foam
Hollow Carbon Microspheres	Urethane Foam				
0	100	70	75	–	Center portion fused
5	95	64	67	10.5	Partially fused
10	90	58	65	12.0	No fused portion
15	85	48	59	12.0	No fused portion
20	80	36	57	11.5	No fused portion
30	70	30	44	10.0	No fused portion

Example 2: Commercially available resin powders of the type used to form novolak phenolic foams (Hi-Lack A-50) were mixed with hollow carbon microspheres of the type used in Example 1. The mixture was poured into a 100 x 100 x 30 mm mold and cured at 140°C for 15 minutes. The cured body was cut to the dimensions of 90 x 30 x 30 mm and baked in a nitrogen atmosphere at 1000°C for 30 hours. The bulk density of the baked body was 0.2 gram per cubic centimeter in average, ranging from 0.18 to 0.22 g/cc.

Table 2 indicates the effect of adding hollow carbon microspheres to the resin precursor. This table shows that the volumetric contraction and weight loss are decreased for phenolic foams in proportion to the amount of carbon microspheres added. The compressive strength was increased by addition of the carbon microspheres, presenting a peak at approximately the composition of 60:40. Reduced hygroscopicity was also noted. The hygroscopicity of the hollow carbon microspheres used was determined to be 13%.

TABLE 2: THE EFFECTS OF THE HOLLOW CARBON MICROSPHERES ON PHENOLIC FOAMS

Composition of Starting Material (ratio by weight).		Volumetric Contraction Rate on Baking* (%)	Weight Loss on Baking* (%)	Compressive Strength of Carbon Foam (kg/cm²)	Hygroscopicity (50% RH)
Hollow Carbon Microspheres	Phenolic Foam				
0	100	46	40	16	16
15	85	40	31	27	16
30	70	23	25	27	15
45	55	14	22	30	15
60	40	9	17	35	15
75	25	5	12	28	14
90	10	3	6	13	14

*at 1000°C

Pyrolyzed Polyurethane Foam

W.R. Adams; U.S. Patent 3,788,938; January 29, 1974; assigned to Scott Paper Company describes a method for treating cellular structures formed from materials which fuse or melt at relatively low temperatures, that is, below about 400°F, such that the finally evolved product will assume substantially the same physical or structural shape of the relatively low temperature material and yet will provide a structure which is relatively stable and possesses greater strength at higher temperatures, that is, temperature on the order of 700°F or even greater.

Briefly, a first basic structure having a desired configuration and comprised of material having a low fusion or deterioration temperature is coated at even lower temperatures with an infusible resin which is capable of being cured to an infusible state at a temperature lower than the fusion or deterioration temperature of the material comprising the first basic structure. The resulting coated structure is then subjected to elevated temperatures so that the basic structure is pyrolyzed and carbonized leaving a stronger, high-temperature-resistant structure substantially in the image of the first basic structure.

One method of the process involves temporarily attaching infusible resin coating material upon the surface of a porous structure formed from material having a given fusion temperature and having open intercommunicating cells or passages. One way of accomplishing this is to wet the structure with a binding liquid and to deposit particulate resin coating material thereon as by dusting techniques, for example.

The coating material has a fusion temperature when in the uncured condition below the fusion temperature of the material comprising the porous structure. The article coated thereby is then heated to a temperature which is above the fusion temperature of the particulate coating material and which is below the decomposition temperature and fusion temperature of the material forming the original porous structure until the coating material is fused over the surface of the structure into a substantially contiguous coating on its surface. The coating material is maintained at the above temperature until the material is cured into its infusible state and into a rigid, self-supporting condition.

The resulting structure is then subjected to a temperature which is below the volatilization temperature of the coating material but above the pyrolyzation temperature of the material forming the basic porous structure. In this manner, the material forming the basic structure is pyrolyzed off thereby reducing the weight of the structure. This results in an article having a shape substantially the same as the shape of a basic structure coated originally but formed of a material, the interior of which is a carbonized product of a material whose melting point is above 200°F, and which at least partially pyrolyzes above 500°F, and the exterior of which is formed of a different material having higher temperature characteristics and which may be at least partially pyrolyzed, but which is infusible. The following examples illustrate the process.

Example 1: A 2" x 2" x 1" piece of reticulated polyurethane foam having approximately 10 pores per lineal inch and weighing 2.0 g was immersed in an ethylene-vinyl acetate copolymer (EVA) emulsion (Flexbond, Grade 150) containing 5% total solids. The sample was agitated in the emulsion to remove all air and insure that the entire surface would be completely contacted by the emulsion. The sample was removed from the emulsion, drained and shaken to remove excess adhesive. It was then air-dried with the aid of a fan at room temperature.

The surface of the foam was observed to have good dry tack. Epoxy resin powder in the "B" stage (Corvel ECA-1283) having a varied particle size of less than 60 mesh was then applied as a coating on the surface of the foam. The sample was placed in a tray coated with the resin powder and then thoroughly dusted by a flooding technique in which the pores of the foam were almost completely filled with resin powder. The sample was then removed and turned in various directions and tapped gently to remove excess powder. The sample was inverted and the process was repeated.

The resin-powder-coated-foam sample was then transferred to a circulating hot air oven and heated for 10 minutes at 428°F to accomplish fusion and curing of the powdered resin upon the surface of the foam structure. After cooling in air, the sample weighed 7.1 g which means it increased in weight by 5.1 g over the polyurethane foam sample uncoated. The coating therefore equaled 250% of the weight based on the weight of the original foam substrate.

The resulting epoxy-coated sample was then sandwiched between two steel plates 0.025 inch thick and 2⅛ inches square with a combined weight of 9.5 g and returned to the oven where the composite sample was subjected to heat treatment for two hours at 572°F, resulting in almost complete pyrolysis of the polyurethane foam from the structure and curing and at least partial pyrolysis of the epoxy coating. After cooling, the assembly weighed 15.1 g (weight loss 21%). The sample between the steel plates was placed in the oven and subjected to a dead weight load of 16 lb (4 psi) while the temperature was slowly elevated. Up to 572°F, no deflection was noted. At 572°F, a deflection of 1.6% of the original thickness was observed with no degradation of its structural integrity. At 626°F, the deflection measured was 4.7% of original thickness and at 644°F, a deflection of 48.0% was measured, representing total collapse for all practical purposes.

Example 2: Polyurethane foam having the same physical characteristics as the sample in Example 1 was coated as in Example 1. In this instance, the adhesive emulsion was air-dried to provide the foam surface with dry tack. The resin was a "B" stage phenolic resin (Union Carbide BRP-4435) having a particle size of 60 mesh. Since phenolic resins yield water upon curing from the "B" stage to the "A" (fully cured) stage, it was necessary to apply the resin in thin multiple coats to allow evaporation of water vapor and to inhibit bubble formation. The resin was applied in

4 successive coats and each coat was heated for 15 minutes at 302°F to accomplish fusion and curing. The resulting phenolic-coated sample was subjected to heat treatment by heating it in an oven gradually to 572°F over a period of one hour and maintaining it at 572°F for two hours. The finished product had a coating amounting to 177% of the original foam substrate weight.

Example 3: A piece of reticulated polyurethane foam having the same physical characteristics that were employed in Example 1 was coated by the same procedure employed in Example 1. However, Alkanex 1003 "B" stage polyester-powdered-resin having a particle size of 60 mesh was used. Because of the fact that a number of the ingredients contained in this resin are chemically sensitive to water, an air-dried adhesive was used as described in Example 2.

Also, as described in Example 2, four multiple coats were applied in order to obtain adequate coverage and each coat was fused and cured upon the foam structure by subjecting the structure to 302°F for 15 minutes. The sample was then heat-treated by the procedure described in Example 2 and the final coating amounted to 266% of the weight of the original substrate.

The following Tables 1 through 5 are self-explanatory and illustrate by relatively comparative figures some of the advantages of the process. The samples in each table were pieces of polyurethane foam having approximately 10 pores per inch and coated with an epoxy resin, unless otherwise noted.

TABLE 1: EFFECT OF CURING CONDITIONS ON COMPRESSIVE STRENGTH

Sample No.	Coating (%)	. . . Curing Conditions . . . Temperature (°F)	Time (min)	Start of Deflection (°F)
1	236	446*	15	203
2	250	572**	120	563

*Normal cure
**Pyrolyzed

TABLE 2: EFFECT OF COATING WEIGHT ON COMPRESSIVE STRENGTH

Sample No.	Coating (%)	. . . Curing Conditions . . . Temperature (°F)	Time (min)	Start of Deflection (°F)
3	250	572*	120	563
4	470	572*	120	635

*Pyrolyzed

TABLE 3: EFFECT OF 10% FIBER GLASS REINFORCEMENT OF COMPRESSIVE STRENGTH

Sample No.	Coating (%)	. Curing Conditions . Temperature (°F)	Time (min)	Start of Deflection (°F)	Fiber Glass as Percent of Coating
5	470	572*	120	635	0
6	365	572*	120	671	10

*Pyrolyzed

TABLE 4: COMPRESSIVE STRENGTH AND DENSITY OF RETICULATED STRUCTURES

	Temperature Test (°F)	Apparent Density (pcf)	Load at Zero Deflection (psi)
10 ppi urethane foam, uncoated	70	1.8	Nil
Cured epoxy coated foam	210	11.4	4
Pyrolyzed epoxy coated foam	662	4.9	4

TABLE 5: LOSS IN WEIGHT OF EPOXY COATED STRUCTURES AFTER PYROLYSIS

	. . . . . . Sample Number		
	7	8	9
Urethane foam, g	2.2	2.0	2.0
Epoxy coated foam, g	8.1	7.1	9.3
Pyrolyzed coated foam, g	6.0	5.6	7.0
Loss on pyrolysis, g	2.1	1.5	2.3

Note: Epoxy coating mixture for samples 7 and 9 contained 10% fiber glass, sample size 2" x 2" x 1".

From the above it can be seen that the process provides a method for preparing a large variety of structures in a manner which allows the fabricating of structures from materials having excellent physical properties in the configuration of and by the use of structures made of very inexpensive materials which have quite undesirable properties for the particular applications involved.

PERLITE TREATMENTS

Milled Perlite

D.L. Ruff; U.S. Patent 3,904,539; September 9, 1975; assigned to Grefco, Inc. describes a thermal insulating composition wherein the composition comprises a mineral aggregate of perlite, a fibrillated organic fibrous material as the primary binder and, optionally, a binder-sizing agent wherein the insulating composition is in the form of insulation and the mineral aggregate has been milled either totally or in part to improve the thermal efficiency by reducing the coefficient of thermal conductivity, K, given in units of $Btu/in/ft^2/hr/°F$.

Additionally, the process for producing the thermal insulating material is provided which involves single or multiple milling of the thermally expanded mineral aggregate prior to or while adequately mixing the aggregate, the organic fibrous binder material and a binder-size prior to depositing a sufficient amount of the composition on a board-forming machine to form a mat of desired thickness and consistency. The mat is then partially dewatered, pressed and dried to produce the thermal insulating material.

Major factors contributing to the value of the thermal conductivity of the resultant composition include: concentration of fiber in the board, since pore size of the resultant product is reduced due to closer packing of fibers and perlite; board density, since the thermal conductivity is generally a linear function of board density over a broad range; and the loose weight density of the thermally expanded perlite utilized in the composition. Additionally, it has been found that milling the perlite prior to or during its incorporation into the board-forming-composition contributes remarkable assistance to reducing the coefficient of thermal conductivity of the resultant insulating board. The milled perlite may be used to completely replace the unmilled perlite or may be used to replace a small portion or substantially all of the unmilled perlite used in prior compositions.

Examples 1 through 6: The following examples illustrate the preparation of the thermal insulating composition. Each of the six examples listed in Tables 1 and 2 below were conducted following the same basic procedure as will be outlined below. The variations in milled perlite density and coefficient of thermal conductivity are indicated in these tables.

In a typical process for preparing the insulating composition of this process, pulped newsprint in such a proportion as to constitute 23% by weight of the final dry board is added to water to make a diluted paper pulp slurry. An appropriate amount of binder-size, in this instance 5% by weight of an asphalt emulsion, is added to the aqueous pulp slurry.

The thermally expanded mineral perlite is milled or crushed and is added to the pulped newsprint in an amount such that it constitutes 72% of the final dry weight of the board. The pulped newsprint-asphalt-perlite is then mixed in accordance with standard procedure for the preparation of insulation board. The composition is then placed on the board former and dewatered by suction, followed by pressing to a standard thickness and drying overnight at a temperature of 230°F. The resulting board in these examples was approximately one inch in thickness.

TABLE 1: PERLITE MILLING CONDITIONS AND PROPERTIES

Ex. No.	Screen Hole Diameter (in)*	Number of Passes	Original Perlite Loose Weight Density (pcf)	Final Milled Perlite Density (pcf)
1	0.094	1	1.46	2.35
2	0.047	1	1.46	2.38
3	0.039	2	1.46	2.92
4	0.094	1	2.53	3.62
5	0.047	1	2.53	3.75
6	0.039	2	2.53	4.64

*Refers to Micro Pulverizer Mill Screens

TABLE 2: PROPERTIES OF BOARDS PREPARED WITH 100% MILLED PERLITE*

Ex. No.	Board Density (pcf)	K Factor of Typical Board with Unmilled Perlite	K Factor of Milled Perlite Board	K Factor Reduction (%)
1	7.7	0.357	0.335	6.6
2	8.8	0.370	0.336	9.2
3	9.4	0.378	0.336	11.1
4	9.6	0.380	0.342	10.0
5	11.8	0.407	0.335	17.7
6	12.4	0.414	0.333	19.6

*Examples as shown in Table 1.

As evidenced by Table 2, there is a pronounced reduction in the coefficient of thermal conductivity following the procedure of the process whereby the thermally expanded mineral aggregate, namely perlite, is milled or crushed prior to incorporation into the board composition.

Aside from the essential ingredients of the insulating materials, other useful ingredients of the finished boards can also be included. For example, termite repellents, materials which prevent the formation of growth of algae, additional sizing materials in cases where the binder component of the boards does not impart sufficient resistance to water absorption may be added as determined by a particular usage of the board.

K factors were obtained using an electronic heat flow meter of the thermal transducer type specified in ASTM C-518. All data were taken at a 75°F nominal mean. Various available board forming machines may be employed for forming the boards described here on a commercial scale, including the Fourdrinier machine, rotary vacuum filters, or cylinder-type board machines.

Acid Treatment

D.K. Dunn; U.S. Patents 3,769,065; October 30, 1973; and 4,000,241; Dec. 28, 1976 describes a method of coating expanded perlite where the material is first sprayed with an acid solution, next sprayed with a water glass solution, and then sprayed with an acid solution.

It has been found that the treatment of expanded perlite particles by moistening the same with an acid prior to the application of the glass coating material produces a coated perlite which substantially eliminates many of the shortcomings of the prior art. Accordingly, this process involves a first step of applying to the expanded perlite particles an acid to moisten the same, a second step of applying a water-soluble silicate, a third step of again applying an acid and a fourth step of baking or curing the coated material to substantially complete dryness. Preferably, the acid comprises an aqueous solution of acetic acid although other mineral and organic acids such as hydrochloric, phosphoric, nitric, sulfuric, boric, formic, propionic, butanoic, malic, citric, and the like, may also be employed. Preferably too, the acid employed in steps one and three will be the same. The resulting coated perlite may be used immediately or conveniently stored for future use.

An acoustical or ceiling tile of general indoor or outdoor applications is formed with a blend of fine and coarse coated perlite particles having the following specifications:

U.S. Screen Size	Blend of Coated Fine and Coarse, %
8	0.0-1.0
16	0.25-5.0
30	60.0-80.0
50	50.0-75.0
100	5.0-15.0
>100	0.0-1.0

After the blend of fine and coarse perlite has been made, the same is used as a starting material and steps one, two and three of the coating method are repeated. The still-moist material is charged into molds where it is subjected to appropriate pressures to form tiles. The tiles are then heat-cured.

Tiles formed according to the process have been found to be extremely stable, physically and chemically, to exposure and weathering and to exceed prior perlite-base tiles in these qualities as well as in breaking strength.

Water-Insensitive Bonded Perlite and Asbestos Structures

S.A. Gerow and V.W. Weldman; U.S. Patent 3,658,564; April 25, 1972; assigned to E.I. du Pont de Nemours and Company describe water-insensitive, birefringent, rigid, low density structural products which consist essentially of expanded perlite bonded by an in situ-produced water-insoluble crystalline reaction product of perlite and sodium or potassium silicate. The process comprises steps (a) to (c) in sequence:

(a)   Mixing 1 to 2 parts filler materials, having at least 75% by weight reactive expanded perlite, with 4 to 1 parts of an aqueous solution of sodium silicate, having an $SiO_2 \cdot Na_2O$ weight ratio of 3:1 to 4:1 and a solids content of 28 to 34 weight percent, or potassium silicate, having an $SiO_2 \cdot K_2O$ weight ratio of 2:1 to 2.6:1 and a solids content of 24 to 30 weight percent.

(b)   Forming the mixture into a coherent article by compression, and

(c)   Curing the coherent article under controlled temperature and humidity conditions, whereby the water content of the article is maintained at a level of not less than 10% of the weight of the water-free solids in the article, for a period sufficient to produce an article capable of withstanding 8 hours of immersion in boiling water.

The most surprising physical property of the bonded perlite-silicate products of this process is the high degree of water-insensitivity such materials possess when compared to the bonded perlite-silicate products known in the art.

According to a related process described by *S.A. Gerow and V.W. Weldman; U.S. Patent 3,663,250; May 16, 1972; assigned to E.I. du Pont de Nemours and Company* water-insensitive insulating structures of fibrous asbestos are bonded by an in situ produced water-insoluble crystalline reaction product of asbestos and sodium or potassium silicate. Structures are obtained by mixing the asbestos with an aqueous solution of the silicate to make a damp granular powder, compacting the powder into a coherent article, and curing the article under controlled temperature and humidity conditions so as to maintain a water level in the article of at least about ten weight percent based on solids for a period of time sufficient to produce the crystalline binder. After curing, the structures are dried, if necessary, to a water content below about 20%, based on solids.

OTHER PROCESSES

Granular, Pourable Siliceous Composition

J.M. Pallo and D.J. Fischer; U.S. Patent 3,950,259; April 13, 1976; assigned to Johns-Manville Corporation describe a dry, free-flowing siliceous insulation material. It has a thermal conductivity significantly lower than most conventional thermal insulations and one which equals or approaches that of dry still air at most temperatures up to about 800°F. The composition consists essentially of

60 to 94 weight percent of a first siliceous material, 1 to 10 weight percent of a second siliceous material, and 5 to 30 weight percent of an opacifier. The first siliceous material is a particulate material having a silica content of at least 85%, a bulk density of 10 to 15 lb/ft^3, an average particle size of 50 to 100 microns and a BET surface area of at least 175 m^2/g. The second siliceous material is also a particulate material having a silica content of at least 85%, a bulk density of 0.5 to 2.5 lb/ft^3, an average particle size of 1 to 10 microns and a BET surface area of 80 to 150 m^2/g.

The opacifier is also a particulate material wherein the individual particles are each of a size in the range of from 10 to 80 microns. In a preferred case, a third siliceous material is also included in an amount of from 0.1 to 10 weight percent to enhance the surface area of the composition. The third siliceous material is a particulate material having a silica content of at least 90%, a bulk density of 0.5 to 3.0 lb/ft^3, an average particle size of 0.005 to 0.1 micron, and a BET surface area of at least 150 m^2/g. The following examples illustrate the process.

Example 1: A composition was prepared containing 70 weight percent of a first siliceous material, 5 weight percent of a second siliceous material and 25 weight percent of a titanium dioxide opacifier. The first siliceous material was a granular siliceous material containing approximately 93% silica, having an average particle size of 80 microns, a BET surface area of 250 m^2/g and a bulk density of approximately 12.5 lb/ft^3. This component was a commercial material, Sipernat 22.

The second siliceous material had a silica content of approximately 92%, an average particle size of 3 microns, a bulk density of approximately 1.6 lb/ft^3 and a BET surface area of 110 m^2/g. This material was a commercial material, Sipernat 17.

The titanium dioxide was a commercial high purity titania containing 97 to 99% titanium dioxide. Its particle size was all minus 325 mesh or less than 43 microns. Minimum particle size was greater than 5 microns.

These materials were mixed thoroughly with a commercial Henschel mixer at 1,800 rpm for a period of 5 minutes. The blades of the Henschel mixer were adjusted to the lowest position in order to prevent any batch material from settling out and remaining at the bottom of the mixer.

The pourability of the formulation was determined using ASTM test method D-1895 (Method A). A 32 gram sample was aerated and then loaded into a funnel. The amount of time required for this 32 grams of material to flow completely out of the funnel was recorded and reported as the pourability rate.

The thermal conductivity of the sample was determined by the heat flow meter method (ASTM test method C-518) at mean temperatures of 300°F, 500°F and 800°F. The results of these tests are all reported in the table below.

Example 2: A composition similar to that of Example 1 was prepared, with this composition containing 68 weight percent of Sipernat 22, 7 weight percent of Sipernat 17 and 25 weight percent of the titania opacifier. The pourability rate and the thermal conductivity at the stated temperatures were determined and also are reported in the table.

Example	Pourability Rate, Sec.	Thermal Conductivity, Btu/in/hr/ft²/°F		
		300°F	500°F	800°F
1	45	0.27	0.30	0.48
2	15	0.26	0.29	0.48

For comparison purposes it will be noted that the thermal conductivity of dry air at 300°F is 0.24, at 500°F is 0.29 and at 800°F is 0.36 Btu/in/hr/ft²/°F. A comparison with the table above will indicate clearly that at the lower temperatures the compositions of this process closely approach and essentially equal the thermal conductivity of the dry still air.

Silica-Alumina Fibers

E.W. Olewinski and L.J. Pluta; U.S. Patent 3,835,054; September 10, 1974; assigned to Nalco Chemical Company describe a method for the preparation of thermal insulation board wherein a ceramic fiber such as Fiberfrax is slurried with silica sol and sequentially with a source of alumina. The solution is adjusted to a slightly acid-neutral pH of 5 to 7 and the slurry is flocculated with a minor amount of an anionic polymer consisting of a latex of acrylamide/acrylic acid. The flocculated mix is subsequently vacuum formed and dried under conventional procedures.

A highly preferred ceramic fiber material usually of equal parts of alumina and silica is Fiberfrax. As preferably utilized in the bulk fiber form, it may be described as a ceramic fiber basically composed of 51.3% of Al_2O_3 and 47.2% SiO_2 where the fibers range in length up to 1½" and have a mean diameter of 2 microns. Such fibers are capable of withstanding continuous use temperatures up to 2300°F and short term exposures to higher temperatures.

Commercially, silica sols are available as Nalcoag, e.g., Nalcoag 1115, and as Ludox. Prefered silica sols contain particles of average diameter of 3 to 150 mμ.

As a flocculating agent a minor amount of an anionic polymer is utilized, which is a latex copolymer of acrylamide and acrylic acid prepared according to the teachings in U.S. Patent 3,624,019. The acrylamide/acrylic acid percentage in the polymer may vary from about 90/10 to 10/90 with the preferred about 70/30. The utilization of this polymer has been found to increase effective strength of the product but not to increase the heat degradation.

The board formed from the slurry mix by vacuum forming and drying has been found to have densities ranging from 8 to 24 lb/ft³ with strengths of 10 to 150 psi. Within the limits of polymer addition, an increase in polymer with constant alumina and silica gives decreased density, strength, and forming time.

Example 1: 140 grams of Fiberfrax were slurried in 5,600 ml tap water. 51.3 grams silica sol (Nalcoag 1115) and 40 grams No. 5 sodium aluminate were added and the slurry was mixed for 15 minutes. The pH was adjusted to 5.5 with H_2SO_4 and 350 ml of 0.1% solution of acrylamide/acrylic acid copolymer were added. Mixing was continued slowly until the system was flocculated. The slurry was then poured into a 10½" x 10½" box with a screen on the bottom and vacuum applied to draw off the water. The pad was then dried overnight at 110°C. The dried pad had a strength of 90 psi and a density of 18.5 lb/ft^3. After firing at 1300°F for one hour, the strength was 100 psi.

Example 2: 140 grams of Fiberfrax were slurried in 5,600 ml tap water. 57.9 grams silica sol (Nalcoag 1115) and 54.9 grams alum were added and the slurry was mixed for 15 minutes. The pH was adjusted to 6.5 with NaOH and 130 milliliters of 0.1% acrylamide/acrylic acid copolymer were added. Mixing was continued slowly until the system was flocculated. The slurry was then poured into a 10½ x 10½ inch box with a screen on the bottom and vacuum applied to draw off the water. The pad was dried overnight at 110°C. The dried pad had a density of 17.7 lb/ft^3 and a strength of 102 psi. After firing at 1300°F for one hour, the strength was 107 psi.

COMPANY INDEX

INVENTOR INDEX

U.S. PATENT NUMBER INDEX

NOTICE

Nothing contained in this Review shall be construed to constitute a permission or recommendation to practice any invention covered by any patent without a license from the patent owners. Further, neither the author nor the publisher assumes any liability with respect to the use of, or for damages resulting from the use of, any information, apparatus, method or process described in this Review.